Springer Optimization and Its Applications

Volume 182

Aims and Scope

Optimization has continued to expand in all directions at an astonishing rate. New algorithmic and theoretical techniques are continually developing and the diffusion into other disciplines is proceeding at a rapid pace, with a spot light on machine learning, artificial intelligence, and quantum computing. Our knowledge of all aspects of the field has grown even more profound. At the same time, one of the most striking trends in optimization is the constantly increasing emphasis on the interdisciplinary nature of the field. Optimization has been a basic tool in areas not limited to applied mathematics, engineering, medicine, economics, computer science, operations research, and other sciences.

The series **Springer Optimization and Its Applications (SOIA)** aims to publish state-of-the-art expository works (monographs, contributed volumes, textbooks, handbooks) that focus on theory, methods, and applications of optimization. Topics covered include, but are not limited to, nonlinear optimization, combinatorial optimization, continuous optimization, stochastic optimization, Bayesian optimization, optimal control, discrete optimization, multi-objective optimization, and more. New to the series portfolio include Works at the intersection of optimization and machine learning, artificial intelligence, and quantum computing.

Volumes from this series are indexed by Web of Science, zbMATH, Mathematical Reviews, and SCOPUS.

More information about this series at http://www.springer.com/series/7393

Dionysis D. Bochtis • Maria Lampridi
George P. Petropoulos • Yiannis Ampatzidis
Panos Pardalos
Editors

Information and Communication Technologies for Agriculture—Theme I: Sensors

Springer

Editors
Dionysis D. Bochtis
Institute for Bio-economy
and Agri-technology (iBO)
Centre for Research and Technology
Hellas (CERTH)
Thessaloniki, Greece

George P. Petropoulos
Department of Geography
Harokopio University of Athens
Athens, Greece

Panos Pardalos
Department of Industrial
and Systems Engineering
University of Florida
Gainesville, FL, USA

Maria Lampridi
Institute for Bio-economy
and Agri-technology (iBO)
Centre for Research and Technology
Hellas (CERTH)
Thessaloniki, Greece

Yiannis Ampatzidis
Department of Agricultural
and Biological Engineering
University of Florida
Immokalee, FL, USA

ISSN 1931-6828 ISSN 1931-6836 (electronic)
Springer Optimization and Its Applications
ISBN 978-3-030-84143-0 ISBN 978-3-030-84144-7 (eBook)
https://doi.org/10.1007/978-3-030-84144-7

This Springer imprint is published by the registered company Springer Nature Switzerland AG
The registered company address is: Gewerbestrasse 11, 6330 Cham, Switzerland

Preface

This book was conceived by a need for an up-to-date knowledge source on sensor design, development, and application for monitoring agricultural production parameters. Agriculture is a multi-parametrical production system, and its production environment is extremely variable. To that effect, there is a necessity for developing cost-effective and high-accuracy sensors that cover diverse needs. Throughout an agricultural production cycle – from planting preparations to harvesting – the use of sensors can offer valuable information to farmers that can be utilized in decision-making, increasing the efficiency of production while reducing the required inputs, with tangible benefits to the economy and the environment. The difficulties of manual data collection processes along with the rapid technological developments have urged the adoption of sensing systems that are able to collect large sets of data at short time. To utilize the data collected from sensors, further processing is required with data analysis and visualization resulting to utility parameters (e.g., crop health, height, size) that can directly be used by farmers in the everyday decision-making process.

The first part of the book (Part I) presents an overview on the state of the art in sensing technologies applied in agricultural production, including remote and proximal sensing, wireless sensors network systems, and IoT. The chapter titled **"Emerging Sensing Technologies for Precision Agriculture"** elaborates on the latest advances of sensing technologies examining various categories of sensing systems that are used for agricultural applications. The chapter also examines the technologies that can be used in combination with sensing. For example, machine vision, combined with artificial intelligence, is utilized in a variety of agriculture-related applications as for example for pest management and precision spraying or yield prediction.

The evolution of the spectral, spatial, and temporal resolution in satellite sensors has enhanced the interest of the scientific community in their use, as they are able to provide accurate estimations with respect to the properties of soil. In fact, during the last decades, the efficiency of soil spectroscopy has been thoroughly examined, while remote and proximal sensing techniques are now widely used to gather primary or secondary data for the assessment of soil properties, with various resolution options depending on the field of application. Soil is considered as an important factor in the implementation of sustainable development goals (SDGs) as it is strongly connected to the efficient management of the Earth's resources and directly connected to environmental degradation, climate change, and other challenges that the Earth is facing. Nevertheless, the assessment of soil condition is still a costly and time-consuming process. Earth observation technologies are a promising potential towards the monitoring and reporting of soil properties, providing valuable information that can be utilized by various sectors. The chapter **"Soil Reflectance Spectroscopy for Supporting Sustainable Development Goals"** elaborates on the significance of soil conditions in realizing the SDGs, demonstrating how soil reflectance spectroscopy can aid towards its evaluation at national to global level. Detailed literature examples are also presented on how soil reflectance spectroscopy is used for the estimation of soil properties including soil moisture, organic carbon/matter, and clay content. Lastly, the barriers and challenges in the utilization of Earth observation techniques for the monitoring of soil are discussed.

Regarding the monitoring of crop growth processes, nowadays the availability of proximal and remote sensing technology for monitoring plant growth is increasing at an unprecedented rate, allowing for faster and cost-effective assessments. The chapter titled "**Proximal Sensing Sensors for Monitoring Crop Growth**" provides a theoretical overview of various monitoring solutions available for precision agriculture with particular emphasis on variable rate fertilization. Different sensor platforms are reviewed, from drone cameras to tractor-mounted and hand-held devices up to sensors carried by robotic platforms. Use-cases from Estonia and Lithuania are also presented to illustrate the most common practices of proximal sensing implementation in the context of precision agriculture.

The second part (Part II) of the book refers to wireless sensor networks dedicated to agricultural production. Wireless sensor networks (WSNs) is the term used to describe a group of, usually large in number, sensors that are distributed over an area for collecting and transmitting information. Transmission can take place either among sensor, or between sensors, and one or more moderators and gateways, depending on the architecture of the network. The sensors can collect a variety of different types of data, for example, from environmental parameters (humidity, temperature, and pressure) up to presence and movement data. The spread of IoT technologies has encouraged the use of WSNs in agriculture. The IoT concept involves the interaction of different technologies within a specific environment and includes all the processes that are used to manage the collected data. These technologies can include, among others, various types of devices, sensors, networks, actuators, routing protocols, as well as databases. The evolving network technology has increased the adoption of IoT systems. This adoption is expected to increase further, with the advent of 5G (fifth generation of mobile networks).

Through the implementation of wireless sensor networks, the capability of monitoring agricultural-related parameters is enabled along with the potential to remotely control the executed activities during the in-field processes. However, in real-life applications, the use of traditional network management techniques should be examined, considering specific characteristics of the sector. As a matter of fact, for their use in agriculture, WSNs should be able to operate efficiently under minimum energy consumption requirements. Considering this, high-end low-power wide-area network (LPWAN) radios, using protocols like LoRa, are constantly gaining ground, among the most promising technologies for the specific purpose. The LoRa protocol exhibits a plethora of benefits, including the ability to operate at long distances, with low energy consumption, also demonstrating immunity to noisy environments. The chapter **"Experimental Performance Evaluation Techniques of LoRa Radio Modules and Exploitation for Agricultural Use"** aims at improving the behavior of LoRa radio modules and presents a low-cost, generic, and flexible methodology that results in an easy-to-use portable equipment enabling fast and satisfactory results with respect to accuracy. Towards that direction, innovative smartphone solutions are used to monitor and/or modify the operational traits of LoRa-equipped sensor nodes through simple commands that safeguard, that the procedures are user-friendly. A series of measurements were performed in realistic agricultural conditions, while the results obtained are in accordance with the underlying characteristics of the LoRa protocol. It is worth noting that the developed testbed infrastructure can be implemented in various scenarios. It can also be expanded to facilitate the communication tasks of typical agricultural applications such as autonomous irrigation systems.

The realization of WSNs in the agricultural sector is a complicated process due to the interaction of technological and biological entities. In the chapter titled **"Evaluating the Performance of a Simulated Softwarized Agricultural Wireless Sensor Network,"** the required specifications for creating a WSN for monitoring the activity of beef cattle are examined. The routing protocols RPL (routing protocol for low-power and lossy networks) and CTP (collection tree protocol) are assessed in the livestock use case. Additionally, proving that softwarization is a trend in the evolution of networks, the use of SDN (software-defined networking) together with the RPL was also examined, giving the best results while achieving the required specifications. The results also encourage the adoption of such systems in various agricultural domains for animal and crop production. However, the introduction of IoT technologies should be accompanied with the use of the most resource-efficient routing protocols. Such routing protocols are essential in the case of smart farming approaches that employ autonomous processes for improved decision-making. Certain limitations of the work presented concern the unavailability of open datasets and the difficulties in the interaction among the relevant stakeholders, including Academia, indicating the importance of farmers' familiarization with such innovative technologies.

Examining the progress of WSNs related to smart agriculture applications, the chapter **"Smart Agriculture: A Low-Cost Wireless Sensor Network Approach"** elaborates on the issues of accountability and accuracy of systems that are developed to monitor agricultural environmental parameters. In the literature review section of the chapter, the different WSN technologies dedicated to smart agriculture applications along with the low-cost equipment that is used in frontier research are discussed. Additionally, various systems for synchronized monitoring are evaluated to select one for further examination. Further examining the potential of modern smart agriculture, the work also elaborates a variety of schemes dedicated to environmental monitoring, which aggregate sensor information from different locations. A relevant scheme includes the cloud/fog computing paradigm, which results in flexible WSNs that interact with a main cloud infrastructure via a fog computing network. Furthermore, the application of this cloud/fog architecture was also examined for its implementation in the prevention of natural hazards, such as wildfires, considering the system's fast response time.

The third part of the book (Part III) is dedicated to remote sensing. Remote sensing is the term used to describe the process in which the required information is collected through satellite images (Earth observation) or aircraft-based sensors. Remote sensing data are very useful in phenological trends monitoring as well as in the assessment of the influence of climate variability. Satellite images can facilitate field scale monitoring, providing information of the crop condition regionally, which is not feasible with the use of hand-held sensors because of their sampling restrictions. Over the last 50 years, the Earth observation technology has evolved and has been optimized with respect to temporal and spatial resolution. Earth observation in the context of monitoring agriculture has recently matured sufficiently, including missions that contribute to the time series processing of information. Several research and commercial satellites are covering the entire planet at various revisit frequencies and with various spatial resolutions (ranging from a few centimeters to hundreds of meters). In addition to the visible spectrum, many of these satellites can also collect data in the red edge, near-infrared, infrared, and thermal spectra, creating valuable data sets with respect to crop monitoring. Moving to lower levels of Earth observation, and when higher resolution than the one obtained by satellites is required, sensors placed on airplanes or unmanned aerial vehicles are used. Such sensors can deliver high spatial resolution measurements and are suitable for monitoring row crops or orchards where crops do not fully cover the surface of the soil.

The chapter titled **"Potential of Sentinel-2 Satellite and Novel Proximal Sensor Data Fusion for Agricultural Applications"** presents a comparative assessment of a ground-based proximal multispectral sensor, called Plant-O-Meter, and the images obtained from the Sentinel-2 satellite mission. The results demonstrated that both instruments offer comparable results at specific growth stages of maize, rendering the Plant-O-Meter proximal sensor an efficient alternative to satellite images that can provide high spatial resolution images. The main advantage of the ground proximal system is that it can perform independently to weather conditions such as cloud coverage. Additionally, data from satellite images can be fused with proximal data collected from the Plant-O-meter sensor, adding value to the derived information, especially since satellite data usually need auxiliary interpretation with ground-truth data. With respect to the spatial resolutions of the measurements, the study demonstrated that the plant development stage plays an important role with respect to the accordance of the indices calculated from Plant-O-Meter and Sentinel-2, especially since the first can provide information at plant level while the later at field or regional level. Over large areas, the vegetation growing stages can be assessed via the optical time series data at high temporal resolutions gathered from Earth observation missions. The chapter **"Trends in Satellite Sensors and Image Time Series Processing Methods for Crop Phenology Monitoring"** elaborates on the latest trends in data collection about land surface phenology (LSP) metrics for the quantification of the key stages of crop growing seasons through image time series processing. The tendency in image time series processing is mostly focused on: (a) increasing spatial and temporal resolution by combining constellations of multiple satellites, (b) fitting multi-year and irregular time series data with machine learning algorithms, (c) time series fusing of various sources of data, and (d) developing the relevant software packages. The rapid increase in the availability of Earth observation data along with new data processing methods has resulted in the advancement of the quantification of LSP metrics. Within the work presented, the DATimeS toolbox is utilized in order to produce LSP for time series of MODIS and S2 data over both homogenous and heterogenous agricultural areas. The results indicate that for homogenous areas, the LSP parameters can be related to the main crop type and be calculated in a consistent manner. In the case of heterogenous areas, LSP parameters can be inconsistent, due to the complexity of the landscape, the temporal resolution variation, and the influence of cloud coverage.

The wide availability of unmanned aerial systems (UAS) technology coupled with the latest discoveries in image segmentation and classification has led to the emerging of new opportunities in agricultural and environmental sciences. Some of UAS's key advantages include the ability for ad-hoc rapid and repetitive collections over small areas without almost any influence from the cloud cover in comparison to satellite solutions. However, here the challenge is the generation and execution of optimized flight plans in terms of minimizing the shadow effects and providing similar environmental conditions for repetitive flights and the best speed over image sharpness ratio. In the chapter "**Drone Imagery in Support of Orchards Trees Vegetation Assessment Based on Spectral Indices and Deep Learning,**" the combination of a deep learning approach for the tree crown delineation and RGB-based vegetation indices generated by drone flights was used for assessing the tree vegetation health. The approach obtained very promising results in the case of the plum, apricot, and walnut trees, mostly because the orientation of the leaves of these trees is towards the camera and the spaces between leaves and branches are relatively small. In contrast, less reliable results were obtained for olive trees due to the specific tree configuration with small leaves and large spaces between branches. Overall, the study demonstrates the real potential of drone applications for small to medium orchards.

The popularity of remote sensing was increased following the release of open-access satellite images from various providers globally. Furthermore, the cost of use of unmanned aerial systems (UAS) has started to decrease, thus their utilization for field monitoring and mapping with the use of the appropriate sensors has become popular. Farmers, on the other hand, have started to familiarize with the potential of such systems, and they are incorporating their use for in-field hotspot insights. However, the most important drawback in the wide use of such systems is the high requirements for pre-processing of the data before they can be used in field. Also, such passive systems are susceptive to exogenous factors such as weather and atmospheric conditions. In parallel, and in many cases complementary to these technologies, on-the-spot sensing technologies have advanced during the last decades, offering both off-line and real-time monitoring solutions. However, there are still many limitations in their use in small- and medium-size farms, related among others to their high cost, the complexity of use, and their accuracy. This leaves room for improvements for the introduction of novel and more efficient proximal sensing solutions in terms of cost and applicability.

The fourth and last part of the book (Part IV) deals with the application of proximal sensing in agriculture. Proximal sensing is the term used for monitoring in short distance from the object of interest usually with the use of sensors placed on tractors or other vehicles. Proximal sensors offer a wide variety of advantages with respect to the spatial resolution (which can be in the range of millimeters to centimeters) and temporal independence since measurements can be scheduled at any time.

The normalized difference vegetation index (NDVI) is one of the most used and standardized indexes in crop production. The relatively simple NDVI formula and the ease of measurement, requiring only a near-infrared (NIR) and visible sensitive instrument, have led to a wide implementation of the index. NDVI can be produced from various data sources, from satellite images to close to canopy platforms and sensors including UASs, close-contact spectral sensors mounted on mobile phenotyping platforms, and active NDVI field sensing. The data processing itself has also been expanded from the simple photosynthesis prediction and change detection to advanced crop analytics, including nutrient, yield, and phenotyping prediction. In parallel, these processes have been enriched with machine learning approaches that are able to handle vast amounts of data generated by high-resolution imaging spectroscopy sensors to relate NDVI or other optical data to crop parameters. However, there is still a gap on bridging the physically based-knowledge together with the data-driven one, in field and plot scales. The chapter **"What Does the NDVI Really Tell Us About Crops? Insight from Proximal Spectral Field Sensors"** covers this issue of relating the historically grounded physically based theory to field-based sensing applications. Authors attempt to answer this question by analyzing proximal data acquired at high temporal and spectral resolution in two agricultural sites in Finland.

One of the key aspects for site-specific management within precision agriculture concept is field heterogeneity. On this aspect and for the lateral heterogeneity of the soil, electrical conductivity data provide reliable information that can be used directly and cost effectively. Apparent electrical conductivity data are correlated with parameters such as soil salinity, soil moisture content, and soil texture. Because soil fertility and yield depend on these parameters, there is also a link between apparent electrical conductivity and yield, while many scientists consider it one of the key features for delineating agricultural management zones. Additional information regarding vertical heterogeneity (e.g., soil layering) helps to understand in more detail soil-moisture relationships, the interaction between soil and plants, as a means for selecting the right plant type in terms of appropriate root depth. To close the gap between two- and three- dimensional soil data (lateral and vertical), several novel techniques including new instruments and inversion routines have been proposed and developed. In the chapter "**Geophysical Sensors for Mapping Soil Layers: A Comparative Case Study Using Different Electrical and Electromagnetic Sensors,**" authors present a field study on the use of geophysical sensors to image lateral and vertical soil heterogeneity within the rooting zone of plants up to a depth range of 1–2 m.

The use of geoinformation tools has improved the management of pests and especially the mapping of pest populations hotspots. In the chapter **"Geoinformation Technologies in Pest Management: Mapping Olive Fruit Fly Population in Olive Trees,"** a novel methodological framework for the monitoring of olive-fruit pest populations is presented. The framework proposes the use of a locally installed network of traps in combination with geoinformation tools such as geographical information systems (GIS) and global positioning systems (GPS). The methodology was practically tested in a typical Mediterranean olive grove in Crete, Greece. The concept presented enables the analysis of population data with the use of the technical tools that the GIS environments offer. The monitoring of the trap network provides valuable information with respect to the allocation and the number of traps in the area under monitoring. One of the main advantages of the methodological framework is related to the mapping of the spatial and temporal variation in the population of pests, in the period when pest management occurs. In this manner, the effectiveness of spraying is increased, since specific spraying zones can be determined, a fact of significant importance especially in ecologically sensitive areas.

Closing this part, and the entire book as well, in the chapter **"In-Field Experiments for Performance Evaluation of a New Low-Cost Active Multispectral Crop Sensor,"** a comparison, in real-conditions, between a newly develop multispectral device (named Plant-O-Meter) and a widely used and accepted commercial hand-held device (GreenSeeker) is attempted. The reference sensor was chosen since it works under the same principles as the tested sensor, while it is of low cost and has comparable measurement characteristics. The two sensors were evaluated in two test maize fields, and nitrogen (N) experiments were designed to simulate variations in canopy development, vigor, and greenness, to facilitate the comparison of the sensors' performance in a wide range of expected readings. The NDVI was estimated for three maize hybrids with the use of both sensors at different growth stages. According to the results, the best measuring time was between the V7 and V8 stage for both sensors. The overall results indicate an almost identical behavior of both sensors with a 1:1 ratio of accordance in measurements, demonstrating the potential of the innovative sensor in plant canopy assessment for on-the-spot variable rate fertilization.

Different aspects of sensors implementation in agricultural production (e.g., types of sensors, parameters monitoring, network types, connectivity, accuracy, reliability, durability, and needs to be covered) are analyzed within the book context providing a variety of up-to-date information and knowledge on the topic. Given the multi-regional distribution of the chapters, we hope that this book provides a "mapping" of the global status in advancement and development of prototyped and applied sensing solutions in agricultural production domain.

Thessaloniki, Greece	Dionysis D. Bochtis
Thessaloniki, Greece	Maria Lampridi
Athens, Greece	George P. Petropoulos
Immokalee, FL, USA	Yiannis Ampatzidis
Gainesville, FL, USA	Panos Pardalos

Contents

Part I
Overview

Emerging Sensing Technologies for Precision Agriculture

Sri Kakarla, Yiannis Ampatzidis, Seonho Park, George Adosoglou, and Panos Pardalos

1 Introduction

Farmers across the world are looking for ways to increase efficiency and productivity of the farms due to increasing demand of agricultural products and decreasing farmland. As the global population continues to rise, the biggest question that arises is how to produce enough food to feed every one of them. To solve this problem, farmers need to find new ways to increase the production on existing farmlands and precision agriculture technologies might be the solution to this challenge.

Precision agriculture uses innovative technologies to increase crop yield while using lesser resources by establishing a decision management system, which uses data from the farm to control and estimate the number of resources required for a particular process with accuracy and precision. Precision agriculture is a rapidly developing area and emerging sensing technologies play an important role in it. Right from planting a plant to packing the end product, sensors are being used in several ways to increase the efficiency of a farm. The whole farming process can be categorized into different areas such as:

1. Planting
2. Soil management
3. Nutrient and water management

S. Kakarla · Y. Ampatzidis
Agricultural and Biological Engineering Department, Southwest Florida Research and Education Center, University of Florida, Immokalee, FL, USA
e-mail: i.ampatzidis@ufl.edu

S. Park · G. Adosoglou · P. Pardalos (✉)
Industrial and Systems Engineering Department, Center of Applied Optimization, University of Florida, Gainesville, FL, USA
e-mail: pardalos@ise.ufl.edu

D. D. Bochtis et al. (eds.), *Information and Communication Technologies for Agriculture—Theme I: Sensors*, Springer Optimization and Its Applications 182,
https://doi.org/10.1007/978-3-030-84144-7_1

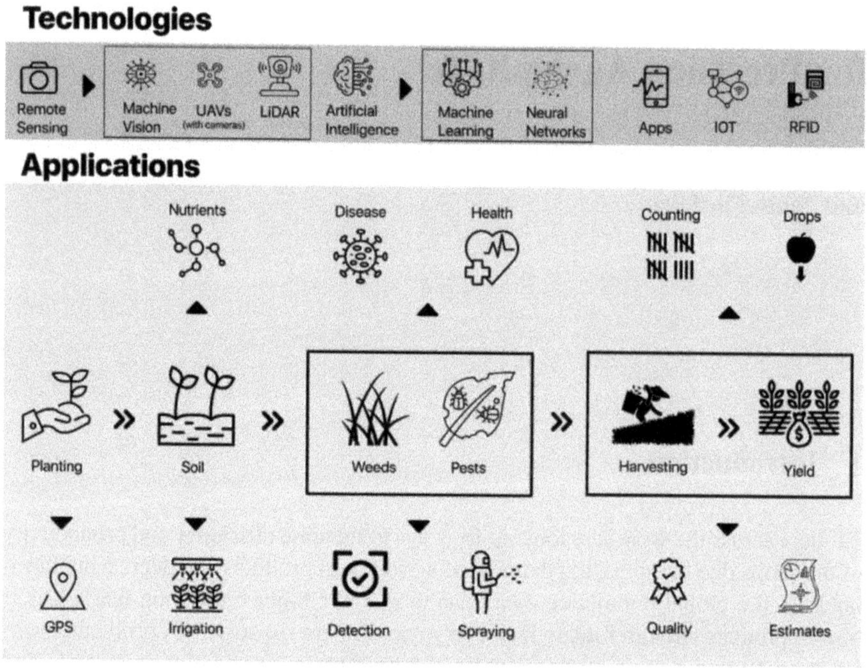

Fig. 1 Sensing technologies and their applications in agriculture

4. Pest and diseases management
5. Yield harvesting and post-harvest processing

There are numerous applications of several sensors in each of the above-mentioned areas (Fig. 1). Some of the most commonly sensors in agriculture include optical sensors, airborne sensors, non-destructive sensors, positioning sensors, quality detection sensors, etc.

1.1 Planting

The most commonly used sensor during planting is the global positioning system (GPS). GPS is used by the planting machines to plant crops in an efficient pattern so as to reduce the amount of fuel and time needed to navigate around the farm. The GPS can also be used for a better inventory management system to identify the location of a specific plant(s).

Airborne sensors (e.g., RGB and multispectral cameras) are being used after planting to identify skips/gaps in the planting pattern so that the planting machine can go back and plant in the gaps (Fig. 2) which helps reduce the loss of farm area being used [1].

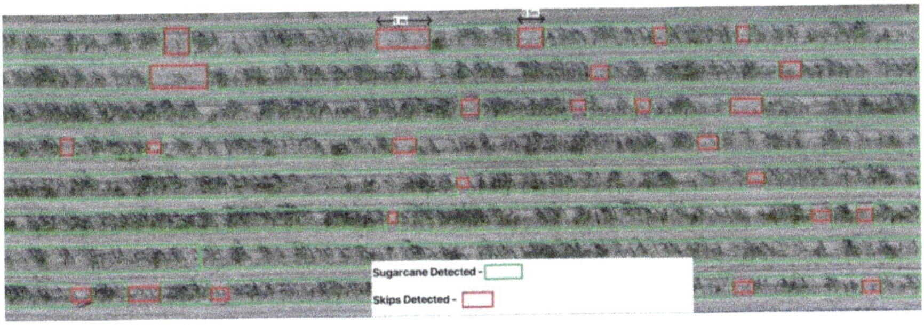

Fig. 2 UAV image showing skips detection in a sugarcane field

1.2 Soil Management

There are various parameters that come into play with soil. Farmers are concerned about the water/moisture levels and nutrient composition of soil layers. There are many geophysical sensors being used to measure moisture levels of the soil in real time, transmitting data into cloud which forms a sophisticated irrigation management system. There are many spectral sensors being used to map different nutrients levels in the soil. These spectral sensors are sometimes integrated into unmanned aerial vehicles (UAVs) to help map the soil nutrient levels across large size farms.

There are also several geophysical sensors such as ground penetrating radar (GPR) and electromagnetic sensors that are being used for soil layer mapping and for creating 3D maps of the plant root system [2]. Other sensors can be used to measure soil water content and help determine various parameters such as evapotranspiration, groundwater recharge, etc. They can help growers efficiently run irrigation systems to maximize yield and also minimize the impacts of farming practices.

1.3 Plant Health Management

Spectral sensors, such as multispectral and hyperspectral cameras, and artificial intelligence (AI) are used in the field for high throughput phenotyping, to identify plant health status and detect plant stress, and to determine plant nutrient concentration [3].

Unmanned aerial vehicles (UAVs) equipped with various types of sensors, such as RGB, multispectral and hyperspectral cameras, combined with AI can be used to scout individual plants and collect information about them such as plant health status, leaf density, plant height and canopy size (Fig. 3), plant nutrient concentration (e.g., development of fertility maps; Fig. 4), etc. [4]. This information can help farmers with maintaining an efficient inventory system and track the progress of the farm and also identify "weak" spots in the farm, which can better manage with a better decision management system.

Fig. 3 Agroview, a cloud- and AI-based application to analyze and visualize UAV collected data [4]. Example of citrus blocks in Florida, USA

Fig. 4 Agroview: Example of a fertility map for discrete N management zones (N ranges based on recommendations from the University of Florida's Institute of Food and Agriculture)

1.4 Pests and Disease Management

Disease diagnosis by visual observation of symptoms can get complex due to the similarity of symptoms between various diseases. This complex nature can even confuse experienced personnel to misidentify a disease which can lead to further problems [5]. Optical sensors combined with machine learning have been utilized to distinguish between diseases using differences in reflectance, shape, color, and textural information that may not be visible to a naked eye [6].

Unmanned aerial vehicles (UAVs) equipped with various types of sensors, such as RGB, multispectral and hyperspectral cameras, combined with AI can be used to identify diseases in large scale farms rapidly and in low-cost compared to manual inspections [7].

Optical sensors like visual RGB cameras and LiDAR sensors are used to detect and identify different types of weeds and insects. These sensors are integrated with existing sprayers to decrease the use of chemicals by gathering data about the amount of weeds and pests present in the farm and there by controlling the amount of resources used according to the gathered data.

Conventional spraying technologies which are not equipped with any sensors spray all along their path irrespective of the presence of weeds/pests. Researchers have been employing new techniques with the help of optical sensors paired with artificial intelligence. For example, these smart sprayers detect and identify the type of weed and controls the spraying system in a way that it only sprays where a weed is present [8]. This will ultimately end up saving a lot of chemicals used by the farmer which in turn reduces the cost required for spraying. Another benefit of these smart sprayers is environmental protection, they reduce the amount of chemicals being used and the chemical drift off that is being carried into the water bodies and into the soil which will help preserve the quality of our natural resources and avoid phenomenon like red tide and other chemical disasters. Similar technologies have also been adopted to estimate the number of pests present in a farm which can support the farm prepare an efficient schedule of pesticide spraying [9].

These variable rate technologies (VRT) use sensors such as LiDAR, RGB or multispectral cameras, depth cameras, and GPS combined with powerful microcontrollers and novel algorithms that enable variable application of chemicals. These technologies automatically control the amount of chemicals applied based on presence of pests and diseases, plant health status, plant or soil nutrient concentration, etc. The sensors used in these technologies can detect several parameters such as the health status of the plants, plant height and canopy size, absence of plants, etc., which are all applied towards the decision making of the VRT system.

1.5 Yield Harvesting and Post-Harvest

Optical sensors like multispectral, thermal, and RGB cameras are used to detect or predict the yield potential of the farms. These are also used to detect the quality, maturity, and the readiness of the product for harvesting purposes. Another application of these sensors can be to count fruit/vegetable drops from plants/trees, which will enable the farmers to estimate the percentage of drops and estimate food wastage.

During the packing and processing of the products, computer vision is used for sorting products (e.g., fruit or vegetable) based on size, shape, and maturity. Smart and automated systems are used for categorizing products for further processing and production, and for disease/bruise detection, which in turn improves efficiency of the logistics and increase food safety. For example, Blasco [10] used computer vision (near infrared-NIR and fluorescent-FL imagery) to detect anthracnose and green mold, which are considered to be serious damage to skins of citrus fruit. This technology was used to sort fruits according to the presence of damages on fruits, which will ensure that only fruits with good quality reach the market. That can increase profits for growers.

2 Types of Sensors

There are a wide range of sensors available in the market and there are various applications for each of these sensors. Below, several categories of sensing systems that are used in agriculture are presented.

2.1 *Remote Sensing*

Remote sensing refers to the use of sensors to acquire information from an object (or area) without any contact with the target object. These sensors collect information by measuring the reflection and radiation being emitted by the objects present in an area. The sensors are usually mounted on ground or aerial vehicles (e.g., UAV, aircraft, and satellite). Recently, unmanned aircraft systems (UAS) are being heavily used by people around the world due to the relatively low cost of the equipment and their ability to cover large areas in shorter time when compared to manual observation. There are various types of spectral sensors that are integrated with UAS or ground based vehicles to collect various farm data by measuring several parameters across the electromagnetic spectrum.

2.2 *Computer Vision*

Computer vision is the process of enabling computers to understand and analyze imagery and therefore process, detect, identify, and classify objects present in an imagery. Computer vision has been made possible due to the advancements in machine learning techniques (e.g., neural networks and deep learning). For example, Costa et al. [11] utilized computer vision and deep learning, a Mask-RCNN algorithm, for object detection and semantic segmentation for area estimation of shuck, shell, and embryo on pecans to better understand the fruit growth curve. This study provided a faster and more reliable methodology to detect and estimate growth of pecan nuts throughout the season. Below, various sensing systems that are used for collecting images in precision agriculture applications are presented.

2.2.1 RGB

Red, green, blue (RGB) sensors are commonly referred to as visual cameras that are widely used in everyday devices such as cellphones, webcams, etc. These sensors measure the reflectance in red, green and blue spectrum and provide users with an image. Unmanned aerial vehicles equipped with RGB cameras are flown over large-scale fields collecting thousands of images. These collected images are thereby stitched together using photogrammetry techniques to produce a map of the entire field. These maps can be used for several precision agriculture applications. For example, they can be used to detect, identify, and count plants in a field, and to estimate leaf density, plant health status [12], canopy size, tree height, etc. Ampatzidis et al. [4] developed a cloud-based application, called Agroview, to process, analyze, and visualize data collected from RGB cameras attached to UAVs (Figs. 3 and 4). This application, upon uploading the user collected images, processes the images and stitches them together into a map and can: (i) detect, count, and geo-locate plants and plant gaps (locations with dead or no plants); (ii) measure plant height and canopy size (plant inventory); (iii) develop plant health (or stress) maps; and (iv) estimate plant nutrient concentration and develop fertility maps. These maps can further be used for applications of variable rate technologies (e.g., variable rate fertilizer applicators). Technologies like Agroview can be used for plant inventory by saving over 70% of data collection cost and 90% of collection time.

2.2.2 Multispectral

Multispectral sensors can provide data that human eyes cannot see, and which cannot be captured by the RGB sensors. They usually provide reflectance data from near infrared spectrum in additional to the red, green, and blue spectrums. This information is very important for the calculation of the most widely used vegetation index called normalized difference vegetation index (NDVI). NDVI is being widely used by researchers across the world to identify the health status of various plants. It is measured on a scale of zero to one, where zero being the most stressed and one being a healthy plant. For example, Guan [13] used small UAVs for calculating high-resolution normalized difference vegetation index (NDVI) values and subsequently correlate these values to fertilizer application levels and yield of rice and wheat crops. The NDVI values helped them differentiate the levels of fertilization levels in the crops which indicated a potential for early detection of nitrogen deficiency. The color NDVI distribution maps of wide rice fields identified differences in stages of ripening and lodging-injury areas, which accorded with practical crop growth status from ground observation.

2.2.3 Hyperspectral

Hyperspectral sensors are one of the most complex spectral sensors that are being used for agricultural applications. They are not being as widely used as the other spectral sensors due to very high cost of the equipment and the complex operating procedures. Contrary to the RGB and multispectral sensors, hyperspectral sensor collects reflectance data in continuous scans along a spectrum usually ranging from 400 to 2400 nm. While the multispectral sensor collects the reflectance data over ranges of spectrum, hyperspectral sensor can collect reflectance data from a wavelength range as narrow as 4 nm. Researchers have been using hyperspectral sensors combined with machine learning algorithms to correlate the reflectance data with various agricultural parameters. For example, hyperspectral sensors are being used to detect, identify, and distinguish diseases with similar visual symptoms, which can be a very complex task for a regular farm worker. For example, Abdulridha et al. [14] used both airborne and benchtop hyperspectral imagery to detect and distinguish bacterial spot (BS) and target spot (TS) diseases in tomato fields. Over 35 vegetation indices were analyzed to select the most optimum wavelengths for disease detection and identification. Classification methods like multilayer perceptron neural network (MLP) and stepwise discriminant analysis (STDA) were compared to distinguish between various stages of the diseases and MLP method had higher classification rates than STDA method in all disease stages. The best vegetation indices to detect both the diseases were also determined. In a follow up study, Abdulridha et al. [7] utilized machine learning models to identify and clasify Tomato yellow leaf curl, BS, and TS in two varieties of tomato using aerial (aka, UAV) and ground (lab-absed) hyperspectral imaging (Fig. 5).

2.2.4 Thermal

Thermal sensors measure the energy radiated by an object at a wavelength corresponding to its surface temperature. They can provide the users with surface temperature of various objects present in a field. Thermal cameras are widely used in agriculture to estimate soil moisture and plant evapotranspiration for precision irrigation applications. For example, Soliman [15] used both airborne and handheld thermal sensors to estimate soil moisture content in vineyards where the soil was covered with grass. In this study, they evaluated the strength of relationships between soil moisture, mechanical resistance, and thermal inertia. They used data from both sensors to explore the effects of different field of views and the high inter-row temperature variability. Despite the low resolution of airborne thermal sensor, it

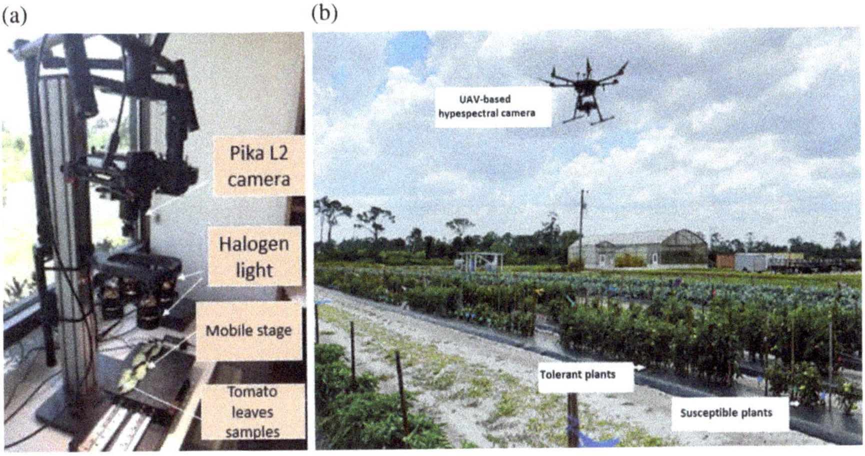

Fig. 5 (**a**) Benchtop Pika L 2.4 camera (Resonon Inc., Bozeman MT, USA) with leaves samples in the laboratory condition; (**b**) Unmanned aerial vehicle using hyperspectral sensing (same Pika L 2.4 camera) to collect spectral reflectance data from susceptible and tolerant tomato plants infected with tomato yellow leaf curl. Ref. [7]

performed better than estimates from the handheld thermal sensor. The results from this study were encouraging for future research on extending thermal remote sensing of soil moisture in heterogeneous land cover regions. Thermal sensors can also be used with machine vision to detect fruits on trees [16].

2.2.5 LiDAR

LiDAR sensors measure the distance to the objects around them by illuminating the target with laser light and calculating the time required for the light to return to the sensor. LiDAR sensors have been used historically to map digital elevation and surface models of the earth's surface. LiDAR sensors are being used in agriculture for 3D modelling of the farms that can help farmers understand the terrains of the farm for an efficient implementation of irrigation system practices. They can also be used to measure various parameters such as crop height, crop density, canopy size, etc. (e.g., Fig. 6) These LiDAR sensors can be used both on ground based and air borne sensors (Fig. 7) for various applications.

2.3 Synthetic Aperture Radar

Synthetic aperture radar (SAR) is a type of radar that is utilized to reconstruct two-dimensional images. The distance the aircraft with the SAR device travels over a target in the time taken for the radar pulses to return to the antenna creates the large

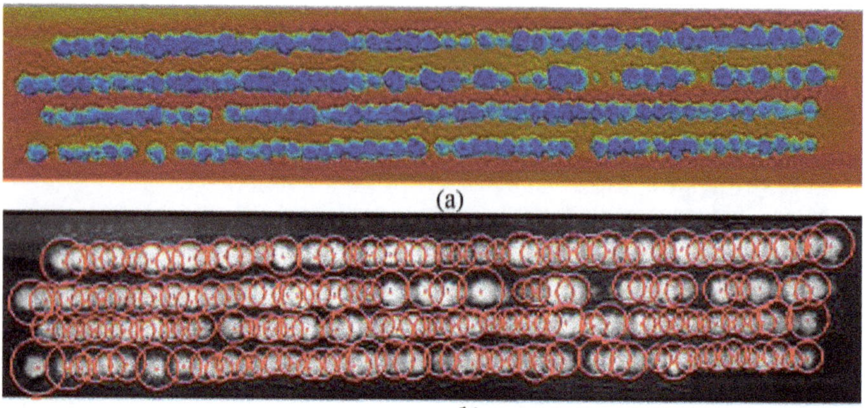

Fig. 6 (a) Point cloud generated using LiDAR; (b) Trees detected using canopy height model (CHM) segmentation post processing

Fig. 7 UAV equipped with a LiDAR airborne sensor

synthetic antenna aperture. Thus, it can have better resolution than comparatively small physical radar systems. This technology has a consistent spatial resolution over a range of views. It is usually attached to UAS and has been utilized in precision agriculture including estimating soil moisture, ground/foliage penetration, biomass. For example, Ndikumana et al. [17] studied the capabilities of multitemporal radar images for rice height and dry biomass retrievals using Sentinel-1 C-band including dual-polarization VV (Vertical receive and Vertical transmit) and VH (Vertical receive and Horizontal transmit) data. To this end, they trained three types of machine learning models (multiple linear regression-MLR, support vector regression-SVR, and random forest-RF) to estimate rice height and dry biomass. Reisi-Gahrouei et al. [18] examined the use of MLR and artificial neural network (ANN) models to estimate the biomass for canola, corn, and soybeans using 14 L-band polarimetric parameters of uninhabited aerial vehicle synthetic aperture radar (UAVSAR) data. Hosseini et al. [19] developed a neural network transfer function between the biomass derived from SAR data and from optimal data. Its purpose is to facilitate the integration of optical and SAR derived biomass estimation.

Synthetic aperture radar is also used to characterize the bare or vegetated soils in agricultural environments. Sekertekin et al. [20] used three models, namely Dubois semi-empirical model, MLR, and water cloud model (WCM) for estimating surface soil moisture in vegetated agricultural fields. They examined the potential of L-band ALOS-2 and C band Sentinel-1 SAR data for estimations in a dry season. Mercier et al. [21] used SAR Sentinel-1 and optical Sentinel-2 times series data to monitor forest-agriculture mosaics. Nasirzadehdizaji et al. [22] aimed to investigate the sensitivity of 10 parameters derived from multi-temporal Sentinel-1 SAR data to estimate crop height and canopy coverage of maize, sunflower, and wheat.

3 Wireless Sensor Networks

Sensors devices can be combined with the Internet of Things (IoT) technologies and Wireless Sensor Networks (WSN) in order to accomplish efficient energy management. A WSN is typically connected to the cloud gathering data, analyzing it, making predictions, and then assisting in intelligent decision making. In precision agriculture, WSNs can be used for better providing crop nutrients and reducing the number of agrochemicals that are needed to control pests and diseases. For irrigation planning purposes, WSNs can be used to optimize water and electricity consumption using state-of-the-art technologies leading to green agriculture while also increasing agricultural yield.

The WSN consists of a set of sensor nodes that collect various data. These data can include air humidity, temperature, rain, light, soil moisture, etc. These nodes are located in strategic spots of the field. The WSN also consists of a coordinator node that is in charge of gathering sensor data from the rest of the nodes and uploading it to the cloud. An analysis of the data received from the network's sensors is then performed on the cloud. Subsequently, the coordinator node determines when and how the actuators (e.g., irrigation system) should be activated.

It is important to note that the communication between sensor nodes is facilitated by merging diverse sensors from simple (e.g., temperature, pressure, humidity) to more complicated (e.g., micro-radars, tracking, and images), thus, WSNs can take advantage of a wide range of surroundings to gather precise data. The cloud platform, which is able to represent the received data and the system's status with graphs and make forecasts, allows the user to easily take decisions and command the nodes. The forecasting can be based on the expected probability of precipitation retrieved from a local weather forecast API service in the case of an irrigation system. The visualization is done in real-time and can be used to control every aspect of the deployed network.

The main challenges in setting up most WSNs is synchronization. In order to effectively monitor the sensor nodes, the acquired measurements must be correlated. Approaches to deal with synchronization challenges include providing the sink node's clock to the entire network as a point of reference. Other challenges include determining optimum deployment schemes, measurement periods, routing protocols, energy efficiency, cost, communication range, scalability, and fault tolerance.

In light of the future arrival of 5G technologies, the research community is turning to alternative solutions such as proposed cloud/fog hybrid models. This way, the system can benefit from the 5G technologies' very high speed and low latency in communications. For further enhancing PA and remote sensing, the systems of remote sensing are requiring an increasing amount of data to be processed from various areas. Consequently, machine learning methods and artificial intelligence algorithms are receiving greater attention. Optimization methods such as genetic algorithms are also employed to achieve a self-organizing, adaptive wireless sensor network design for energy management that takes into account communication and energy-conservation constraints while also guaranteeing maximum life span.

4 Sensor Fusion

Sensor fusion can be defined as the ability to combine the inputs from multiple sensors (usually low-cost sensors) to collect data that result in an increased accuracy; compared to the data being collected by individual sensor. For example, stereoscopic vision, the combination of two or more optical sensors, can be used to detect the distance between a sensor and an object. This technology is being used in harvesters to calculate the distance from the sensor to the detected object for accurate movement of robotic arms to pick the product(s). Barrero [23] used a fusion of RGB and multispectral sensors to detect Gramineae weeds in rice fields. This technology combines the texture information given by a high resolution RGB image and the reflectance information given by low resolution multispectral images to obtain a fused RGB-multispectral image with better weed discrimination features. The normalized green red difference index (NGRDI) generated using multispectral imagery was fused with the RGB image that was able to accurately detect the weeds. Sankey et al. [24] used LiDAR and hyperspectral fusion to monitor the landscape changes in southwestern forests of USA using topography modelling.

5 Conclusions

From planting a seed to harvesting the yield, sensors can help growers with providing critical information in every stage of the production. This information can be used by growers to make key decisions to increase application efficiencies and optimize inputs usage. Remote sensing systems can provide growers with large sets of data in a very short time, compared to manual data collection processes. This can save growers both time and money. The data acquired from the sensors undergoes post processing and with the help of data analysis and visualization software, and hence growers can get information about various parameters like crop health, crop size, crop height with just a few clicks. For example, advanced optical sensors are being used to detect and differentiate between diseases that have similar visual symptoms. Although an expensive process, scientists are working towards developing low-cost sensor models using information gathered from currently undergoing research.

Machine vision technologies combined with artificial intelligence have been also used in various applications such as pest detection, variable rate and precision agrochemical applications, and yield estimation and forecasting. Variable rate spraying technologies can detect the presence of pests and diseases, and adjust the amount of chemicals being sprayed according to the information processed by the sensing system. This can help growers save chemicals, which in turn will save money and also minimize the negative environmental impact. Sensor devices can be merged with wireless sensor networks (WSN) and Internet of Things (IoT) technologies to achieve efficient energy management, real-time data collection, and precision decision-making.

Acknowledgments This material was made possible, in part, by a Cooperative Agreement from the U.S. Department of Agriculture's National Institute of Food and Agriculture (NIFA), award #2020-67021-30761. Its contents are solely the responsibility of the authors and do not necessarily represent the official views of the USDA.

References

1. Ampatzidis Y., and Partel V. (2019). UAV-based high throughput phenotyping in citrus utilizing multispectral imaging and artificial intelligence. *Remote Sensing*, 11(4), 410; https://doi.org/10.3390/rs11040410.
2. Zhang X., Derival M., Albrecht U., and Ampatzidis Y. (2019). Evaluation of a ground penetrating radar to map root architecture of HLB-infected citrus trees. *Agronomy (Special Issue: Precision Agr.)*, 9(7), 354. https://doi.org/10.3390/agronomy9070354.
3. Ampatzidis Y., Partel V., Meyering B., and Albrecht U. (2019). Citrus rootstock evaluation utilizing UAV-based remote sensing and artificial intelligence. *Computers and Electronics in Agriculture*, 164, 104900, doi.org/https://doi.org/10.1016/j.compag.2019.104900.
4. Ampatzidis Y., Partel V., Costa L. (2020). Agroview: Cloud-based application to process, analyze and visualize UAV-collected data for precision agriculture applications utilizing artificial intelligence. *Computers and Electronics in Agriculture*, 0168-1699.
5. Cruz A., Ampatzidis Y., Pierro R., Materazzi A., Panattoni A., De Bellis L., Luvisi A. (2019). Detection of grapevine yellows symptoms in Vitis vinifera L. with artificial intelligence. *Computers and Electronics in Agriculture*, 157 (February 2019), 63–76, https://doi.org/10.1016/j.compag.2018.12.028.
6. Abdulridha J., Ampatzidis Y., Roberts P., Kakarla S.C. (2020a). Detecting powdery mildew disease in squash at different stages using UAV-based hyperspectral imaging and artificial intelligence. *Biosystems Engineering* 197, 135-148. https://doi.org/10.1016/j.biosystemseng.2020.07.001.
7. Abdulridha J., Ampatzidis Y., Qureshi J., Roberts P. (2020b). Laboratory and UAV-based identification and classification of tomato yellow leaf curl, bacterial spot, and target spot diseases in tomato utilizing hyperspectral imaging and machine learning. *Remote Sensing*, 12(17), 2732; https://doi.org/10.3390/rs12172732.

8. Partel V., Kakarla S.C., and Ampatzidis Y. (2019a). Development and evaluation of a low-cost and smart technology for precision weed management utilizing artificial intelligence. *Computers and Electronics in Agriculture*, 157 (February 2019), 339-350.
9. Partel V., Nunes L., Stansley P., and Ampatzidis Y. (2019b). Automated vision-based system for monitoring Asian citrus psyllid in orchards utilizing artificial intelligence. *Computers and Electronics in Agriculture*, 162, July 2019, 328-336.
10. Blasco N.A.J. (2007). Citrus sorting by identification of the most common defects using multispectral computer vision. *Journal of Food Engineering*, 384-393.
11. Costa L., Nunes L., Ampatzidis Y. (2020). A new visible band index (vNDVI) for estimating NDVI values on RGB images utilizing genetic algorithms. *Computers and Electronics in Agriculture*, 172, 105334, https://doi.org/10.1016/j.compag.2020.105334.
12. Costa L., Ampatzidis Y., Rohla C., Maness N., Cheary B., Zhang L. (2021). Measuring pecan nut growth utilizing machine vision and deep learning for the better understanding of the fruit growth curve. *Computers and Electronics in Agriculture*, 181, 105964, https://doi.org/10.1016/j.compag.2020.105964.
13. Guan S, F. K. (2019). Assessing Correlation of High-Resolution NDVI with Fertilizer Application Level and Yield of Rice and Wheat Crops Using Small UAVs. *Remote Sensing*, 112
14. Abdulridha J., Ampatzidis Y., Kakarla S.C., Roberts P., (2020c). Detection of target spot and bacterial spot diseases in tomato using UAV-based and benchtop-based hyperspectral imaging techniques. *Precision Agriculture*, 21, 955-978. https://doi.org/10.1007/s11119-019-09703-4.
15. Soliman A, H. R. (2013). Remote Sensing of Soil Moisture in Vineyards Using Airborne and Ground-Based Thermal Inertia Data. *Remote Sensing*, 3729-3748.
16. Gan H, Lee, W. S., Alchanatis V., Abd-Elrahman A. (2020). Active thermal imaging for immature citrus fruit detection. *Biosystems Engineering*. 198. 291-303. https://doi.org/10.1016/j.biosystemseng.2020.08.015.
17. Ndikumana E., Ho Tong Minh, D., Dang Nguyen H. T., Baghdadi N., Courault D., Hossard L., El Moussawi I. (2018). Estimation of rice height and biomass using multitemporal SAR Sentinel-1 for Camargue, Southern France. *Remote Sensing*, 10(9), 1394.
18. Reisi-Gahrouei, O., Homayouni, S., McNairn, H., Hosseini, M., & Safari, A. (2019). Crop biomass estimation using multi regression analysis and neural networks from multitemporal L-band polarimetric synthetic aperture radar data. *International Journal of Remote Sensing*, 40(17), 6822-6840.
19. Hosseini M., McNairn H., Mitchell S., Robertson L.D., Davidson A., Homayouni S. (2019). Synthetic aperture radar and optical satellite data for estimating the biomass of corn. *International Journal of Applied Earth Observation and Geoinformation*, 83, 101933.
20. Sekertekin, A., Marangoz, A. M., & Abdikan, S. (2020). ALOS-2 and Sentinel-1 SAR data sensitivity analysis to surface soil moisture over bare and vegetated agricultural fields. *Computers and Electronics in Agriculture*, 171, 105303.
21. Mercier A., Betbeder J., Rumiano F., Baudry, J., Gond V., Blanc L., … Hubert-Moy L. (2019). Evaluation of sentinel-1 and 2 time series for land cover classification of forest–agriculture mosaics in temperate and tropical landscapes. *Remote Sensing*, 11(8), 979.
22. Nasirzadehdizaji R., Balik Sanli F., Abdikan S., Cakir Z., Sekertekin A., Ustuner M. (2019). Sensitivity analysis of multi-temporal Sentinel-1 SAR parameters to crop height and canopy coverage. *Applied Sciences*, 9(4), 655.
23. Oscar Barrero, S. A. (2018). RGB and multispectral UAV image fusion for Gramineae weed detection in rice fields. *Precision Agriculture*, 809–822.
24. Sankey T., Donager J., McVay J., Sankey J., (2017). UAV lidar and hyperspectral fusion for forest monitoring in the southwestern USA. *Remote Sensing of Environment*. 195, 30-43. https://doi.org/10.1016/j.rse.2017.04.007.

Soil Reflectance Spectroscopy for Supporting Sustainable Development Goals

Theodora Angelopoulou

1 Introduction

Global economy depends on the goods and services provided by the natural environment. Soil as a natural resource is considered one of the most important components of the natural environment because it is our life support system and largely non-renewable [1, 2]. It also performs a number of key environmental, social and economic functions such as: biomass production, including agriculture and forestry; storing, filtering and transforming nutrients, substances and water; biodiversity pool, for habitats, species and genes; it is the physical and cultural environment for humans and human activities; it is a source of raw materials; acts as a carbon pool; and is the archive of geological and archaeological heritage [3]. Despite soil's significance, few initiatives have been adopted for its protection and conservation in the past years i.e., the Global Assessment of Soil Degradation (GLASOD) project that produced a world map of human-induced soil degradation [4] and the EU Thematic Strategy for Soil Protection in 2006 [5]. In 2012 at the United Nations conference in Rio de Janeiro member states decided to develop a set of Sustainable Development Goals (SDGs)[1] to build upon the previous Millennium Development Goals in order to establish the UN High – level Political Forum on Sustainable

[1] https://sustainabledevelopment.un.org/sdgs

T. Angelopoulou (✉)
Institute for Bio-economy and Agri-technology (iBO), Centre for Research and Technology Hellas (CERTH), Thessaloniki, Greece

Laboratory of Remote Sensing, Spectroscopy, and GIS, Department of Agriculture, Aristotle University of Thessaloniki, Thessaloniki, Greece
e-mail: d.angelopoulou@certh.gr; agtheodor@agro.auth.gr

D. D. Bochtis et al. (eds.), *Information and Communication Technologies for Agriculture—Theme I: Sensors*, Springer Optimization and Its Applications 182,
https://doi.org/10.1007/978-3-030-84144-7_2

Development. As a result, in 2015, the 2030 Agenda for Sustainable Development was adopted with 17 SDGs and 169 targets at its core [6]. Agriculture plays a very important role considering end of poverty and hunger, climate change adaptation and mitigation while sustaining natural resources and therefore lies in the heart of the 2030 Agenda for Sustainable development [7]. Although for most SDGs there is no direct link with soils, they can contribute in the realization of the SDGs. Specifically, SDGs two and three concern food security that could be addressed by promoting sustainable agriculture and SDG six (clean water and sanitation) is related with soil's function of storing and filtering water. One of the most important SDG is about climate action and as it was highlighted by the "four per 1000 initiative" an increase of 0.4% soil carbon stocks per year would significantly reduce the CO_2 concentration in the atmosphere related to human activities [8] while SDG 15 (Protect, restore and promote sustainable use of terrestrial ecosystems, sustainably manage forests, combat desertification, and halt and reverse land degradation and halt biodiversity loss) is directly related with soil conservation and sustainable management, hence soil protection.

Thus, monitoring soil's condition is an imperative need to evaluate the progress made towards achieving soil protection and SDGs in particular. However, soil analysis is a time consuming and particularly costly procedure that could be characterized inefficient for large scale estimations. Therefore, an extensive need for optimizing the monitoring and recording capacity using big data analytics and accurate measuring methods at low cost, needs to be developed [9]. Prediction of soil properties in a fast and accurate way is considered a necessity in order to overcome the lack of consistent soil property information. Advances in the past three decades, have contributed to significant progress in such predictions. With the development of geostatistics and the evolution of sensing techniques (satellite, airborne, in-situ and laboratory spectroscopy), soil properties can now be predicted with accuracy using novel tools and approaches such as digital soil mapping [10]. These techniques are most frequently used and evaluated by soil scientists in the developed world. However, they could be also applied in regions where there is lack of knowledge and infrastructure providing beneficial information for the specific areas. To this end, efforts have been established to address the collective need for more coordinated Earth Observation (EO) based interoperable data and systems. One of these efforts is the Group on Earth Observations (GEO), created in 2005 to address the necessity for more comprehensive global data sets. GEO is a voluntary partnership of governments and international organizations coordinating efforts to build a Global Earth Observation System of Systems, (GEOSS) [11]. In addition to that, the European Union's Earth Observation Programme Copernicus offers information services based on satellite Earth Observation and in situ (non-space) data for the benefit of European citizens. Specifically, the Copernicus Land Monitoring Service (CLMS) provides geographical, reliable and up-to-date information in six thematic areas: land, marine, atmosphere, climate change, emergency management and security to a broad range of users in Europe and across the World in the field of environmental terrestrial applications [12].

The advantages of remote and proximal sensing EO techniques refer to the potential of large scale, cost effective and repeatable measurements that make them attractive for soil monitoring. The main principle underlying these techniques is energy – matter interactions [13, 14]. Particularly, the reflected electromagnetic radiation of a material is measured and the reflectance is related to the wavelength, the materials' surface and also to reflectance and emissivity characteristics, [15]. Specifically, in the Visible and Near Infra-Red (VNIR) region the distinct spectral signatures are caused by electronic transitions that are the result of overtones and combinations of fundamental vibrational bands due to the stretching and bending of N-H, O-H, and C-H bonds [16]. For that reason soil reflectance spectroscopy could be used for quantitative and qualitative analysis of soil properties and therefore after [17] several efforts for building soil spectral libraries in global, international and national level have initiated [18–21]. Data from spectral libraries could be eventually used as reference data for classes identification in multispectral and hyperspectral images using classification algorithms. An SSL could be utilized to label unknown spectra by matching spectral signatures derived from satellite imagery enabling automatic identification and extraction of various objects and assist in proper selection of training pixels of the image [22].

2 The Important Role of Soil in Supporting SDGs

Europe's land is considered to be one of the most intensively used in the world. The way land is used has crucial impacts on the environment. Land resources are also affected by forest fires, extreme weather conditions, floods and droughts as a consequence of climate change. Therefore, monitoring soil's condition is essential for soil conservation that allows governments and society to identify problems, provide solutions and assess the progress made towards meeting the 2030 Agenda for Sustainable Development. Soil services are very important for supporting life on planet as they provide water and nutrients for plant growth, regulate the water cycle and are the second largest terrestrial carbon pool. It is estimated that in the EU around 75 billion tons of carbon is stored in soil. Hence, it plays a significant role in climate change mitigation and adaptation (Schils et al., 2008). The European Environmental Agency's (EEA) estimates reported that in 2017, EU's CO_2 emissions were about 4.5 million tons and although is reported that there is an increase of Soil Organic Carbon (SOC) in EU, these estimates are considered highly uncertain. The complexity of identifying the amount of SOC stocks is due to its constant change as it is a dynamic parameter that depends on the relationship between plants that capture carbon dioxide and release it back in the atmosphere. To prevent climate change consequences global warming should be below 2 °C and the Intergovernmental Panel on Climate Change (IPCC) suggests that greenhouse gas emissions from every sector should be reduced. A potential solution that is cost effective is improving soil's quality. The Food and Agriculture Organization (FAO) reported that restoring degraded soils could remove up to 62 billion tons of carbon.

Though defining the relationships between climate change and soil quality is very complex and for that reason the results of hypothetical scenarios are more qualitative than quantitative. Initiatives are starting to emphasize soil's importance in climate change mitigation such as the EU thematic strategy for soil protection and the Paris agreement. While a recent EU regulation suggests that Member States should fully offset the sectors Greenhouse Gases (GHGs) from 2021 to 2030. To do so a consistent monitoring and reporting agenda should be implemented with the support of EEA which continues to develop knowledge for sustainable land management and for that purpose uses data acquired from earth observation from the Copernicus Land Monitoring Service.

2.1 *Monitoring Soils for Optimization of Precision Agriculture*

Precision agriculture (PA) involves the application of practices for better management of the farm inputs such as herbicides, fertilizers, fuel and seeds by determining the right place and time [23]. The main goal is to increase the productivity and profitability of a farm system in the long term while preventing the excessive use of chemicals with targeted applications by providing information on the spatiotemporal variability of soil and crop properties including environmental parameters [24]. PA bases its applicability on the use of technologies i.e., the use of sensors for mapping, monitoring and prediction of specific essential variables regarding crop yield, soil parameters, diseases and pests [25]. There is a global research interest during the past decades towards the use and development of the right technologies to support PA applications. Among these, remote and proximal sensing techniques in combination with geographical information systems (GIS) and global positioning systems (GPS) are evaluated for their efficiency to be used as tools for data acquisition and data analysis for their further implementation in decision support systems [26]. In the field of agriculture these techniques are increasingly used showing leveraging potentials due to the development of spectral, spatial and temporal resolution of the sensors that has provided unpresented information about spectral characteristics of soil properties.

Soils and plants are the main receivers of inputs in agriculture. Site specific management does not only have environmental but also economic impacts and therefore it could potentially be the main method of agricultural practices. Determining quantitatively and qualitative soil properties is of great importance for correct site-specific applications and soil variability maps will be the way to determine soil inputs while also identifying possible issues apart from nutrient deficiencies such as soil erosion, soil salinity etc. at field, regional, national and global scales. Although some soil properties like soil texture could be considered stable over time, PA applications need continuous monitoring, therefore conventionally soil analysis is inadequate. For that reason, remote and proximal sensing techniques could be used as an alternative method for acquiring soil information. However, PA practices, have not been largely accepted from agricultural producers. Their adoption depends on the provision of tangible examples of the economic and environmental benefits, clarification that remote and proximal sensing techniques do not replace traditional methods but act supplementary, there must be consistent commitment for several growing seasons to lead to conclusion [27].

3 Earth Observation Supporting SDGs

EO technologies can find application in various environmental research topics as biodiversity and wildlife trends; land use change monitoring; mitigating, and managing the impact of natural disasters; natural resources sustainability, such as energy, freshwater, and agriculture; address emerging diseases and other health risks; and climate change adaptation and mitigation [28]. The World Summit on Sustainable Development Implementation in 2002, highlighted the need for coordinated observations relating to the state of the Earth. Acknowledging EO contribution for monitoring purposes the Group on Earth Observations (GEO) was established in February 2005 by the Third Earth Observation Summit in Brussels. GEO is a global network that connects government institutions, academic and research institutions, data providers, businesses, engineers, scientists and experts to create innovative solutions to address global challenges at a time of data plethora [29].

The GEO community is creating a Global Earth Observation System of Systems (GEOSS) in order to integrate observing systems in a better manner while using common standards for data sharing; it also links the existing and forthcoming Earth observation systems while supporting the development of new ones. The amount of open data resources excides 400 million and are contributed from more than 150 national and regional providers such as NASA and ESA; international organizations such as WMO and the commercial sector such as Digital Globe.

In September 2015, the United Nations (UN) General Assembly adopted the 2030 Agenda for Sustainable Development that entails a plan of action for people the planet and prosperity. The 2030 Agenda features 17 Seventeen Sustainable Development Goals (SDGs) with 169 related targets and indicators which among others calls for public-private cooperation to exploit data from EO technologies [30]. Many of these goals either directly or indirectly concern land and soil and therefore their achievement is related to healthy soils and a sustainable land use. Specifically, SDG two – "*End hunger, achieve food security and improved nutrition and promote sustainable agriculture*" - is closely related to soil preservation. Restoring soil productivity and ecosystem functions in these contexts requires new ways of managing soil fertility such as innovative forms of precision agriculture that consider the diversity, heterogeneity and dynamics of smallholder farming systems. SDGs three and six are related to health issues. By protecting and enhancing the ability of the Earth's soils to provide clean water in sufficient quantities for humanity, ecosystems and agriculture will be a key element for full filling these SDGs. The importance of soils in sullying, storing and filtering water is explained in Sect. 2. SDG 13 – "*Take urgent action to combat climate change and its impacts*" – is directly linked with soils as they are both affect and affected by climate change. It is suggested that improvements in soil management could result in reducing greenhouse gas (GHG) emissions or increase soil carbon stock. Soil management can therefore be used as a climate mitigation option [31]. A step forward is to propose new SDG indicators based on new Essential Variables (EV). For example regarding Target 15.3 "*By 2030, combat desertification, restore degraded land and soil, including land affected by desertification, drought and floods, and strive to achieve a land degradation-neutral world directly references degraded soil*" has a single indicator 15.3.*1* – "*Proportion of land that is degraded over the total land area*" that does not mention the quality of the soil in terms of soil variables. For that reason, the Essential Agriculture Variables mentioned in Sect. 3.2, should include soil organic carbon and soil moisture together with soil type, soil degradation and soil productivity [32].

3.1 Extracting Soil Information from Earth Observation

Accurate and timely information on soil's condition at global or even local scale is a challenging task for soil scientists. Despite the importance of such data there are few countries with the required monitoring programs that are needed to achieve current needs while many existing data sets are out of date. For example, the Harmonized World Soil Database [33] that provides information on world soil resources at approximately 1 km scale was last updated in 2013 and notes that the reliability of the information in the database is variable. In the same line, the best available estimation of SOC stocks at the country level, The Global Soil Organic Carbon Map [34] has limitations as data provided by some countries are outdated. Therefore, specific issues that arise concern the accuracy of the data, the date the data were acquired and their spatial reference. The use of remote sensing technologies will provide information for inaccessible areas to improve incomplete spatial and thematic coverage of current regional and global soil databases. In this regard, optical remote sensing observations and in particular reflectance spectroscopy at the remote sensing scale, referred to as imaging spectroscopy (IS), or hyperspectral imaging, have shown to be powerful techniques for the quantitative determination and modelling of a range of soil properties. IS has been used since more than 20 years in various soil applications such as evaluation and monitoring of soil quality and soil function (e.g., soil moisture and carbon storage), soil fertility and soil threats (e.g., acidification and erosion) and soil pedogenesis (i.e., soil formation and evolution). Further, soil degradation (salinity, erosion and deposition), soil mapping and classification, soil genesis and formation, soil contamination and soil hazards (swelling soils) are also important soil science issues nowadays examined with IS, enlarging the soil spectroscopy into the spatial domain from airborne to satellite platforms.

Soil analysis is very important as it provides valuable information for agriculture and environmental monitoring. However, conventionally is performed by wet chemical analysis that is time consuming and costly while consumes a lot of chemical products. As [35] refer in their study approximately 1.5 billion hectares are used globally as agriculture area and if two samples in two depths were gathered for every 5 ha this will resulted in 600 million soil samples. If for example these samples were measured for Soil Organic Matter (SOM), by the wet combustion method, the reagents used for such analysis would have a great ecological impact without proper disposal. In addition to that, the time needed for soil analysis is prohibitive for precision agriculture applications. What is more is that soil analysis is not free from errors as it could be affected by various factors regarding the reagents used and the process itself [36], while it was also observed that the accuracy of the results could be different among different laboratories or even at the same laboratory [37]. For that reason, alternatives of soil routine analysis started to be investigated such as remote and proximal sensing techniques in the VNIR- SWIR (Sort Wave Infrared) spectral region that are time and cost effective while environmental friendly and can predict several soil properties from one measurement [38].

Earth observation according to the Group on Earth Observations (GEO 2016), is the gathering of information about planet Earth's physical, chemical and biological systems with the use of remote sensing instruments (e.g., onboard unmanned aerial vehicles (UAVs), drones, aircraft, and satellites), among others [39]. Because soil is characterized as a complex system with various constituents which contribute to variability in its physical and chemical composition over time and space, earth observation techniques could provide the ability of clustering several soil properties with a single measurement by utilizing data mining techniques [40]. The analysis is based on searching the interaction between the electromagnetic radiation and the active chemical groups also referred as chromophores [13]. For example different clay types in soil have shown distinct spectral signatures particularly in the short-wave infrared region due to strong absorption of the SO_4^{2-},CO_3^{2-} and OH^- and combinations of fundamental features of H_2O and CO_2 for example [41]. In general electronic absorptions are highly associated with Fe-minerals such as hematite, biotite, goethite, and olivine that also give distinctive colours in soil [42]. Despite the strong relationship between soil chromophores and the chemical and physical characteristics of a material, this relationship is not straightforward due to the multivariate and overlapping behaviour of spectral data [43].

Soil reflectance data can be acquired in various conditions i.e., in the laboratory, in the field and from the air and space. Laboratory measurements provide the most controlled conditions not being affected by variations in the viewing angle, illumination changes, soil moisture, soil roughness and ambient conditions that occur with in situ measurements. Spectral data acquisition from air and space has added hindrances as measurements are also affected by the low signal to noise ration and atmospheric attenuations [44]. Laboratory based measurements enable to understand soil spectra physical and chemical principles while field spectroscopy is evolving to become a common tool for soil monitoring. Regarding the development of air and space born sensors bandwidth has significantly decreased with the advent of hyperspectral remote sensing and the forthcoming imagers will provide a lot of potentials in soil mapping [45]. The accuracy and field of application of these platforms depends on their spectral, spatial and temporal resolution. Concisely, satellites offer high spatial coverage but with lower spectral resolution than laboratory and in situ measurements that are also affected from external factors. However despite the poorer quantitative interpretation of satellite data they provide significant assistance for environmental monitoring in large scales [46].

3.2 Estimating Essential Agricultural Variables with EO Techniques the Cases of Soil Organic Carbon and Soil Moisture

The remote sensing community has recently been introduced to the concept of essential variables that refer to a minimum set of variables that are necessary for determining a systems state and development. In the light of the need for monitoring and reporting about earth's ecosystems condition, the adoption of a commonly accepted essential set of parameters to succeed on achieving national and global sustainable goals is of great importance. In addition to that, the number of sensors that share the same characteristics is large and therefore there is a need for data harmonization and standardization in order to provide consistent and comparable information in a more comprehensive manner [47]. Regarding the agriculture sector the GEOGLAM initiative strengthens the ability to produce and distribute relevant, timely and accurate information using EO data at regional, national, and global level. In order to track essential aspects of agricultural production GEOGLAM suggested the adoption of Essential Agricultural Variables (EAVs) [48]. Among these soil moisture content and soil organic carbon/matter are considered very important and hence are the most studied.

Soil moisture (SM) is a critical environmental parameter that is considered essential for agriculture monitoring because it links the energy and water exchange between the atmosphere and land surface and also provides information about groundwater conditions. SM is used in regional water cycle studies, irrigation management in agriculture, climate change and environmental monitoring and therefore it is an indispensable variable in terrestrial research [49]. Many approaches have been developed based on optical and thermal infrared remote sensing, by establishing relationships between SM and soil reflectance or surface temperature, vegetation coverage and soil thermal properties [50]. As it was noticed by [51], the reflectance of water absorption bands is different from that of non-absorption bands and is used to establish the relationship between soil moisture. The general observation after experiments is that reflectance decreases with increasing SM content. Since then, many studies have focused in the specific matter and have established empirical relationships between soil reflectance and SM. For example, [52] found that the albedos of all dry soils were approximately two times greater than those of wet soils, except for sandy soils [53]. Analyzed the water absorption values in the near-infrared region and accurately estimated the SM of a large number of soil samples. Estimating soil moisture from the reflectance measurements could also be used prior estimating a soil property, to reduce its effect in the spectral signature.

Soil organic matter consists of organic constituents found in soil in various stages of decomposition such as tissues from dead plants and animals, materials less than 2 mm in size, and soil organisms. SOM turnover plays a critical role in soil ecosystem functioning and global warming. SOM is crucial for soil structure stabilization, retention and release of plant nutrients and the maintenance of water-holding capacity. For these reasons it is characterized as a key indicator not only for agricultural productivity, but also for environmental resilience [1]. It was observed that soils' reflectance at specific wavelengths could be correlated with organic components (cellulose, lignin, starch). SOC is a constituent of SOM and generally affects spectral reflectance in the visible region, mainly at 550–700 nm due to change in soil colour, as the concentration increases the soil appears darker [54] and narrow peaks between 2100–2300 nm [55]. Specifically, [40] found that at the regions of 1400–1900 nm the OH^- groups have strong absorption features mainly due to soil water content, hydroxyls and clay content.

3.2.1 Approaches for Soil Moisture and Soil Organic Carbon/Matter Estimation Using EO Techniques

An empirical method involves the use of statistical data generated from the EO measured data and the ground-based variable without being a well understood causal relationship. However these methods may not be reliable when they are used outside the conditions under which the relationship was established [56]. Soils reflectance in the VNIR- SWIR region has many overlapping absorptions caused by its constituents or by the soil structure itself and for that reason is characterized largely non-specific. In order to interpret these complex absorption patterns and correlate them with soil properties we need to use various multivariate calibrations [57]. Primary, the scientific community uses linear regression to explain the relationships between soil reflectance and soil properties, while the most commonly used were and still are partial least squares regression (PLSR), principal component analysis (PCA) and multiple linear regression (MLR) among others [40, 53]. These techniques can deal with data comprised of large numbers of predictor variables that are highly collinear.

The first studies were performed in controlled laboratory conditions and therefore the main concerning issue, apart from system configuration, was data processing. In addition to that and as already mentioned, spectral data are also affected by various components in the soil, some directly connected to the estimated soil property, and some not. Thus, the application of preprocessing algorithms on the spectra prior to developing the model can amplify relevant spectral features and can remove variability from light scattering effects. These techniques can be categorized in scatter correction methods and spectral derivatives [58]. Authors like [59–61] evaluated the effect of several pre-processing techniques and concluded that they can improve the obtained prediction accuracy. The most commonly used are the Savitzky-Golay first derivative, standard normal variates (SNV) and the continuum removed reflectance (CRR) among others. Feature selection is also a very common practice to reduce the complexity of the spectral data and correlate soil properties with specific important wavelengths [62]. The selection of the most suitable pre-processing technique should be related to the subsequent modelling procedure. However, including additional factors to a model could have adverse effects by increasing its complexity and hence its robustness for future applications. There is also the case of applying the wrong type or too severe preprocessing technique that could remove important spectral information. However, it is difficult to determine the suitability of a preprocessing technique before model validation [58]. Nevertheless spectroscopic laboratory measurements, similar to conventional analytical methods, need to be applied in a commonly accepted manner by the scientific community to allow comparison between them and for that reason [63] suggested the use of common standards and protocols.

Calibration methods are evaluated not only for their performance but also for their interpretability and computational time. When referring to regression in general we include all these methods that attempt to fit a model to observed data to quantify the relationship between the response and predictor variables or to predict new values. To that end, PLSR is the most frequently used linear multivariate statistical method mainly in the early years of soil spectroscopy [64–67] because it relates the response and predictor variables so that the model explains more of the variance in the response with fewer components, it is more interpretable, and the algorithm is computationally faster [68]. Meanwhile, the use of non-linear data mining techniques has emerged in the last decade due to the rapid development in information and computer technology that has subsequently led to increase in data volume. Despite these advances there are still efforts to describe the complex relationships between these data, while the final goal is to use these models in remote sensing applications [69]. Neural Networks (NN) [70, 71], Multivariate Adaptive Regression Splines (MARS) [72, 73], Support Vector Machines (SVM), Random Forests [74, 75], and the Cubist [76] are some of the most frequently used.

Numerous studies from early years have attempted to model soil's reflectance with soil moisture content. Laboratory based methods have been developed either to estimate or correct soil moisture from soil reflectance. However, there are various approaches found in the literature considering the artificial wetting procedure in order to acquire different values of soil moisture content. A first approach is to rewet soil samples until they reach a near to saturation point (gravimetric soil moisture content $\theta g \approx 50\%$), and perform the measurements each day until the soil is in dry state ($\theta g < 10\%$) [77]. A second approach is to apply known volumes of water in the soil samples and then perform the measurements [78] and the last and most time consuming approach is with the use of a pressure plate extractor to adjust soil moisture to different levels under different designed potential levels [79]. There are also approaches that used spectral indices for SM estimation. Specifically, [80] developed the Normalized Difference Soil Moisture Index (NSMI) that was based on two reflectance values 1800 nm and 2119 nm because it was found that the SWIR spectral region is the most promising for deducing soil moisture quantities from a combination of surface reflectance. Spectral indices could pave the path for rapid in field and remote sensing applications for soil moisture estimation [81]. Proposed two new spectral indices that are less sensitive to the quality of atmospheric correction like NSMI and WISOIL [82]; the Normalized Index of NSWIR domain for SM estimation from Linear correlation (NINSOL) and Normalized Index of NSWIR domain for SM estimation from Nonlinear correlation (NINSON). These indices are based in the 2000–2200 nm spectral region which is sensitive to soil texture variations. Recently, [83] have developed, a multilayer radiative transfer model of soil reflectance (MARMIT) to estimate soil water content which is an improvement of the Bach model [84], that could be applied on imaging spectroscopy data in the laboratory, but not yet in remote sensing data.

The use of airborne sensors has also been evaluated for SM estimation from early years for example [85] aimed to predict soil water content of a partially irrigated field using the AVIRIS airborne sensor with sufficient accuracy ($R^2 = 0.88$) [86]. used the Airborne Prism Experiment (APEX) to collect airborne hyperspectral data by developing a new SM index similar to the existing NINSOL index that has application to airborne and in field data. They concluded that an index requires wavelength dependent, non-linear compensation which could be based on an external spectral SM dataset.

Satellite EO sensors for SM retrieval can be categorized to optical and thermal EO techniques, passive and active microwave – based methods and there is also the synergistic use of different types of EO data. Although these methods have proven the direct relationship of reflectance with SM or the indirect with vegetation spectral indices [87] optical remote sensing has limitations due to its inability to penetrate clouds and vegetation canopy while it is also affected by atmospheric conditions [88]. The use of thermal infrared based methods is more attractive as there are several studies that relate thermal inertia with SM [89–91]. These methods rely on land surface temperature (LST) that is sensitive to SM content due to its impact on the surface heating process mainly on bare soil or sparse vegetation cover conditions [92]. Soil moisture retrieval using microwave sensors is based on the principle that the dielectric properties of soil-water mixture strongly affect the soil scattering and emission in the respective frequency range. Therefore a change in soil moisture will result in a measurable variation of the soil scattering and emission because the dielectric constant of the water is much higher than that of dry soil at the microwave bands [93]. In addition to that, microwave sensors are not limited by cloud cover and observations can be made any time of the day as they are independent of solar illumination. The influence of soil moisture is most prominent at low frequencies (~10–1 GHz) while longer wavelengths allow for deeper penetration into the soil and reduce the influence of vegetation in attenuating soil's moisture signal. Although active sensors provide higher spatial resolution than passive sensors, they are more affected by the local topography, soil roughness and vegetation cover [94]. The high spatial and temporal resolution of the spaceborne Synthetic Aperture Radar systems (SAR) has made them attractive for hydrological applications at regional scale [95]. suggested that L-band backscattering data could provide information on the spatial distribution of SM content, while the C- band data could be proven grossly inadequate.

In the absence of detailed data regarding soil roughness, empirical models could be a viable approach for mapping SM content in flat areas with low vegetation cover. RADARSAT −1 satellite has the ability to acquire data at different incident angles that provides a better solution for separating SM and roughness signals [96]. exploited this multi angular approach using existing theoretical and empirical backscatter models and concluded that the Modified Dubois Model gave better results than the Oh Model. The Soil Moisture and Ocean Salinity (SMOS) proved to be very satisfying for global soil moisture retrieval based on L – band measurements for cases with low vegetation and low radio frequency interferences [97]. However, the spatial resolution of the abovementioned satellites is not adequate for SM estimation at field scale. Sentinel 1 was launched in 2014 and is the first Earth observation satellite to be built for Europe's Global Monitoring for Environment and Security (GMES) programme with a spatial resolution of 10 m. Within the framework of the ESA-funded Sentinel-1 Soil Moisture Algorithm Development (S1-SMAD) project, an algorithm has been proposed and validated for its practicability for deriving near-operational soil moisture content estimates. The algorithm was based on the inversion of an analytical electromagnetic model through an ANN and intensively validated considering several configurations taking into account the S1 acquisition modes. NDVI was also considered to be a key factor in determining a reliable SM content estimation, when only VV polarization is available [98].

To explore the potential of different sensors, [99] suggested that the synergist use of S1 and Sentinel −2 could reduce the uncertainty that is caused by vegetation cover and proposed multitemporal change detection approaches to address soil roughness effects. Similarly, [100] presented an innovative synergistic method combining S1 microwave and Landsat-7/8 (L7/8) thermal data. The methodology was based on the thermal-derived soil evaporation efficiency to calibrate the radar backscatter and SM relationship. The comparison of S1 VV- and VH-polarized data, showed that the VV backscatter coefficient is more sensitive to SM variation, and the combined use of the two sensors provided better results for SM estimations, paving the path for further evaluation of the synergies between radar and optical data. The Sentinel-1 C band backscatter coefficient (σ^{o}) is complementary to Soil Moisture Active and Passive (SMAP) L band, brightness temperature (TB) observations. Taking into account that SMAP observations provide higher sensitivity in soil moisture estimation on large scales while Sentinel-1 has better spatial resolution, joint Sentinel-1 and SMAP data assimilation was found to moderately improve soil moisture estimates [101].

The effects of different SOM content in spectral signature were observed over 40 years ago [102] and since then many studies have moved towards SOM/SOC estimations using laboratory soil spectroscopy [53, 71, 103]. These efforts have tried to find the appropriate multivariate statistical method to model soil properties [104] using mostly the PLSR technique due to its low computational time and interpretability. However studies showed that the relationship between soil properties and soil reflectance was not always linear and therefore the progress made in machine learning algorithms has enabled their use in the respective field of application showing most of the times better accuracy from linear methods [60, 105–107].

After the proven ability of the use of soil reflectance spectroscopy for soil properties estimation, in situ soil sensors have been created to be used as for handheld measurements or on board of vehicles. The various prototypes that have been developed are either now used commercially [108–110] or are still under development. Studies have shown that, specifically for SOC/SOM estimations, the measurements are affected by various factors. Apart from the uncontrolled external factors that were tried to be addressed by utilizing a closed chamber [111] there was also found that the presence of vegetation cover, straws [112], stones and soil moisture content [113] also affect the measurements. For that reason, it was preferred and suggested to make the measurements at seed bed level and with low moisture content and vegetation cover. The results are also promising, still external factors should be addressed either with the use of algorithms such as external parameter orthogonalization (EPO) and direct standardization (DS) [114] or with better sensor configuration.

Remote sensing applications either from air or space have also been evaluated for SOC/SOM estimation. The use of airborne hyperspectral sensors provide the ability for large areas coverage because a single flight can have adequate duration and the flight could be also be scheduled according to weather conditions [115]. However, the flight cost does not allow very frequent operations. Although these measurements are affected by ambient conditions and need atmospheric corrections the results from their application are quite promising for digital soil mapping [116]. It was also reported that soil heterogeneity also affects measurements because the results could vary among different soil types: therefore it was suggested that splitting the dataset according to soil type, image number and region could be a possible solution [117]. One of the main factors that affect prediction models from airborne data is the vegetation cover. Since it is difficult to find large areas with bare soils, many efforts have been made to address this issue [118]. tried to estimate the bare soil fractional cover of an area however, the results were poor and significant information about some areas' SOC content was lost due to the high vegetation cover. Another proposed technique to remove vegetation influence from mixed pixel was the Residual Spectral Unmixing (RSU) [119]. This technique gave the ability to identify SOC field variations though in some cases SOC was over or underestimated. To increase the bare soil sample area, [120] utilized the Iterative Spectral Mixture Approach. Despite the 45.4% increase of the sample area results were still poor. Considering that certain soil properties do not change over a short period of time, [121] used crop rotation to create multitemporal composites of airborne data again the low prediction of SOM strengthen the fact that there are multiple parameters that affect soil properties estimation like soil roughness which was also reported from [122]. When it comes to utilizing in situ spectral measurements combined with airborne hyperspectral data, the acquisitions should be performed in close dates as suggested by [123]. However, to develop the prediction models soil sampling and analytical laboratory measurements are needed. Hence [55] aimed to make the most of the already developed soil spectral libraries and proposed a bottom up approach. The concept of this approach is that instead of using chemically analyzed soil samples as independent variables a representative of the studied area subset from a large soil spectral library is utilized and the results were relative good.

The use of spaceborne EO data for soil properties estimation could be enhanced with the advent of the new (PRISMA) and forthcoming hyperspectral satellite sensors (EnMAP, HyspIRI) [45]. Nonetheless there are efforts utilizing satellite data for SOM/SOC estimations [16]. was one of the first studies that evaluated the potential of Hyperion hyperspectral sensor for SOC estimation and concluded that the predictions were similar to those used in field spectral measurements. They also reported that the percentage of SOC content affected the model's accuracy that was reduced at levels of SOC below 1% [124]. also investigated the potential of the multispectral Advanced Land Imager, (ALI) and Hyperion satellite sensors for SOM estimation with the latter presenting slightly lower RMSE. Vegetation cover should also be addressed because the spatial resolution of satellite sensors results in mixed pixels. To increase the bare soil area, [125] proposed the Geospatial Soil Sensing System (GEOS3) to develop a bare soil composite from Landsat 5 TM time series. Multi-temporal data from two optical sensors, RapidEye and Landsat 8, were also used in the study of [126] who reported a correlation between SOC content and elevation while June was the month were better accuracy was observed.

3.2.2 Estimating Other Agronomic Variables (pH, Clay, and Others)

In the VNIR-SWIR region, many minerals show diagnostic absorption bands due to vibrational overtones, electronic transitions, charge transfer, and conduction processes [127]. As already mentioned, different clay types have distinct spectral signatures mainly in the SWIR region therefore soil clay content estimation has been well studied. Extracting reflectance values from remote sensing approaches is a more complex process than laboratory measurement; however, if the sensor is sensitive enough and the atmospheric effects can be properly removed from the original data, this technique could be of great use for rapid quantitative mapping over large areas [128]. Conducted one of the first studies that evaluated the capability of the airborne DAIS-7915 (0.4–14 μm) sensor for soil properties estimation and reported that electrical conductivity (EC), although a featureless property, could be spectrally explained from field moisture content. The HyMAP sensor (400–2500 nm) was used by [129] for clay and $CaCO_3$ estimation. They reported that scaling up laboratory measurements to airborne observations poses uncertainties due to lack of inconsistency of airborne spectral measurements and errors in correcting the atmospheric effects. It was also reported that the presence of calcareous pebbles affected the estimates of $CaCO_3$ [130]. Suggested that transferring a model built in laboratory conditions to airborne images could generate better prediction models than those obtained by the conventional model built from only HyMap spectra. In the same line, [131] explored the potential of transfer learning for soil clay mapping using hyperspectral imagery and a pre-trained CNN model developed from a large number of spectra measured in the laboratory and provided results ($R^2 = 0.60$) suggesting a new way to utilize large scale soil spectral libraries and hyperspectral data. Predictions of soil properties from the previous sensor were also computed by [132] for eight soil properties using PLSR.

The results showed the following: four out of the eight soil properties ($CaCO_3$, iron, clay and CEC) were suitable for mapping using hyperspectral data, and both accurate local predictions and good representations of spatial structures were observed; and the application of prediction models using hyperspectral data over the study area provided statistical characterizations within field variations and variograms that describe in details the short range soil variations [133]. assessed the effect of spectral resolution on clay topsoil property estimation by simulating artificial sensors (EnMap. HYPXIM, HyspIRI and HYPERION) and existing multispectral sensors (ASTER, SENTINEL-2 MSI, LANDSAT-7 ETM+ and LANDSAT-8 OLI). All these sensor simulations were based on real airborne hyperspectral VNIR-SWIR data acquired over landscapes at a 5-m spatial resolution (AISA-DUAL hyperspectral sensor). The simulation of sensors allowed to assess the influence of the spectral resolution on the estimated soil property, independently to other specifications (e.g., spatial resolutions, acquisition dates, signal to noise ratio). Analysis of the PLSR model performances highlighted the ASTER spectral configuration may allow for the estimation of clay content and pedological patterns of estimated clay content are preserved with spectral configurations up to 100/100 nm (the first number indicates the spectral resolution in the VNIR region and the second the spectral resolution in the SWIR region) and with the ASTER spectral configuration.

3.2.3 Methods Exploiting Ancillary Information

In most cases measurements are very few and scattered in space and time, therefore estimating a targeted essential variable could be improved by deriving auxiliary information from other related categorical or continuous variables such as land use maps, a digital elevation model and/or remote and proximal sensing data [134]. Similarly, remote sensing data could be also used as auxiliary variables for soil properties prediction from proximal sensing and by this way, the advantage of high spectral resolution of proximal sensing can be combined with the high spatial resolution of remote sensing [135].

Mapping soil properties in densely vegetated areas could, for example, be achieved by exploiting the indirect relations between vegetation and soil attributes. There are several vegetation indices as well as the use of time series to determine possible soil patterns. For that reason, data acquired from satellite sensors could also be used as auxiliary variables for mapping soil properties. Subsequently, the complimentary use of various geostatistical methods combined with spectral data was proven to be more accurate than for example using only ordinary kriging in predicting SOC spatial variability and led to the development of high-quality maps [76, 136]. Biochemistry spectral indices based on wavelengths related to cellulose, starch and lignin could be used for SOC prediction [137]. Mondal et al. [138] also reported the correlation of SOC distribution with indices and other remotely sensed variables i.e., brightness index (BI), greenness index (GI), wetness index (WI), normalized difference vegetation index (NDVI), vegetation temperature condition index (VTCI), digital elevation model (DEM), and slope and compound topographic index (CTI). A direct link was observed by [139] between NDVI and SOC which had a positive linear relationship [140]. Evaluated the potential of Sentinel −2 for SOC estimation using several spectral indices and band combinations and good correlations were found between SOC and B4, B5, B11 and B12 bands for a specific region.

4 Remote Sensing for Soil Monitoring: Limitations and Ways Forward

Despite the several advantages remote sensing has i.e., being a nondestructive method, providing data systematically over large geographical areas, obtaining information about areas that are inaccessible, providing information that can be used in other domains and having large temporal resolution, there are still some limitations towards their establishment as fully acceptable methods [141]. A significant hindrance is the need for consistent evaluation of the accuracy of the acquired data which is difficult since the evaluation depends on reliable and accurate ground measurements which are unavailable in many parts of the world. In addition to that, soil monitoring with conventional methods also entails errors that makes the correlation between physical and spectral measurements more doubtful. There has also been described by many authors an evident decrease in accuracy when laboratory spectral measurements are compared to airborne and spaceborne data acquisitions. The decrease of the accuracy comes as result of the conditions the measurements are being conducted (controlled laboratory conditions versus measurements affected by environmental and atmospheric conditions), sample condition (dried, sieved samples versus undisturbed soil samples with different roughness and soil moisture content, crust, vegetation cover), and differences in spectral and spatial resolution [115, 128, 142].

Despite that, there is an increasing availability of handheld optical sensors, unmanned aerial vehicles, and high-resolution satellite imagery for precision agriculture applications that present new opportunities for soil monitoring. Handheld reflectance sensors have been used to show differences in growth trajectories under different tillage and irrigation schemes in wheat [143], a method that could be adapted to monitor resistance and recovery for other disturbances and systems. Furthermore, the increasing popularity of thermal imaging and multispectral or infrared sensors in high-throughput phenotyping for crop breeding suggests similar applications for near-real-time detection of crop stress and recovery. Integrated use of multiple remote sensing sources and increased remote sensing capacity can help overcome many of these known challenges, as long as data and product requirements are clearly identified. The prioritization of new satellite missions associated with freely accessible data for scientific use might indeed be facilitated by the formulation of clear, consensual demands from ecosystem researchers [144].

5 Conclusions and Recommendations

The efficiency of soil spectroscopy has been largely investigated in the last decades with promising results. Remote and proximal sensing techniques could be used as primary or auxiliary data sources for soil properties estimation, depending on their spectral, spatial and temporal resolution and field of application. Advances in spectral, spatial and temporal resolution of satellite sensors has turned the scientific interest towards their utilization and provided the opportunity for more accurate soil properties estimation. There are many studies that developed different methodologies for estimating soil parameters. Soil moisture estimation could be derived from the direct relationship between soli reflectance and different soil moisture contents or indirectly evaluating vegetation traits related to water stress. Additionally, there is the use of the thermal region of the spectrum which correlates soil moisture with the thermal properties of soil.

The most frequently used approaches are the thermal inertia and the temperature index method. However, the synergistic use of the VNIR-SWIR and thermal region could potentially provide more accurate results for SM estimations. Specifically, for satellite observations, the use of spectral indices such as NDVI have been used as proxies in vegetated areas either for direct estimation of SOC or to derive bare soil areas using time series. There are several efforts to estimate soil properties directly from their spectral signature due to their characteristic correlations with specific parts of the spectrum. These correlations are commonly described with linear statistical techniques such as PLSR. As another option, data mining techniques that have the ability to explain non-linear relationships, have also been evaluated for soil properties estimation (SVM, ANN, Random Forests, and Cubist). Although these techniques have shown better performance from conventional statistical methods, they have the drawback of needing specific knowledge for their implementation and interpretation while needing a significant amount of data that are usually unavailable. Therefore, simpler models are recommended for small datasets. Feature selection should also be performed based on reliable predictors that will provide accurate information about a model's interpretability in order to understand soils relationship with electromagnetic radiation. Machine learning techniques have still lower interpretability compared to linear models for the reason that the latter have been well studied for longer period of time. Therefore, it is recommended, studies using machine learning techniques focus on interpreting the results rather than only presenting their accuracy.

Technological advances in the field of agriculture include the development and use of different soil sensors. These innovations have given farmers information that allowed them to optimize their management capabilities, enhance their knowledge about a fields condition while applying sustainable practices that lead to soil preservation and environmental protection. It could increase farmer's power by allowing site specific applications according to crop type, soil type and environmental conditions with the ability to reduce inputs while maintain their yield production. Due to various spatial resolutions soil sensing offers, the acquired information could also be used by different scientific disciplines towards the development of new products and solutions while help in the implementation of new policies for soil protection, sustainable development, food security, nutrient and water managements and climate change mitigation and adaptation. Governments could also use data from remote and proximal sensing technologies in order to make important decisions about the policies they will adopt and how to work on national issues concerning agriculture.

Currently the main issues regarding the use of hyperspectral satellite sensors for soil applications concentrate on atmospheric corrections and sensor errors, the low signal to noise ratio, lower spectral resolution compared to laboratory spectroradiometers and the absence of commonly accepted methodologies for soil properties estimations. To that end, in order to address these issues, there is a need for accurate and reliable data acquisition both for the sensors and for ground truth data, and international collaboration and development of commonly accepted protocols for processing the respected data. The volume of remote and proximal sensed data is large and therefore big data technologies should be further developed for the specific purposes. By these it is hoped to achieve quasi real time predictions that could have global, national, regional and farm applications. Towards systems automation there is a need for tools development that will enable rapid and accurate processing and visualization of the data with an easily interpretable manner that will in turn provide information directly to the farmer or policy makers. Building networks specifically for agricultural purposes will enable a constant data flow that will help to coordinate efforts for solving problems related to model's accuracy and further spatial analysis of agricultural parameters. For that reason, data interoperability will provide systems and services to have a clearer view about the context and meaning of the data.

To achieve the use of EO data in the field of agriculture for soil properties estimations in a more efficient and manner interdisciplinary collaboration between soil scientists and the earth observation community is suggested as these scientific fields are inextricably linked and should work complementary. Knowledge must be shared through the education of scientists towards the advances in each field. Data should be easily accessible and at low cost or even freely available. Data processing and integration should be established for communities to benefit the most of EO technologies for achieving the sustainable development goals.

References

1. FAO (2017) Soil Organic Carbon the Hidden Potential
2. Lampridi M, Sørensen C, Bochtis D (2019) Agricultural Sustainability: A Review of Concepts and Methods. Sustainability 11:5120. https://doi.org/10.3390/su11185120

3. Blum WEH (2005) Functions of Soil for Society and the Environment. Rev Environ Sci Bio/Technology 4:75–79. https://doi.org/10.1007/s11157-005-2236-x
4. Oldeman LR, Hakkeling RTA, Sombroek WG (1991) World map of the status of human-induced soil degradation : an explanatory note, 2nd. rev. ed. Wageningen [etc.]:ISRIC [etc.]
5. European Commission (2006) Thematic Strategy for Soil Protection IMPACT ASSESSMENT OF THE THEMATIC STRATEGY ON SOIL PROTECTION
6. United Nations (2016) Transforming Our World: The 2030 Agenda for Sustainable Development
7. Fao FOOD AND AGRICULTURE Key to achieving the 2030 Agenda for Sustainable Development 2
8. Singh BP, Setia R, Wiesmeier M, Kunhikrishnan A (2018) Agricultural Management Practices and Soil Organic Carbon Storage. Soil Carbon Storage 207–244. https://doi.org/10.1016/B978-0-12-812766-7.00007-X
9. Paustian K, Lehmann J, Ogle S, et al (2016) Climate-smart soils. Nature 532:49–57. https://doi.org/10.1038/nature17174
10. McBratney AB, Mendonça Santos ML, Minasny B (2003) On digital soil mapping. Geoderma 117:3–52. https://doi.org/10.1016/S0016-7061(03)00223-4
11. Borzacchiello MT, Craglia M (2011) Socio-Economic Benefits from the Use of Earth Observation
12. Land Monitoring Service-Copernicus
13. Hunt GR (1980) Electromagnetic radiation: The communication link in remote sensing. Remote Sens Geol 5–45
14. Angelopoulou T, Balafoutis A, Zalidis G, Bochtis D (2020) From Laboratory to Proximal Sensing Spectroscopy for Soil Organic Carbon Estimation—A Review. Sustainability 12:443. https://doi.org/10.3390/su12020443
15. Stuart BH (2004) Infrared Spectroscopy: Fundamentals and Applications
16. Gomez C, Viscarra RA, Mcbratney AB (2008) Soil organic carbon prediction by hyperspectral remote sensing and fi eld vis-NIR spectroscopy : An Australian case study. Geoderma 146:403–411. https://doi.org/10.1016/j.geoderma.2008.06.011
17. Stoner ER (1980) Atlas of soil reflectance properties. Agricultural Experiment Station, Purdue University
18. Viscarra Rossel R (2009) The Soil Spectroscopy Group and the development of a global soil spectral library. NIR news 20:14. https://doi.org/10.1255/nirn.1131
19. Tóth G, Jones A, Montanarella L (2013) The LUCAS topsoil database and derived information on the regional variability of cropland topsoil properties in the European Union. Environ Monit Assess 185:7409–7425. https://doi.org/10.1007/s10661-013-3109-3
20. Brodský L, Klement A, Penížek V, et al (2011) Building soil spectral library of the czech soils for quantitative digital soil mapping. Soil Water Res 6:165–172
21. Romero DJ, Ben-Dor E, Demattê JAM, et al (2018) Internal soil standard method for the Brazilian soil spectral library: Performance and proximate analysis. Geoderma 312:95–103. https://doi.org/10.1016/J.GEODERMA.2017.09.014
22. Piekarczyk J, Kazmierowski C, Krolewicz S, Cierniewski J (2016) Effects of Soil Surface Roughness on Soil Reflectance Measured in Laboratory and Outdoor Conditions. IEEE J Sel Top Appl Earth Obs Remote Sens 9:827–834. https://doi.org/10.1109/JSTARS.2015.2450775
23. Pierpaoli E, Carli G, Pignatti E, Canavari M (2013) Drivers of Precision Agriculture Technologies Adoption: A Literature Review. Procedia Technol 8:61–69. https://doi.org/10.1016/j.protcy.2013.11.010
24. Liaghat S, Balasundram SK (2010) A Review: The Role of Remote Sensing in Precision Agriculture. Am J Agric Biol Sci 5:50–55
25. Lee WS, Alchanatis V, Yang C, et al (2010) Sensing technologies for precision specialty crop production. Comput Electron Agric 74:2–33. https://doi.org/10.1016/j.compag.2010.08.005
26. Neményi M, Mesterházi PÁ, Pecze Z, Stépán Z (2003) The role of GIS and GPS in precision farming. In: Computers and Electronics in Agriculture. Elsevier, pp 45–55

27. Seelan SK, Laguette S, Casady GM, Seielstad GA (2003) Remote sensing applications for precision agriculture: A learning community approach. Remote Sens Environ 88:157–169. https://doi.org/10.1016/J.RSE.2003.04.007
28. Anderson K, Ryan B, Sonntag W, et al (2017) Earth observation in service of the 2030 Agenda for Sustainable Development. Geo-Spatial Inf Sci 20:77–96. https://doi.org/10.1080/10095020.2017.1333230
29. GEO
30. United Nations (2015) Transforming our world: The 2030 Agenda for Sustainable Development. https://doi.org/10.1163/157180910X12665776638740
31. Keesstra SD, Bouma J, Wallinga J, et al (2016) The significance of soils and soil science towards realization of the United Nations sustainable development goals. SOIL 2:111–128. https://doi.org/10.5194/soil-2-111-2016
32. Masó J, Serral I, Domingo-Marimon C, Zabala A (2019) Earth observations for sustainable development goals monitoring based on essential variables and driver-pressure-state-impact-response indicators. Int J Digit Earth 1–19. https://doi.org/10.1080/17538947.2019.1576787
33. Nachtergaele FA, van Velthuizen HB, Batjes NC, et al (2010) The harmonized world soil database
34. FAO, ITPS (2018) Global Soil Organic Carbon Map (GSOCmap) Technical Report
35. Demattê JAM, Dotto AC, Bedin LG, et al (2018) Soil analytical quality control by traditional and spectroscopy techniques: Constructing the future of a hybrid laboratory for low environmental impact. Geoderma 337:111–121. https://doi.org/10.1016/J.GEODERMA.2018.09.010
36. O' Rourke SM, Holden NM (2011) Optical sensing and chemometric analysis of soil organic carbon - a cost effective alternative to conventional laboratory methods? Soil Use Manag 27:143–155. https://doi.org/10.1111/j.1475-2743.2011.00337.x
37. Cantarella H, Quaggio JA, Van Raij B, De Abreu MF (2006) Variability of Soil Analysis in Commercial Laboratories: Implications for Lime and Fertilizer Recommendations. Commun Soil Sci Plant Anal 37:2213–2225. https://doi.org/10.1080/00103620600817523
38. Minasny B, McBratney AB (2008) Regression rules as a tool for predicting soil properties from infrared reflectance spectroscopy. Chemom Intell Lab Syst 94:72–79. https://doi.org/10.1016/j.chemolab.2008.06.003
39. Politi E, Paterson SK, Scarrott R, et al (2019) Earth observation applications for coastal sustainability: potential and challenges for implementation. Anthr Coasts 2:306–329. https://doi.org/10.1139/anc-2018-0015
40. Ben-Dor E, Banin A (1995) Near-Infrared Analysis as a Rapid Method to Simultaneously Evaluate Several Soil Properties. Soil Sci Soc Am J 59:364–372. https://doi.org/10.2136/sssaj1995.03615995005900020014x
41. Hunt GR (1977) Spectral signatures of particulate minerals in the visible and near infrared. Geophysics 42:501–513
42. Brown DJ, Shepherd KD, Walsh MG, et al (2006) Global soil characterization with VNIR diffuse reflectance spectroscopy. Geoderma 132:273–290. https://doi.org/10.1016/j.geoderma.2005.04.025
43. Schwartz G, Ben-Dor E, Eshel G (2012) Quantitative analysis of total petroleum hydrocarbons in soils: Comparison between reflectance spectroscopy and solvent extraction by 3 certified laboratories. Appl Environ Soil Sci 2012:. https://doi.org/10.1155/2012/751956
44. Rencz AN, Ryerson RA (1999) Manual of Remote Sensing, Remote Sensing for the Earth Sciences. Wiley
45. Castaldi F, Palombo A, Santini F, et al (2016) Evaluation of the potential of the current and forthcoming multispectral and hyperspectral imagers to estimate soil texture and organic carbon. Remote Sens Environ 179:54–65. https://doi.org/10.1016/j.rse.2016.03.025
46. Cécillon L, Barthès BG, Gomez C, et al (2009) Assessment and monitoring of soil quality using near-infrared reflectance spectroscopy (NIRS). Eur J Soil Sci 60:770–784. https://doi.org/10.1111/j.1365-2389.2009.01178.x

47. Patias P, Verde N, Tassopoulou M, et al (2019) Essential variables: describing the context, progress, and opportunities for the remote sensing community. In: Papadavid G, Themistocleous K, Michaelides S, et al (eds) Seventh International Conference on Remote Sensing and Geoinformation of the Environment (RSCy2019). SPIE, p 40
48. Whitcraft AK, Becker-Reshef I, Justice CO, et al (2019) No pixel left behind: Toward integrating Earth Observations for agriculture into the United Nations Sustainable Development Goals framework. Remote Sens Environ 235:111470. https://doi.org/10.1016/j.rse.2019.111470
49. Robinson DA, Campbell CS, Hopmans JW, et al (2008) Soil Moisture Measurement for Ecological and Hydrological Watershed-Scale Observatories: A Review. Vadose Zo J 7:358–389. https://doi.org/10.2136/vzj2007.0143
50. Zhang D, Zhou G (2016) Estimation of soil moisture from optical and thermal remote sensing: A review. Sensors (Switzerland) 16:1308
51. Ångström A (1925) The Albedo of Various Surfaces of Ground. Geogr Ann 7:323–342. https://doi.org/10.1080/20014422.1925.11881121
52. Jackson RD, Idso SB, Reginato RJ (1976) Calculation of evaporation rates during the transition from energy-limiting to soil-limiting phases using albedo data. Water Resour Res 12:23–26. https://doi.org/10.1029/WR012i001p00023
53. Dalal RC, Henry RJ (1986) Simultaneous Determination of Moisture, Organic Carbon, and Total Nitrogen by Near Infrared Reflectance Spectrophotometry1. Soil Sci Soc Am J 50:120. https://doi.org/10.2136/sssaj1986.03615995005000010023x
54. Ladoni M, Bahrami HA, Alavipanah SK, Norouzi AA (2010) Estimating soil organic carbon from soil reflectance: A review. Precis. Agric. 11:82–99
55. Castaldi F, Chabrillat S, Jones A, et al (2018) Soil Organic Carbon Estimation in Croplands by Hyperspectral Remote APEX Data Using the LUCAS Topsoil Database. Remote Sens 10:153. https://doi.org/10.3390/rs10020153
56. Andries A, Morse S, Murphy R, et al (2019) Translation of Earth observation data into sustainable development indicators: An analytical framework. Sustain Dev 27:366–376. https://doi.org/10.1002/sd.1908
57. Geladi P (2003) Chemometrics in spectroscopy. Part 1. Classical chemometrics. Spectrochim Acta Part B At Spectrosc 58:767–782. https://doi.org/10.1016/S0584-8547(03)00037-5
58. Rinnan Å, Berg F van den, Engelsen SB (2009) Review of the most common pre-processing techniques for near-infrared spectra. TrAC Trends Anal Chem 28:1201–1222. https://doi.org/10.1016/J.TRAC.2009.07.007
59. Dotto AC, Dalmolin RSD, ten Caten A, Grunwald S (2018) A systematic study on the application of scatter-corrective and spectral-derivative preprocessing for multivariate prediction of soil organic carbon by Vis-NIR spectra. Geoderma 314:262–274. https://doi.org/10.1016/J.GEODERMA.2017.11.006
60. Peng X, Shi T, Song A, et al (2014) Estimating soil organic carbon using VIS/NIR spectroscopy with SVMR and SPA methods. Remote Sens 6:2699–2717. https://doi.org/10.3390/rs6042699
61. Gholizadeh A., Boruvka L., Saberioon MM., et al (2015) Comparing different data preprocessing methods for monitoring soil heavy metals based on soil spectral features. Soil Water Res 10:218–227. https://doi.org/10.17221/113/2015-SWR
62. Shi T, Chen Y, Liu H, et al (2014) Soil organic carbon content estimation with laboratory-based visible-near-infrared reflectance spectroscopy: Feature selection. Appl Spectrosc 698:831–837. https://doi.org/10.1366/13-07294
63. Ben Dor E, Ong C, Lau IC (2015) Reflectance measurements of soils in the laboratory: Standards and protocols. Geoderma 245–246:112–124. https://doi.org/10.1016/j.geoderma.2015.01.002
64. Bikindou FDA, Gomat HY, Deleporte P, et al (2012) Are NIR spectra useful for predicting site indices in sandy soils under Eucalyptus stands in Republic of Congo? For Ecol Manage 266:126–137. https://doi.org/10.1016/J.FORECO.2011.11.012

65. Rodionov A, Pätzold S, Welp G, et al (2014) Sensing of Soil Organic Carbon Using Visible and Near-Infrared Spectroscopy at Variable Moisture and Surface Roughness. Soil Sci Soc Am J 78:949. https://doi.org/10.2136/sssaj2013.07.0264
66. Jiang Q, Li Q, Wang X, et al (2017) Estimation of soil organic carbon and total nitrogen in different soil layers using VNIR spectroscopy: Effects of spiking on model applicability. Geoderma 293:54–63. https://doi.org/10.1016/J.GEODERMA.2017.01.030
67. Gupta A, Vasava HB, Das BS, Choubey AK (2018) Local modeling approaches for estimating soil properties in selected Indian soils using diffuse reflectance data over visible to near-infrared region. Geoderma 325:59–71. https://doi.org/10.1016/J.GEODERMA.2018.03.025
68. Stenberg B, Viscarra Rossel RA, Mouazen AM, Wetterlind J (2010) Visible and near infrared spectroscopy in soil science. 107:163–215. https://doi.org/10.1016/S0065-2113(10)07005-7
69. Fang Q, Hong H, Zhao L, et al (2018) Visible and Near-Infrared Reflectance Spectroscopy for Investigating Soil Mineralogy: A Review. J Spectrosc 2018:1–14. https://doi.org/10.1155/2018/3168974
70. Goldshleger N, Chudnovsky A, Ben-Dor E (2012) Using reflectance spectroscopy and artificial neural network to assess water infiltration rate into the soil profile. Appl Environ Soil Sci 2012:1–9. https://doi.org/10.1155/2012/439567
71. Rossel RAV, Behrens T (2010) Using data mining to model and interpret soil diffuse reflectance spectra. Geoderma 158:46–54. https://doi.org/10.1016/j.geoderma.2009.12.025
72. Nawar S, Buddenbaum H, Hill J (2015) Estimation of soil salinity using three quantitative methods based on visible and near-infrared reflectance spectroscopy: a case study from Egypt. Arab J Geosci 8:5127–5140. https://doi.org/10.1007/s12517-014-1580-y
73. Zhang W, Goh ATC (2016) Multivariate adaptive regression splines and neural network models for prediction of pile drivability. Geosci Front 7:45–52. https://doi.org/10.1016/J.GSF.2014.10.003
74. Gholizadeh A, Borùvka L, Vašát R, et al (2015) Estimation of potentially toxic elements contamination in anthropogenic soils on a brown coal mining dumpsite by reflectance spectroscopy: A case study. PLoS One 10:1–14. https://doi.org/10.1371/journal.pone.0117457
75. Chen S, She D, Zhang L, et al (2019) Spatial Downscaling Methods of Soil Moisture Based on Multisource Remote Sensing Data and Its Application. Water 11:1401. https://doi.org/10.3390/w11071401
76. Wang B, Waters C, Orgill S, et al (2018) High resolution mapping of soil organic carbon stocks using remote sensing variables in the semi-arid rangelands of eastern Australia. Sci Total Environ 630:367–378. https://doi.org/10.1016/J.SCITOTENV.2018.02.204
77. Peng Y, Xiong X, Adhikari K, et al (2015) Modeling soil organic carbon at regional scale by combining multi-spectral images with laboratory spectra. PLoS One 10:. https://doi.org/10.1371/journal.pone.0142295
78. Roudier P, Hedley CB, Lobsey CR, et al (2017) Evaluation of two methods to eliminate the effect of water from soil vis–NIR spectra for predictions of organic carbon. Geoderma 296:98–107. https://doi.org/10.1016/j.geoderma.2017.02.014
79. Nocita M, Stevens A, Noon C, van Wesemael B (2013) Prediction of soil organic carbon for different levels of soil moisture using Vis-NIR spectroscopy. Geoderma 199:37–42. https://doi.org/10.1016/J.GEODERMA.2012.07.020
80. Yin Z, Lei T, Yan Q, et al (2013) A near-infrared reflectance sensor for soil surface moisture measurement. Comput Electron Agric 99:101–107. https://doi.org/10.1016/J.COMPAG.2013.08.029
81. Haubrock SN, Chabrillat S, Lemmnitz C, Kaufmann H (2008) Surface soil moisture quantification models from reflectance data under field conditions. Int J Remote Sens 29:3–29. https://doi.org/10.1080/01431160701294695
82. Fabre S, Briottet X, Lesaignoux A, et al (2015) Estimation of Soil Moisture Content from the Spectral Reflectance of Bare Soils in the 0.4–2.5 μm Domain. Sensors 15:3262–3281. https://doi.org/10.3390/s150203262

83. Whalley WR, Leeds-Harrison PB, Bowman GE (1991) Estimation of soil moisture status using near infrared reflectance. Hydrol Process 5:321–327. https://doi.org/10.1002/hyp.3360050312
84. Bablet A, Vu PVH, Jacquemoud S, et al (2018) MARMIT: A multilayer radiative transfer model of soil reflectance to estimate surface soil moisture content in the solar domain (400–2500 nm). Remote Sens Environ 217:1–17. https://doi.org/10.1016/j.rse.2018.07.031
85. Bach H, Mauser W (1994) Modelling and model verification of the spectral reflectance of soils under varying moisture conditions. In: International Geoscience and Remote Sensing Symposium (IGARSS). IEEE, pp 2354–2356
86. Diek S, Chabrillat S, Nocita M, et al (2019) Minimizing soil moisture variations in multi-temporal airborne imaging spectrometer data for digital soil mapping. Geoderma 337:607–621. https://doi.org/10.1016/J.GEODERMA.2018.09.052
87. Wang H, Li X, Long H, et al (2010) Monitoring the effects of land use and cover type changes on soil moisture using remote-sensing data: A case study in China's Yongding River basin. Catena 82:135–145. https://doi.org/10.1016/j.catena.2010.05.008
88. Petropoulos GP, Ireland G, Barrett B (2015) Surface soil moisture retrievals from remote sensing: Current status, products & future trends. Phys. Chem. Earth 83–84:36–56
89. Minacapilli M, Iovino M, Blanda F (2009) High resolution remote estimation of soil surface water content by a thermal inertia approach. J Hydrol 379:229–238. https://doi.org/10.1016/j.jhydrol.2009.09.055
90. Lu S, Ju Z, Ren T, Horton R (2009) A general approach to estimate soil water content from thermal inertia. Agric For Meteorol 149:1693–1698. https://doi.org/10.1016/j.agrformet.2009.05.011
91. Taktikou E, Bourazanis G, Papaioannou G, Kerkides P (2016) Prediction of Soil Moisture from Remote Sensing Data. Procedia Eng 162:309–316. https://doi.org/10.1016/J.PROENG.2016.11.066
92. Zhao W, Li Z-L (2013) Sensitivity study of soil moisture on the temporal evolution of surface temperature over bare surfaces. Int J Remote Sens 34:3314–3331. https://doi.org/10.1080/01431161.2012.716532
93. Jackson TJ, Schmugge J, Engman ET (1996) Remote sensing applications to hydrology: soil moisture. Hydrol Sci J 41:517–530. https://doi.org/10.1080/02626669609491523
94. Kumar S V., Dirmeyer PA, Peters-Lidard CD, et al (2018) Information theoretic evaluation of satellite soil moisture retrievals. Remote Sens Environ 204:392–400. https://doi.org/10.1016/j.rse.2017.10.016
95. D'Urso G, Minacapilli M (2006) A semi-empirical approach for surface soil water content estimation from radar data without a-priori information on surface roughness. J Hydrol 321:297–310. https://doi.org/10.1016/j.jhydrol.2005.08.013
96. Sahebi MR, Angles J (2010) An inversion method based on multi-angular approaches for estimating bare soil surface parameters from RADARSAT-1. Hydrol Earth Syst Sci 14:2355–2366. https://doi.org/10.5194/hess-14-2355-2010
97. Kerr YH, Waldteufel P, Richaume P, et al (2012) The SMOS soil moisture retrieval algorithm. IEEE Trans Geosci Remote Sens 50:1384–1403. https://doi.org/10.1109/TGRS.2012.2184548
98. Paloscia S, Pettinato S, Santi E, et al (2013) Soil moisture mapping using Sentinel-1 images: Algorithm and preliminary validation. Remote Sens Environ 134:234–248. https://doi.org/10.1016/j.rse.2013.02.027
99. Hajj M El, Baghdadi N, Zribi M, Bazzi H (2017) Synergic use of Sentinel-1 and Sentinel-2 images for operational soil moisture mapping at high spatial resolution over agricultural areas. Remote Sens 9:. https://doi.org/10.3390/rs9121292
100. Amazirh A, Merlin O, Er-Raki S, et al (2018) Retrieving surface soil moisture at high spatio-temporal resolution from a synergy between Sentinel-1 radar and Landsat thermal data: A study case over bare soil. Remote Sens. Environ. 211:321–337

101. Lievens H, Reichle RH, Liu Q, et al (2017) Joint Sentinel-1 and SMAP data assimilation to improve soil moisture estimates. Geophys Res Lett 44:6145–6153. https://doi.org/10.1002/2017GL073904
102. Krishnan P, Alexander JD, Butler BJ, Hummel JW (1980) Reflectance Technique for Predicting Soil Organic Matter1. Soil Sci Soc Am J 44:1282. https://doi.org/10.2136/sssaj1980.03615995004400060030x
103. Shepherd KD, Walsh MG (2002) Development of reflectance spectral libraries for characterization of soil properties. Soil Sci Soc Am J 66:988–998. https://doi.org/10.2136/sssaj2002.0988
104. Stevens A, Nocita M, Tóth G, et al (2013) Prediction of Soil Organic Carbon at the European Scale by Visible and Near InfraRed Reflectance Spectroscopy. PLoS One 8:. https://doi.org/10.1371/journal.pone.0066409
105. Vohland M, Besold J, Hill J, Fründ H-C (2011) Comparing different multivariate calibration methods for the determination of soil organic carbon pools with visible to near infrared spectroscopy. Geoderma 166:198–205. https://doi.org/10.1016/J.GEODERMA.2011.08.001
106. Liakos K, Busato P, Moshou D, et al (2018) Machine Learning in Agriculture: A Review. Sensors 18:2674. https://doi.org/10.3390/s18082674
107. Benos L, Tagarakis AC, Dolias G, et al (2021) Machine Learning in Agriculture: A Comprehensive Updated Review. Sensors 21:3758. https://doi.org/10.3390/s21113758
108. Knadel M, Thomsen A, Schelde K, Greve MH (2015) Soil organic carbon and particle sizes mapping using vis-NIR, EC and temperature mobile sensor platform. Comput Electron Agric 114:134–144. https://doi.org/10.1016/j.compag.2015.03.013
109. Pei X, Sudduth K, Veum K, Li M (2019) Improving In-Situ Estimation of Soil Profile Properties Using a Multi-Sensor Probe. Sensors 19:1011. https://doi.org/10.3390/s19051011
110. Bricklemyer RS, Brown DJ (2010) On-the-go VisNIR: Potential and limitations for mapping soil clay and organic carbon. Comput Electron Agric 70:209–216. https://doi.org/10.1016/J.COMPAG.2009.10.006
111. Rodionov A, Welp G, Damerow L, et al (2015) Towards on-the-go field assessment of soil organic carbon using Vis-NIR diffuse reflectance spectroscopy: Developing and testing a novel tractor-driven measuring chamber. Soil Tillage Res 145:93–102. https://doi.org/10.1016/j.still.2014.08.007
112. Rodionov A, Pätzold S, Welp G, et al (2016) Proximal field Vis-NIR spectroscopy of soil organic carbon: A solution to clear obstacles related to vegetation and straw cover. Soil Tillage Res 163:89–98. https://doi.org/10.1016/j.still.2016.05.008
113. Veum KS, Parker PA, Sudduth KA, et al (2018) Predicting Profile Soil Properties with Reflectance Spectra via Bayesian Covariate-Assisted External Parameter Orthogonalization. Sensors 18:3869. https://doi.org/10.3390/s18113869
114. Viscarra Rossel RA, Lobsey CR, Sharman C, et al (2017) Novel Proximal Sensing for Monitoring Soil Organic C Stocks and Condition. Environ Sci Technol 51:5630–5641. https://doi.org/10.1021/acs.est.7b00889
115. Usha K, Singh B (2013) Potential applications of remote sensing in horticulture—A review. Sci Hortic (Amsterdam) 153:71–83. https://doi.org/10.1016/J.SCIENTA.2013.01.008
116. Stevens A, van Wesemael B, Bartholomeus H, et al (2008) Laboratory, field and airborne spectroscopy for monitoring organic carbon content in agricultural soils. Geoderma 144:395–404. https://doi.org/10.1016/J.GEODERMA.2007.12.009
117. Stevens A, Udelhoven T, Denis A, et al (2010) Measuring soil organic carbon in croplands at regional scale using airborne imaging spectroscopy. Geoderma 158:32–45. https://doi.org/10.1016/j.geoderma.2009.11.032
118. Franceschini MHD, Demattê JAM, da Silva Terra F, et al (2015) Prediction of soil properties using imaging spectroscopy: Considering fractional vegetation cover to improve accuracy. Int J Appl Earth Obs Geoinf 38:358–370. https://doi.org/10.1016/J.JAG.2015.01.019
119. Bartholomeus H, Kooistra L, Stevens A, et al (2011) Soil Organic Carbon mapping of partially vegetated agricultural fields with imaging spectroscopy. Int J Appl Earth Obs Geoinf 13:81–88. https://doi.org/10.1016/J.JAG.2010.06.009

120. Bayer AD, Bachmann M, Rogge D, et al (2016) Combining Field and Imaging Spectroscopy to Map Soil Organic Carbon in a Semiarid Environment. IEEE J Sel Top Appl Earth Obs Remote Sens 9:3997–4010. https://doi.org/10.1109/JSTARS.2016.2585674
121. Diek S, Schaepman M, de Jong R (2016) Creating Multi-Temporal Composites of Airborne Imaging Spectroscopy Data in Support of Digital Soil Mapping. Remote Sens 8:906. https://doi.org/10.3390/rs8110906
122. Hbirkou C, Pätzold S, Mahlein AK, Welp G (2012) Airborne hyperspectral imaging of spatial soil organic carbon heterogeneity at the field-scale. Geoderma 175–176:21–28. https://doi.org/10.1016/j.geoderma.2012.01.017
123. Vaudour E, Gilliot JM, Bel L, et al (2016) Regional prediction of soil organic carbon content over temperate croplands using visible near-infrared airborne hyperspectral imagery and synchronous field spectra. Int J Appl Earth Obs Geoinf 49:24–38. https://doi.org/10.1016/j.jag.2016.01.005
124. Castaldi F, Casa R, Castrignanò A, et al (2014) Estimation of soil properties at the field scale from satellite data: a comparison between spatial and non-spatial techniques. Eur J Soil Sci 65:842–851. https://doi.org/10.1111/ejss.12202
125. Demattê JAM, Fongaro CT, Rizzo R, Safanelli JL (2018) Geospatial Soil Sensing System (GEOS3): A powerful data mining procedure to retrieve soil spectral reflectance from satellite images. Remote Sens Environ 212:161–175. https://doi.org/10.1016/j.rse.2018.04.047
126. Forkuor G, Hounkpatin OKL, Welp G, Thiel M (2017) High resolution mapping of soil properties using Remote Sensing variables in south-western Burkina Faso: A comparison of machine learning and multiple linear regression models. PLoS One 12:. https://doi.org/10.1371/journal.pone.0170478
127. Clark RN, King TVV, Klejwa M, et al (1990) High spectral resolution reflectance spectroscopy of minerals. J Geophys Res 95:. https://doi.org/10.1029/jb095ib08p12653
128. Ben-Dor E, Patkin K, Banin A, Karnieli A (2002) Mapping of several soil properties using DAIS-7915 hyperspectral scanner data - A case study over soils in Israel. Int J Remote Sens 23:1043–1062. https://doi.org/10.1080/01431160010006962
129. Lagacherie P, Baret F, Feret J-B, et al (2008) Estimation of soil clay and calcium carbonate using laboratory, field and airborne hyperspectral measurements. Remote Sens Environ 112:825–835. https://doi.org/10.1016/J.RSE.2007.06.014
130. Nouri M, Gomez C, Gorretta N, Roger JM (2017) Clay content mapping from airborne hyperspectral Vis-NIR data by transferring a laboratory regression model. Geoderma 298:54–66. https://doi.org/10.1016/j.geoderma.2017.03.011
131. Liu L, Ji M, Buchroithner M, et al (2018) Transfer Learning for Soil Spectroscopy Based on Convolutional Neural Networks and Its Application in Soil Clay Content Mapping Using Hyperspectral Imagery. Sensors 18:3169. https://doi.org/10.3390/s18093169
132. Gomez C, Lagacherie P, Coulouma G (2012) Regional predictions of eight common soil properties and their spatial structures from hyperspectral Vis-NIR data. Geoderma 189–190:176–185. https://doi.org/10.1016/j.geoderma.2012.05.023
133. Gomez C, Adeline K, Bacha S, et al (2018) Sensitivity of clay content prediction to spectral configuration of VNIR/SWIR imaging data, from multispectral to hyperspectral scenarios. Remote Sens Environ 204:18–30. https://doi.org/10.1016/J.RSE.2017.10.047
134. Mulder VL, de Bruin S, Schaepman ME, Mayr TR (2011) The use of remote sensing in soil and terrain mapping — A review. Geoderma 162:1–19. https://doi.org/10.1016/J.GEODERMA.2010.12.018
135. Lagacherie P, McBratney A, Voltz M (2006) Digital soil mapping: an introductory perspective. Elsevier
136. Mirzaee S, Ghorbani-Dashtaki S, Mohammadi J, et al (2016) Spatial variability of soil organic matter using remote sensing data. CATENA 145:118–127. https://doi.org/10.1016/J.CATENA.2016.05.023
137. Bartholomeus HM, Schaepman ME, Kooistra L, et al (2008) Spectral reflectance based indices for soil organic carbon quantification. Geoderma 145:28–36. https://doi.org/10.1016/J.GEODERMA.2008.01.010

138. Mondal A, Khare D, Kundu S, et al (2017) Spatial soil organic carbon (SOC) prediction by regression kriging using remote sensing data. Egypt J Remote Sens Sp Sci 20:61–70. https://doi.org/10.1016/J.EJRS.2016.06.004
139. Sankar Bhunia G, Kumar Shit P, Reza Pourghasemi H (2017) Geocarto International Soil organic carbon mapping using remote sensing techniques and multivariate regression model Soil organic carbon mapping using remote sensing techniques and multivariate regression model. https://doi.org/10.1080/10106049.2017.1381179
140. Gholizadeh A, Žižala D, Saberioon M, Borůvka L (2018) Soil organic carbon and texture retrieving and mapping using proximal, airborne and Sentinel-2 spectral imaging. Remote Sens Environ 218:89–103. https://doi.org/10.1016/J.RSE.2018.09.015
141. Angelopoulou T, Tziolas N, Balafoutis A, et al (2019) Remote Sensing Techniques for Soil Organic Carbon Estimation: A Review. Remote Sens 11:676. https://doi.org/10.3390/rs11060676
142. Ben-Dor E (2011) Characterization of Soil Properties Using Reflectance Spectroscopy. Hyperspectral Remote Sens Veg 513–558. https://doi.org/10.1201/b11222-31
143. Verhulst N, Carrillo-García A, Moeller C, et al (2011) Conservation agriculture for wheat-based cropping systems under gravity irrigation: Increasing resilience through improved soil quality. Plant Soil. https://doi.org/10.1007/s11104-010-0620-y
144. Paganini M, Leidner AK, Geller G, et al (2016) The role of space agencies in remotely sensed essential biodiversity variables. Remote Sens Ecol Conserv. https://doi.org/10.1002/rse2.29

Proximal Sensing Sensors for Monitoring Crop Growth

Lea Hallik, Egidijus Šarauskis, Marius Kazlauskas, Indrė Bručienė, Gintautas Mozgeris, Dainius Steponavičius, and Toomas Tõrra

1 Introduction

Remote and proximal sensing technologies, coupled with global satellite positioning systems, provide the foundation for the digital innovation of agriculture. Seventeen (17) Sustainable Development Goals (SDGs) are at the heart of the 2030 Agenda for Sustainable Development, which have been adopted by all United Nations Member States [1]. Precision farming in combination with proximal sensing offers solutions to achieve the number of goals of the 2030 Agenda for Sustainable Development, such as SDG 2: Zero Hunger ("End hunger, achieve food security and improved nutrition, and promote sustainable agriculture"), SDG 6: Clean Water ("Ensure availability and sustainable management of water and sanitation for all"), SDG 9: Industries, Innovation & Infrastructure ("Build resilient infrastructure, promote inclusive and sustainable industrialization, and foster

L. Hallik (✉)
University of Tartu, Tartu Observatory, Vegetation Remote Sensing Group, Nõo Parish, Tartu County, Estonia
e-mail: lea.hallik@ut.ee

E. Šarauskis · M. Kazlauskas · I. Bručienė · D. Steponavičius
Vytautas Magnus University, Agriculture Academy, Akademija, Kaunas, Lithuania
e-mail: egidijus.sarauskis@vdu.lt; marius.kazlauskas@vdu.lt; indre.bruciene@vdu.lt; dainius.steponavicius@vdu.lt

G. Mozgeris
Vytautas Magnus University, Agriculture Academy, Akademija, Kaunas, Lithuania
e-mail: Gintautas.mozgeris@vdu.lt

T. Tõrra
Estonian University of Life Sciences, Institute of Agricultural and Environmental Science, Tartu, Estonia
e-mail: toomas.torra@emu.ee

D. D. Bochtis et al. (eds.), *Information and Communication Technologies for Agriculture—Theme I: Sensors*, Springer Optimization and Its Applications 182,
https://doi.org/10.1007/978-3-030-84144-7_3

innovation"), and SDG 15: Life on Land ("Protect, restore and promote sustainable use of terrestrial ecosystems, sustainably manage forests, combat desertification, and halt and reverse land degradation, and halt biodiversity loss") [2]. Digital Innovation Hubs (DIHs) play an essential role in providing relevant services for digital innovations in agriculture within the EU [3]. The Farm Sustainability Tool for Nutrients (FaST) will enable all EU farmers to use fertilizers sustainably and to better support them in implementing the green and digital agriculture transition [4].

1.1 General Scope

The need to monitor and measure plants during growth, stems from the visual differences between the plants growing in the field, which are observable by "naked eye". In the beginning of this Chapter, we will give a brief overview of problems in the field, that can be detected by remote and proximal sensing. We will demonstrate that by looking at the differences in the field, it can be perceived that certain parts of the field have more suitable conditions for plant growth than other parts of the field. However, one cannot simply determine the magnitude of the difference in measurable units by visual observation. Also, in many cases, it is not possible to explain the causes of the problems simply by visual inspection.

In order to obtain unit-comparable results for differences in the field, appropriate measurement methods as well as different measuring devices need to be used. Today, there are many contact-, proximal-, and remote monitoring solutions for measuring the growth processes of plants in the field. The causes of plant growth limitations can be investigated by collecting discrete soil samples from the field in the different areas and determining the content of different nutrients in the soil in the laboratory. The same procedure can be conducted with plant samples to determine the uptake of nutrients. However, such measurements are very resource-intensive, time-consuming, and costly. Now-a-days the availability of proximal and remote sensing technology for monitoring plant growth is increasing at an unprecedented rate, allowing for faster and cheaper assessments. This Chapter will provide an overview of contact and proximal monitoring solutions currently available.

Proximal sensors can be hand-held, mounted on tractors, or attached to drones. Geo-spatial positioning is important for all proximal sensing applications to save results by location. Sensors mounted on agricultural machinery allow simultaneous measurements during specific management activities such as fertilization. It can be a cost-effective and time-savings solution to perform vegetation measurements while executing an ordinary field work of the cropping cycle. If the goal is not to collect information simultaneously with other field management activities, then it makes sense to use drone or satellite platforms for monitoring. Airborne and spaceborne data collection methods are more sensitive to weather conditions such as clouds, rain, and wind. Therefore, all methods have their strengths and weaknesses, and the most optimal solutions can be achieved by combining different sensors and measurement methodologies. This Chapter will provide an overview of different monitoring platforms for precision agriculture, including autonomous platforms and robots. Use-cases from Estonia and Lithuania illustrate the most common usage.

2 Problems in Fields That Can Be Detected by Proximal Sensing

Looking from the edge of the field, it often seems to the observer that everything is fine with the field and there are no major problems. Such a visual assessment from the edge of the field does not usually give an objective overview of the overall condition of the field. In order to get an overview of the whole field, one should drive through this field closely with a tractor or a car, but this is again quite a time-consuming and resource-intensive activity and also may cause undesirable compaction of the field soil. Of course, if such a field monitoring activity is carried out in conjunction with a planned operation in the field, such as harrowing, crop protection, or fertilization, then it is beneficial because the farmer can carry out two necessary operations at the same time. If this is not possible, then it is wise to use a drone, or satellite images to get an overview of the field. When the farmer gets a complete picture of the field from above, he can often be surprised that a field that he thought was in fine conditions, does not look so promising in every field spot. But what do these remote sensing images show and what can be done in the field with this information?

2.1 Problems in Crop Emergence

One can get an idea of the first problems that have arisen in the field, already after the emergence of the crop. When taking an orthophoto or satellite image of a field and analyzing it, it can be noticed that there are areas in the field where there are no plants or only few individual plants have sprouted (Fig. 1). The reasons can, of course, be various; seed material of poor quality, incorrect sowing depth, drought, or due to varying soils.

2.2 Agrotechnical Mistakes

Agrotechnical mistakes can also be detected after the emergence of the crop using drone or satellite images for monitoring. One can consider agrotechnical errors as problems in the field that the farmer could have avoided by choosing the right practices. For example, inadequate tillage, which results in many tops and straw of previous crops

Fig. 1 Crop emergence disorders due to soil variations. (Photo by Toomas Tõrra)

remaining in the topsoil. In the presence of excess straw in the seed layer, the seeds cannot come into contact with the soil and are not able to germinate due to low moisture and insufficient soil contact. Similar problems are encountered when sowing excessively compacted and water-saturated soil, where the plant does not receive enough oxygen. Weeds can be also a major problem. If weed control has not been sufficient in previous years, weeds can displace a large proportion of the crop sown (Fig. 2).

2.3 *Overwintering Damage to the Crop Field*

After sowing winter crops, the field can be in good condition and all sown seeds may germinate well. Problems can occur with water puddles accumulating in the field with autumn rains, where water does not seep into the soil due to the compaction (Fig. 3). The same problem can occur with spring precipitation or snowmelt water that does not seep into the soil but remains in large puddles on the field surface for a long time. In addition, plant diseases can affect large areas, e.g., snow mold, which in the spring may destroy plants within its area of occurrence. If such areas with damaged vegetation are abundant in the field, and their extent and location have been identified by remote sensing, then it is worth considering differentiated fertilization in the field. Differentiated fertilization saves on the use of fertilizers

Fig. 2 Weeds have destroyed a large part of the winter turnip field. (Photo by Toomas Tõrra)

Fig. 3 Germination problems caused by excessive soil moisture in winter wheat. (Photo by Toomas Tõrra)

by applying fertilizer to areas with an optimal number of plants and reducing or not applying fertilizer to areas where there are no or few plants. By providing less fertilizer to an area where there are few plants, we protect the environment because there is no over-fertilization and no leaching of unused nutrients.

3 Precision Agriculture

Precision agriculture has been used for more than 30 years to optimize production costs, increase profitability, and reduce negative impacts on soil and the environment [5]. The main principle of precision agriculture is the adaptation of technological operations to the conditions of individual field locations. Precision agriculture is based on the ability to determine the spatial variability of a field and use this information for more targeted crop management. Some researchers describe precision farming as a complex set of farmers' practices that can ensure the sustainability of agriculture based on four Rs: Right amount, Right application, Right place, and Right time [6–9].

Precision agriculture is a set of modern methods and technologies, first introduced in North American soils and later spread to other countries [6]. Diacono et al. [6] review on the literature of the last two decades on precise agriculture methods for nitrogen fertilizer management in wheat crops, showed that the advantages of precise agricultural practices used in the production of cereals outweigh the disadvantages in order to improve crop yields and reduce agricultural risks to the environment. It has been found that the use of a fixed nitrogen rate for the whole field cannot be economically or ecologically sustainable. Precision fertilization can help solve excess or improper use of fertilizer.

Germanas [10] describes precision agriculture as a crop system managed by autonomous and operational information based on new or improved technologies. The most important are location systems and sensor technology for data acquisition. The collected and processed data is transmitted by special systems for plant fertilization, spraying, etc. agricultural machinery, so that the necessary measures for the plants are precisely adapted to each area. Depending on the relationship between data acquisition, the decision made, and the impact measures used, precision farming relies on fundamentally different autonomous, operational, and mixed (autonomous and operational) data acquisition methods [10].

More and more manufacturers of tillage, sowing, fertilizing, crop protection, and harvesting equipment are offering machines with a variety of precision agricultural accessories and sensor systems. Optoelectronic sensors mounted on the sprayers are able to detect weeds by separating them from the main plants, and sensors that determine the amount of biomass per unit area allow to automatically increase or decrease the rate of plant protection measures [11]. Soil samples can be taken automatically, and plant nutrient levels can be analyzed with multifunction sensors. Nitrogen sensors allow farmers to adjust the fertilizer rate depending on the color of the plant leaves. The lighter the color of the leaves compared to the standard leaf color of "saturated" plants, the more fertilizer, "hungry" plants receive. Fertilizer spreaders are guided by wind speed meters, which allow to compensate the wind resistance so that the fertilizer is spread evenly, according to the set rate. Using field map databases, the actual position of the machine in the field is recorded and the intensity of the spreading fertilizer flow at the field edges, inserts and protrusions is automatically reduced or increased without operator intervention [11]. By using resources more efficiently, precision agriculture can make agriculture more productive, sustainable, and reduce its negative impact on the environment [12].

3.1 Fertilization Effect on Environment

According to the United Nations (UN), in 2050 the population worldwide can reach up to 10 billion [13]. The rapidly growing population leads to a constantly growing demand for agricultural products. Plant nutrition is one of the most important factors in controlling the yield and quality of agricultural products [14, 15]. Therefore, organic and mineral fertilizers are used to enrich the soil with nutrients and make the most of the soil fertility. Typically, in temperate climates (USA and England), almost 40–60% of agricultural crops are grown using a variety of fertilizers [16, 17].

In recent years, fertilizer use has been growing exponentially around the world. In 2017, the global consumption of chemical and mineral fertilizers amounted to 123 kg ha^{-1}, of which nitrogen (N) fertilizers accounted for an average of 70, P_2O_5–29 and K_2O – 24 kg ha^{-1} [18]. Consumption of chemical fertilizers per hectare from 2002 to 2017 increased by about 30%. In Europe, the application of chemical and mineral fertilizers amounted to 79 kg ha^{-1} (52 N, 13 P_2O_5, 14 K_2O) [18]. Intensive use of mineral fertilizers and improper application of fertilization technologies in agriculture have negative effects on both the environment and human health [6, 15, 19]. The main consequences of improper use of chemical fertilizers may be groundwater and surface water pollution, eutrophication of water bodies, accumulation of heavy metals and nitrates in plants, air pollution, imbalance of soil elements, salinity, deterioration of soil fertility and degradation [15, 20].

Although N fertilizer has increased crop yields, the over-application of N can have unintended negative economic and environmental consequences [19]. Concerning energy consumption by inputs, it is evident that among all inputs, fertilizing (with nitrogen N, phosphorus and potassium) is the most energy consuming one [21]. Management strategies to reduce N loss and increase crop N recovery have been studied extensively, including N fertilizer source, application method, timing of fertilizer application, tillage, N loss inhibitors (fertilizer additives), and, more recently, site-specific management to account for within-field crop N needs [22, 23]. Fertilization rates can be refined by using various monitoring tools. Plant spectral reflectance may provide the information needed to assess N supply [24, 25].

The recommended crop fertilization rates, for example, 150 to 180 kg N ha^{-1} for wheat, are often disregarded in intensive agriculture [26]. During the past half of the century, the continued increase in the use of nitrogen fertilizer and the intensity of production has exceeded the population growth [27], which led to nitrogen fertilizer accounting for 1.2% of greenhouse gas (GHG) emissions and of all the energy used worldwide [28–30]. In the future, GHG emissions from agriculture are expected to increase all over the world. However, because GHG emissions per 1 ha or 1 kg of produce largely depend on the nitrogen fertilizer application rates and crop yields, these amounts may be controlled, but it is necessary to research the impact of agricultural technologies on greenhouse gas emissions [31].

In agriculture, farming systems should be more sustainable in order to achieve economic and social benefits while preserving the environment [6]. Agricultural production with lower resources, optimal yields, increased re-source efficiency and, at the same time, low environmental pollution can be achieved through precision farming, including precision fertilization technologies. The willingness to adopt a precision agriculture variable rate fertilization technology is strongly affected by its cost, which can include a perception of both a high monetary cost and cost in the difficulty in the use of technology [32].

3.2 *Variable Rate Fertilization (VRF)*

Fertilization is one of the most important technological operations in agricultural production. Traditionally, fertilizers are spread over the entire soil surface, regardless of changes in soil properties [33]. Variable rate fertilization is essential for the implementation of precision farming and ensuring the efficient use of fertilizers and the management of nutrients in the soil adapted to the conditions of individual field sites [34]. Precise variable rate fertilization technology allows for the application of different amounts of fertilizer, depending on soil fertility conditions and the needs of individual crop sites. The application of such a fertilization method allows to improve the efficiency of fertilizer use and reduce fertilizer leaching [35]. The goal of variable rate fertilization technology is to maximize crop yields by reducing the amount of fertilizer used, achieving an optimal cost-benefit ratio. The VRF method works well to take advantage of crop and soil variability, but this method has no greater value if the field is completely uniform. Variable rate fertilization is suitable for both liquid and granular fertilizers.

The first variable rate fertilization was implemented on a farm owned by the University of Minnesota in 1993–1994. The results of experimental studies showed that using variable rate fertilization increased plant yield by about 30% compared to traditional fertilization [20].

3.3 *Sensors for Precision Fertilization*

Recent developments in observation technologies, geostatistical analysis, data aggregation, and interpolation methods have improved the accuracy and reliability of the determination of soil management zones, making it a viable strategy for commercial agriculture [36]. For precision fertilization, three main methods are distinguished for estimating field variability: measurement of crop properties, measurement of soil properties, and measurement of crop yield [8]. Precision agriculture and variable rate fertilization are strongly linked to smart technologies: global satellite navigation system, geographic information systems, remote sensing,

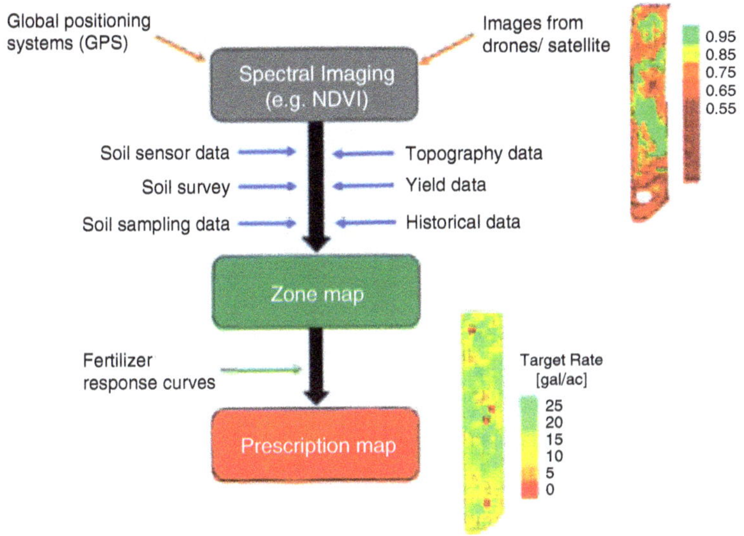

Fig. 4 Creating of prescription map [5]

harvest or soil mapping tools, machine vision, remote and proximal sensors that can measure or even predict crop and soil properties in real time [6, 34, 37–43].

Sensors and mapping are the basis for the application of basic variable rate technologies [40]. Each of these methods has its advantages and disadvantages. Therefore, a combination of both methods would be the best. Map-based variable rate application relies on the use of electronic maps that generate accurate information about input rates applicable in individual field zones [40]. The mapping approach is usually based on the following information: soil type, soil color and structure, topography, crop yield, field monitoring data, and many other sources of information that can be adapted to a particular plant and location.

Models are developed in which prescription maps are applied for field application of symbiotic nitrogen fixation (Fig. 4). Global Positioning System (GPS) spectral images can be generated using different software algorithms, vegetative reflectance (e.g., normalized difference vegetation index-NDVI), other images obtained during plant vegetation, captured from satellites or drones. Such spectral images can be used to create control areas in the field, including soil properties and other data such as sensor data, soil sample analyzes, topography data, yield data, and other available historical data. Prescription maps of different plant fertilizer needs are created based on field area maps [5].

Thilakarathna and Raizada [5] presented a conceptual model (Fig. 5) that integrates different precision farming tools together to improve symbiotic nitrogen fixation of legumes at field level using on-the-go variable management. Under this model, soil chemical and physical properties would be collected and analyzed using soil monitoring sensors mounted on tractors. Crop properties (e.g., plant N status)

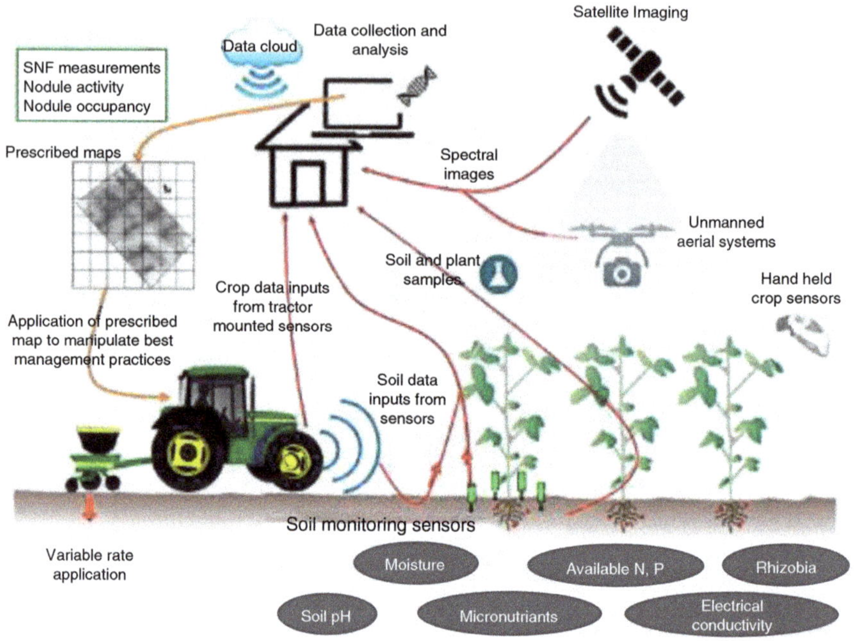

Fig. 5 Conceptual model combining different tools for data collection, processing and use of agricultural information using on-the-go variable management [5]

would be collected using tractor-mounted sensors, similar to Green-Seeker NDVI, or satellite/Unmanned Aerial System (UAS) images. Legume measurements by sensors would be combined with spatial maps that would be generated using satellite or airborne systems to program variable rates for a variety of field conditions.

Mozgeris et al. [24] stated that, the strategy of using ultralight aircraft as a sensor platform for precision agriculture aimed aerial imaging projects, outperforms the UAV-based and professional photogrammetric aerial photography solutions in terms of potential costs associated with the current service demand and supply conditions in Lithuania. This study investigates an imaging system based on a Rikola hyperspectral (HSI) and Nikon D800E (CIR) cameras installed on a manned ultralight aircraft Bekas Ch-32 for applications involving precision agriculture. The efficiency of this technical solution is compared with that of using Canon PowerShot SX260HS camera images acquired from helicopter-type unmanned aerial vehicle (UAV) to accomplish similar tasks. The criteria for comparison were the suitability of acquired images for modelling chlorophyll concentration in spring wheat and for estimating the normalized difference red edge (NDRE) index, which is conventionally obtained using OptRx proximal sensors.

Maps or positioning systems are usually not required for the application of variable rates based on sensors. Sensors measure the properties of soil or plants as they pass through the crop in real time [40]. However, in order to use the information obtained for the management of future crop areas for specific crops, sensor data should be recorded and presented under a geographical indication. Plant property sensors have the greatest potential because of their ability to detect crop needs (e.g., nitrogen) in real time during variable rate fertilization.

Sensor-based nitrogen management systems, compared to conventional farming practices, showed that the efficiency of nitrogen fertilizer use increased to 3.68 times. Sensor systems saved between 10% and 80% of nitrogen fertilizer, and the residual nitrogen fertilizer in the soil was reduced by 30–50% without affecting wheat grain yield and quality [6].

4 Proximal Measurement Sensors

4.1 Soil Characteristics

With precision spreading technology, perhaps the most important data set is maps depicting changes in soil properties that affect plant growth and yield. The most commonly used properties for this purpose are the following [12, 44]:

- soil organic matter content, fractions labile and stable;
- soil organic carbon content;
- soil particle size (clay, loam and sand content);
- soil pH;
- soil structure;
- soil moisture;
- soil bulk density;
- soil compaction;
- soil porosity;
- soil macronutrient level (nitrogen, phosphorus, potassium);
- soil microelements content;
- soil temperature; and
- depth of any root restricting layers.

4.2 Proximal Soil Sensors

Adamchuk et al. [45] divided various real-time soil sensors into main categories based on their design concepts, including electrical and electromagnetic, optical and radiometric data, mechanical, acoustic, pneumatic, and electrochemical soil sensors. The authors added that most soil sensors are affected by more than one agronomic properties of the soil. To explain the methods of the successful measurement of soil properties by sensors, the following five categories of laboratory, stationary and real-time measurement sensors are named:

1. Reflection-based soil sensors;
2. Conductivity and resistance-based soil sensors;
3. Passive radiometric soil sensors;
4. Soil sensors based on force/strength;
5. Electrochemical soil sensors.

One of the most promising and widely used research methods in precision agriculture is the measurement of apparent soil electrical conductivity (EC), i.e., a measurement of how much electrical current soil can conduct. The EC has become one of the most common tools for characterizing the spatial variability of a field because it is reliable and easy to measure [46]. In agricultural practice soil EC sensors are used to devise management zones, set soil sampling locations, create variable rate seeding prescriptions, better manage nitrogen applications, and improve irrigation prescriptions [47]. Soil EC enables to identify soil physic-chemical properties determining patterns of agricultural crop yield [48]. Friedman [49] divides the factors affecting EC to three main categories. The first category describes the bulk soil and defines the respective volumetric fractions occupied by the three phases and possible secondary structural configurations (aggregation): porosity, moisture content and structure. Factors in the second and third categories are the important solid particle quantifiers (particle shape and orientation, particle-size distribution, cations exchange capacity) and soil solution attributes, respectively.

The principle of determining the electrical conductivity of soil is widely applied to assess soil properties. One way to assess soil electrical conductivity is electromagnetic induction using a commercially available Geonics Ltd EM38 meter [44]. The transmitting coil induces a magnetic field, the strength of which varies with the depth of the soil. The ratio of magnetic field strength to soil depth can be altered to measure various soil depths up to 1.5 m. The take-up coil measures the currents in the soil and relates them to the electrical conductivity of the soil [12, 44]. Another commercially available soil electrical conductivity measuring device is the Veris MSP, which measures electrical conductivity while driving. A set of coulter electrodes is used for this, transmitting an electrical signal through the soil [12, 44, 50].

The electrical conductivity of the soil is measured indirectly, i.e., the specific resistance of the soil is measured and converted into electrical conductivity EC (mS m^{-1}). Using the specific electrical resistance measurement method, an electric current is applied to the soil by current electrodes located on the soil surface, and a potential difference is measured in potential measurement electrodes located close to the current electrodes [51, 52].

With the same Veris MSP machine, pH and the amount of organic matter in the soil can be determined in the same run. An optical sensor built into the coulter of the machine measures the reflectance and absorption characteristics of the soil at a depth of 5 cm. The machine is equipped with a navigation system that records the location of the machine. The measurement values, together with the GPS

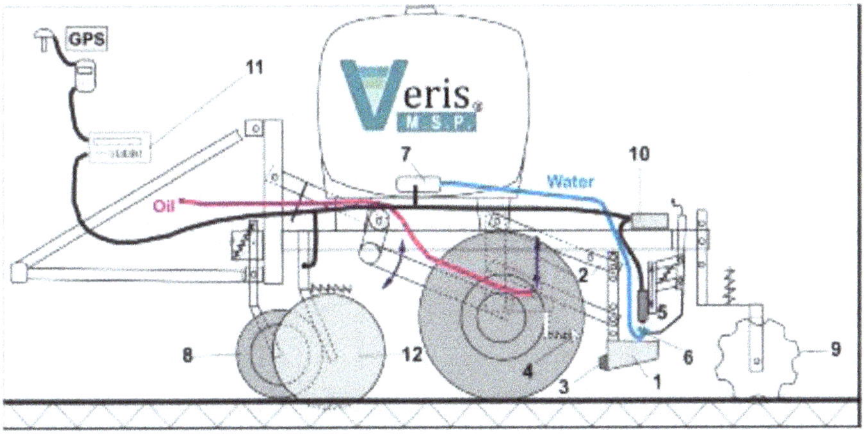

Fig. 6 Veris MSP machine with soil sensing system: 1 – scoop; 2 – mechanism of scoop lifting; 3 – adapter; 4 – hog; 5 – pH sensors; 6 – water supply with nozzles; 7 – water tank; 8 – plant residues removal; 9 – furrow filling hoes; 10 – controller; 11 – data recorder; 12 – sensor of soil electrical conductivity [50]

coordinates, are transmitted to the machine computer, where the field survey data medium is generated. Based on the obtained data, a preliminary map of soil organic matter is generated using the Veris FieldFusion platform [51, 53].

The soil pH value is a predictor of various chemical activities and a rough indica-tor of the plant availability of nutrients within the soil. Having the correct pH is crucial for the healthy plant's growth as it will affect the amount of nutrients avail-able to plants. In Lithuania the study was carried out using the mobile unit Veris 3150 MSP equipped with the Soil pH Manager system. The Soil pH Manager consisted of three main components: a hydraulic soil sampling system, two antimony pH electrodes, and a water wash system (Fig. 6). Selecting the lime rates according to the maps created by using the data of the unit, lime savings was about 40% [54].

The amount of soil organic matter affects the efficiency of crop maintenance work. A device with optical sensors (Fig. 7) capable of automatically adjusting the amount of herbicides without a pre-prepared map or other input data was used to determine the amount of soil organic matter. This device with sensors is pulled or pushed through the soil [40, 44].

Due to different soil types, raw materials, soil and environmental factors such as water content, temperature, humidity, organic matter, topography and soil color, the performance of different sensors varies greatly due to the different results presented.

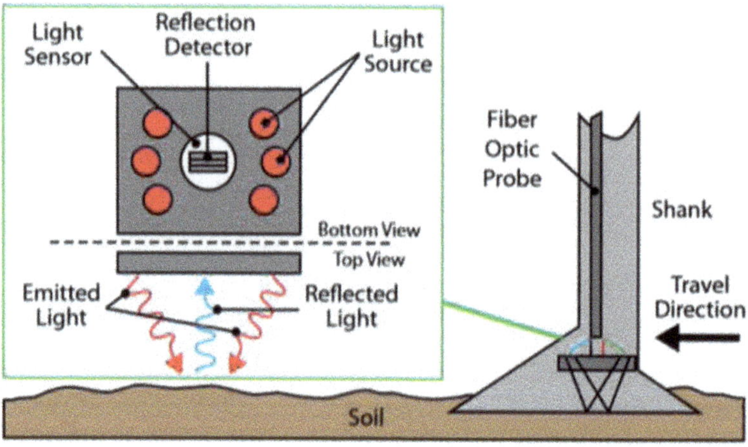

Fig. 7 Scheme of an optical sensor for measuring soil organic matter [40, 44]

4.3 Proximal Crop Sensors

There are many different crop characteristics that can benefit from the use of proximal sensors. From the farmer's point of view, the features that affect the yield as well as the quality of the crop are of the greatest interest. Another important aspect is the ability to solve problems where they can be identified and managed during the growing season. Biomass accumulation, crop water status, nutrient (especially nitrogen) deficiency, disease onset, and weed and insect infestation are all factors that crop growers may want to monitor throughout the season. A system of proximal sensors, in contrast to remote sensing, has the potential advantage that sensors on satellite and antenna platforms are more exposed to weather and clouds, which may limit the benefits of remote sensing during the critical plant growth season [12].

Passive and active light proximal sensors are distinguished. Passive sensors are dependent on sunlight, while active sensors, which have their own light sources, al-low the condition of crops to be assessed independently of ambient light conditions [6, 55]. Commonly used commercial proximal sensors include the Yara N-Sensor® /FieldScan passive and FieldSpec® portable spectroradiometer, as well as the GreenSeeker® and "Crop Circle™" active sensors [12, 56, 57].

Active sensors such as the Trimble® GreenSeeker® crop sensing systems emit brief bursts of red and infrared light, and then measure the amount of each type of light that is reflected back to the sensor. This information is used to calculate the Normalized Difference Vegetation Index (NDVI), which is a direct indicator of the density of the foliage in the sensor's view. Green plants absorb strongly red light and scatter infrared light so that in vegetated areas the reflectance of red light is reduced compared to bare soil while the reflectance of infrared light is increased. Trimble offers both GreenSeeker® crop sensing systems which can be mounted on agricultural machinery for a variable rate application and crop vigor mapping as well as handheld version (GreenSeeker, website assessed 11th June 2020 [56]. The handheld version is a very easy-to-use device with which NDVI measurements of vegetation can be made quickly and conveniently only by walking in the field. The results of using GreenSeeker Handheld Crop Sensor in a fertilization field trial experiment will be presented later in this book chapter in the Estonian use case.

The aim of the Aranguren et al. [58] study was to evaluate the usefulness of the proximal sensors Yara N-TesterTM and RapidScan CS-45 in diagnosing the N nutritional status of wheat after application of manure and sowing. The authors performed 3-year field trials using five different rates of mineral N fertilization (fertilization doses) (0, 40, 80, 120 and 160 g N ha^{-1}). Proximal sensors were used during plant stem elongation before fertilization with mineral N. When any proximal sensor reading was 60–65% N, the optimal N rate for maximum yields was 118 to 128 kg N ha^{-1}. When the sensor readings were 85–90%, the optimal N rate was reduced to 100–110 kg N ha^{-1}. Previous studies have shown that the Yara N-Tester™ (chlorophyll meter) and RapidScan CS-45 (ground-based active-light proximal sensor) readings were valuable indicators of N control in individual crop zones when assessing plant nutritional status, and measurements with these sensors should be performed periodically to effectively monitor the N status of the crop [12, 58].

Padilla et al. [59] performed an analysis of three different types of proximal optical sensors (chlorophyll meters, reflectance sensors, and fluorescence-based flavonols meters) for their use in N-state control in vegetable crops. Thanks to this analysis, the authors highlighted the possibilities of proximal optical sensors that can help control N in plants and the practical problems of their use. The choice of the optimal type of sensor in a particular situation, may depend on the crops grown, the condition of the crop, the specialization and management capabilities of the farm, the available agricultural machinery, and the expertise of the operator. Proximal reflectance sensors typically measure a large area of plant surface and can significantly reduce variability compared to sensors that typically measure small areas of individual leaves. Most reflection sensors and some fluorescence-based flavonol meters can perform continuous measurements while moving, thus connecting large measurement areas. This is especially useful on large farms. Compared to passive plant surface reflection sensors, active sensors are not sensitive to exposure conditions, which is a noticeable practical advantage. Chlorophyll and fluorescent flavonol meters do not combine spurious signals from soil uncovered by plants or plant residues (unlike some plant surface reflection sensors), making them more suitable for early crop growth stages and widely distributed crops [59].

4.4 Chlorophyll Meters

Chlorophyll meters belong to a group of optical sensors that undestructively estimate the relative amount of chlorophyll per unit area of leaf surface area [60]. Chlorophyll absorbs visible light in 400 nm – 700 nm spectral region but

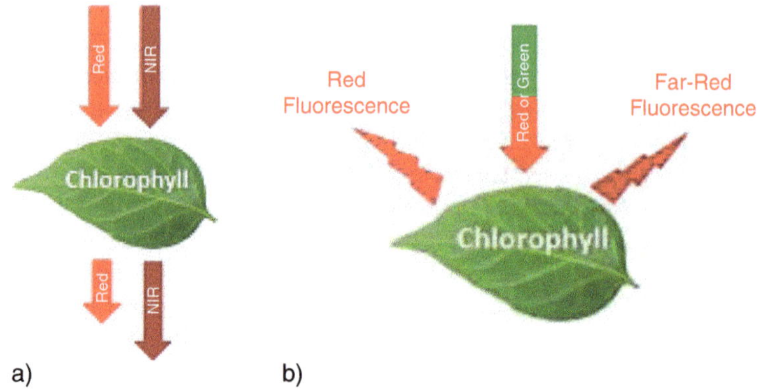

Fig. 8 Chlorophyll meters: (**a**) – transmittance-based; (**b**) – fluorescence-based [59]

chlorophyll content is also strongly associated with leaf N content, as most leaf N is present in the photosynthetic apparatus and enzymes involved in photosynthesis [55, 61–63]. Most chlorophyll meters are hand-held devices that are pressed to the surface of the leaves or measured right next to them. Chlorophyll meters provide a dimensionless value when measured that correlates strongly with the actual chlorophyll content [59, 60, 64].

Most chlorophyll meters determine the relative chlorophyll content of leaves by measuring leaf absorbance and transmittance of radiation. Red rays are absorbed by chlorophyll (Fig. 8a) and near-infrared (NIR) rays by chlorophyll transmits [59, 65]. The chlorophyll content increases as the absorption of red rays increases. The measurement values displayed by the chlorophyll meter then increase [63, 66, 67]. Optical sensors of this type are called transmittance-based chlorophyll meters [59].

Another method for measuring leaf chlorophyll content is to determine the ratio of chlorophyll fluorescence emission using red and far-red radiation [59, 68–70] (Fig. 8b). The fluorescence ratio of red to very red chlorophyll is highly dependent on the chlorophyll content. Due to repeated reabsorption of chlorophyll on the leaf, this ratio decreases with increasing chlorophyll content [68]. Sensors using this measurement method are called fluorescence-based chlorophyll meters [59, 69].

4.5 *Reflectance Sensors for Nitrogen (N)*

Sensors of this type assess the N state of the crop usually indirectly by measuring the chlorophyll sensitive wavelengths of absorbed and reflected rays from the crop foliage [59, 71–73]. When measured, proximal canopy reflection sensors are located close enough to the plant crop (e.g., 0.4–3.0 m). Green plant tissue typically absorbs

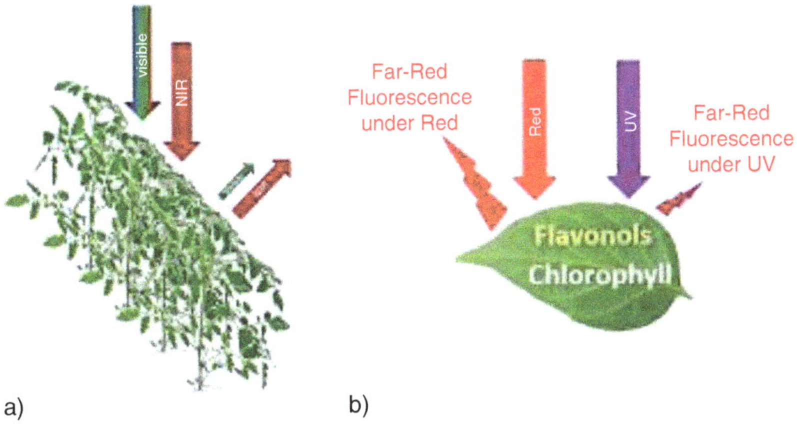

Fig. 9 Optical crop meters: (**a**) – canopy reflectance sensor; (**b**) – fluorescence-based flavonols meter [59]

about 90% of visible rays with wave lengths between 390 and 750 nm due to pigments and reflects rays with wavelengths around 750 to 1300 nm [74]. Due to the strong association between chlorophyll and N content the degree of absorbance and reflectance in the visible and NIR parts of the spectrum varies with the amount of N in the crop (Fig. 9a) [59]. Crops with N deficiency generally reflect more visible and reflect less NIR than crops with sufficient N [59, 63, 73].

Another method used to monitor the N state of the crop with proximal sensors is based on the relative flavonol content from fluorescence measurements [59, 69]. Flavonols are a class of polyphenolic compounds that are carbon-based secondary metabolites that increase in leaf levels under lower N supply conditions [75, 76]. Leaf flavonol content is generally inversely related to leaf chlorophyll content [61]. The relative amount of flavonols is calculated with flavonol meters using the chlorophyll fluorescence screening method [77, 78]. The measurement principle is based on fluorescence radiation from the red to the outermost red-light spectrum [68, 79]. Flavonols accumulating in the epidermis of the leaf absorb significant amounts of UV radiation and reduce the radiation of far red chlorophyll fluorescence during UV radiation, but do not affect the radiation of far red chlorophyll fluorescence under red excitation (Fig. 9b). Flavonol content is calculated by comparing the fluorescence of distant red chlorophyll under red and ultraviolet light [61, 78, 80]. Flavonol meters show a dimensionless value that is very strongly related to the actual amount of flavonols [59, 65, 69, 81]

Deeper knowledge of leaf and canopy reflection has given impetus to the wider application of remote sensing in agriculture. The application of remote sensing in agriculture starts with the observations of plant leaves, and then the obtained connections and dependencies are applied to the canopy. Remote sensing tools have been designed so that the amount of pigment strongly affects the absorption

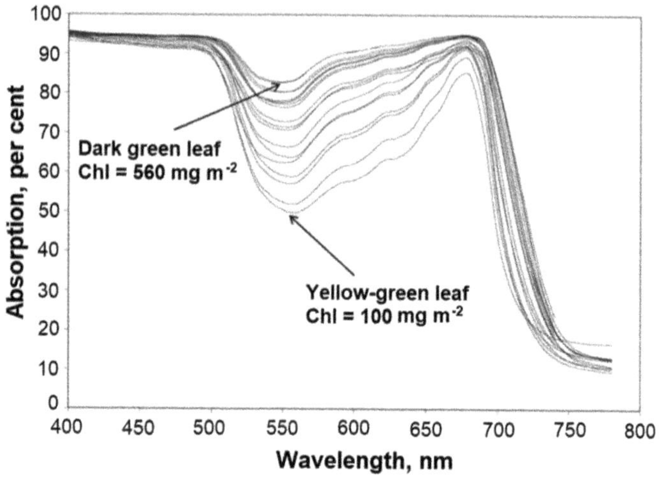

Fig. 10 Absorption spectra of maize leaves [71]

spectrum of leaves [71]. Figure 10 shows the absorption spectra of maize in the presence of dark green leaf and yellow green leaf.

4.6 High-Resolution Spectrometers

Many crop sensors using the high-resolution electromagnetic spectrum have recently been used to describe crop properties. Such sensors may be passive, using reflected, scattered, or radiated energy, or active, using an internal pulsed or modulated energy source and measuring the reflection or fluorescence of that source. The spectra of most interest to researchers and practitioners range from visible (400–700 nm), near-infrared (700–1300 nm), medium-infrared (1300–2500 nm) and thermal infrared regions (usually the main region of interest is 8000–14000 nm) [12]. The most important factors influencing light absorption and reflectance in the visible part of the spectrum are plant pigments such as chlorophyll, carotenoids, and anthocyanins. Cell structure, water content, and crop architecture influence crop reflection in the near-infrared region, and water content in leaves affects reflection in the medium-infrared region (Fig. 11) [12].

Often, absolute reflection at a given wavelength is not a very informative indicator of plant stress. Reflection is related to various plant properties. The vegetation index provides an opportunity to draw conclusions about specific plant characteristics and sources of stress. To determine plant properties, the vegetation index uses a reflection ratio in two or more spectral regions joined by an equation. More than 150 vegetation indices are currently known [82], but the most widely known is the NDVI.

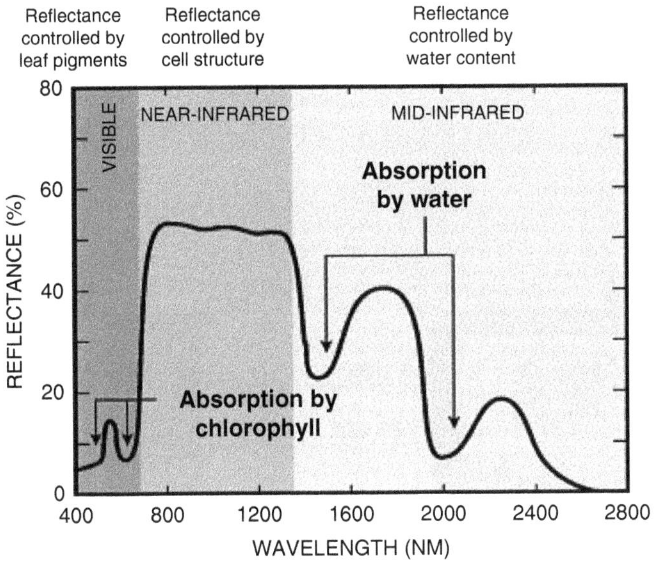

Fig. 11 Crop canopy reflectance factors in the region of visible to medium infrared wavelengths [12]

4.7 *Yara N-Sensor*

The electromagnetic sensors for commercial assessment of the N condition of field crops, the Yara N-Sensor, which is mounted on tractor and consists of two spectrometers, one for scanning the crop on the side of the tractor (Fig. 12) and the other for measuring ambient light, is widely used, in real-time to correct the reflected signal at a selected wavelength from 450 to 900 nm [12, 55, 83]. Reflection is used to calculate NDVI or other vegetation indices of interest. The tractor-mounted system uses an algorithm to calculate the optimal rate of N fertilizer in the scanned region and transmits this information to the controller to change the rate of N fertilizer use in real-time fertilizer application [12, 57]. Depending on the height of the plants, the crop can be scanned with a Yara N-Sensor approximately 3–4 m wide stripes to the left and right of the tramline. The sensors can be easily mounted on almost all sprayers and spreaders that can be adjusted electronically, regardless of manufacturer and type. The connection takes place via serial interfaces or via an Isobus adapter. The Yara N sensor is controlled from the driver's cab using a multifunction display.

For research and commercial purposes, OptRx® integrated crop monitoring sensor technology is used, using a three-spectrum sensor system to measure reflectance at 670, 730, and 780 nm [12]. One of the advantages of this sensor system using the red-edge NDRE spectrum (730 nm) instead of the red-spectrum in the NDVI equation is the higher sensitivity of the red-edge reflection to the crop chlorophyll and thus to the N content in the crop with high biomass content. The most competitive

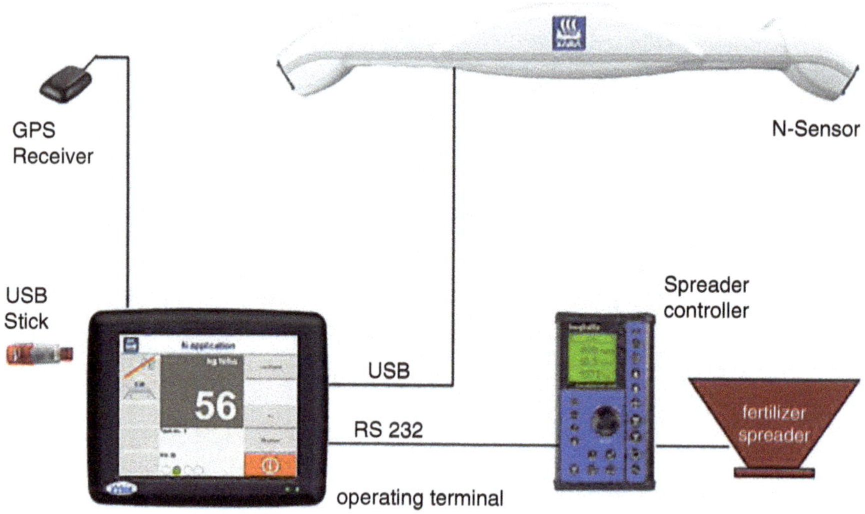

Fig. 12 Schematic of the devices and its connections using Yara N-sensor ALS [84]

Fig. 13 OptRx crop sensing for real-time variable rate applications [50, 85]

products typically use only the red wavelength, which is not sufficiently resistant to the high density of plant biomass [12, 85].

OptRx crop sensors (Fig. 13) measure and record plant data in real time using the reflection of the illuminating light of the plants. The sensors can be mounted on the entire beam of the unit to collect information when driving through the field. Data is recorded, systematized, and used for further analysis or real-time variable rate programs.

5 Autonomous Platforms in Precision Agriculture

One of the key elements of smart farming is farm management information systems that support the management of data collection and processing, monitoring, planning, decision making, documentation, and automation of farm operations. An increasing number of farm management information systems are now using the Internet of Things (IoT) technology to further optimize targeted business goals [86]. The IoT is the result of technological advances in many parallel and often overlapping areas, including the results of ubiquitous and comprehensive computing, mobile telephony, telemetry and communication between machines, wireless sensor networks, mobile computers, and computer networks. The IoT complements current information and communication technologies that already provide "anytime" and "anywhere" communication [86].

According to scientists [87], in order for modern agriculture to be sustainable, it is necessary to automate farming processes by combining sensors, robotics and artificial intelligence. New technological systems and materials, rapid progress of artificial intelligence and machine learning, the right speed and costs have increasing applicability in a commercial context. The diagram below (Fig. 14), in terms of intelligent technologies, illustrates the connections and subsystems for wide application in agriculture.

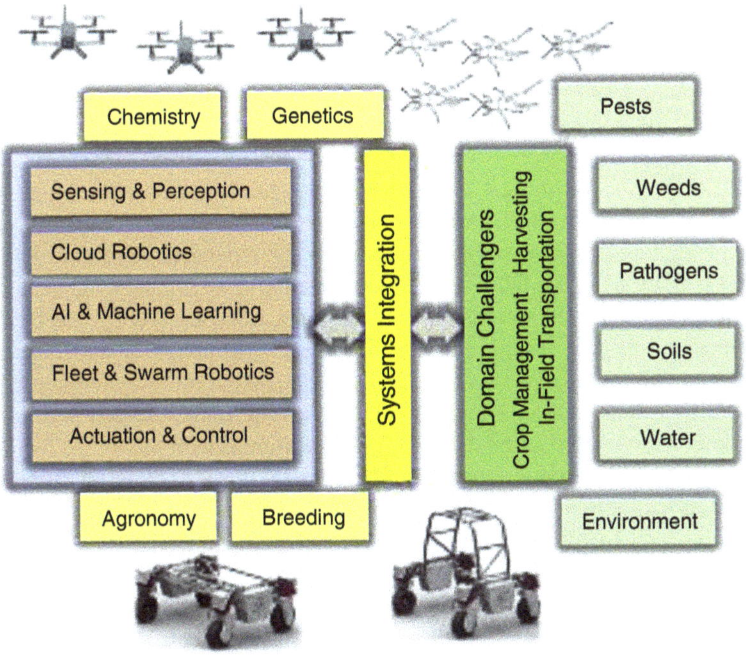

Fig. 14 Smart Technology for broadacre agriculture subsystems [87]

Fig. 15 Universal mobile robot platform for soil sampling [91]

Commercial agricultural robots can already be seen running in the fields on both European and North American farms. The companies offer modern farmers innovative agricultural machines that increase the efficiency of technological processes and reduce labor costs and negative environmental impact. Some robot manufacturers focus on the robot itself, while others focus on its functionality, such as collecting data on plant condition, measuring soil properties, weed recognition, or improving interaction with the tools used.

The use of robots in precision agriculture is mostly focused on soil sampling and mapping [88, 89]. Generating a sampling model is a very important task, so automation of the sampling process has advantages. In order to achieve higher efficiency and quality, automation of the whole soil sampling process is required [90]. A mobile robot platform has been developed at the Estonian University of Life Sciences (Fig. 15). Weighing 470 kg, this equipment is ideal for automating repetitive light tasks commonly performed by humans, such as automatic measurements and soil sampling [89, 91]. Soil samples are highly dependent on the current situation and conditions in the field, which requires a higher frequency and accuracy of data collection. This automated and remotely controlled technology enables more frequent sampling than traditional human-operated manual methods. Experiments have shown that the current system is 50% faster than the traditional method [89].

Recently, robots in agriculture have been an important research and development topic, which will play an increasingly important role in the future for various agricultural purposes and will be particularly useful in the application of precision agricultural technologies. Robots can be fitted with a very wide range of sensors that can be used for functions such as guidance, navigation, obstacle avoidance, crop and row detection, crop condition detection, and weed monitoring and control [38, 92]. The authors [38, 93] reviewed sensors mounted on agricultural machines and robots for monitoring and mapping various basic soil parameters. Sensors can be used to measure parameters such as soil quality and chemical

Fig. 16 Solar-powered robot developed in the Australian Center for Fields Robotics, University of Sydney [38]

composition (e.g., moisture content, nitrogen, and pH) and meteorological variables that affect crop growth, such as air temperature, relative humidity, precipitation, sunlight, wind speed, and direction.

The Australian Center for Field Robotics at the University of Sydney has developed a Ladybird robot suitable for precision farming (Fig. 16) [38, 93]. This solar-powered autonomous robot has an integrated GPS navigation system, forward and rear-facing LIDARs sensors, and a high-resolution digital video camera (Fig. 17). Thanks to the sensors mounted on the robot, it can independently drive through crops, identify and avoid obstacles, and detect crop rows. The height of the crop is determined by a laser sensor. The hyperspectral video camera captures infrared and ultraviolet data in the 400–900 nm wavelength range. The resulting spectral information allows the system to determine the shape and color of the crop, and their spectral traces allow some conclusions to be drawn about crop health [38, 93].

The data obtained with the help of sensors can be used to spread the optimized amount of fertilizer. Vougioukas [94], notes that robotic precision spraying and fertilization operations are often performed using sensors rather than relying on existing maps. Bogue [38], draws attention to robotic agricultural machines and notes their important role in precision agriculture, as these and similar systems have been shown to increase yields while significantly reducing the amount of agrochemicals used.

A team of researchers [95] from the Free University of Bozen- Bolzano, Faculty of Science and Technology, Italy, presented a mobile robot sui-table for plants sensing, that meets the following basic technical requirements: the ability to move quickly, off-road and turn manoeuvres on even or steep hills; is easily transported; individual crop monitoring sensors can be easily attached to it (Fig. 18). In addition to meet these requirements, the robot must ensure a high level of reliability and

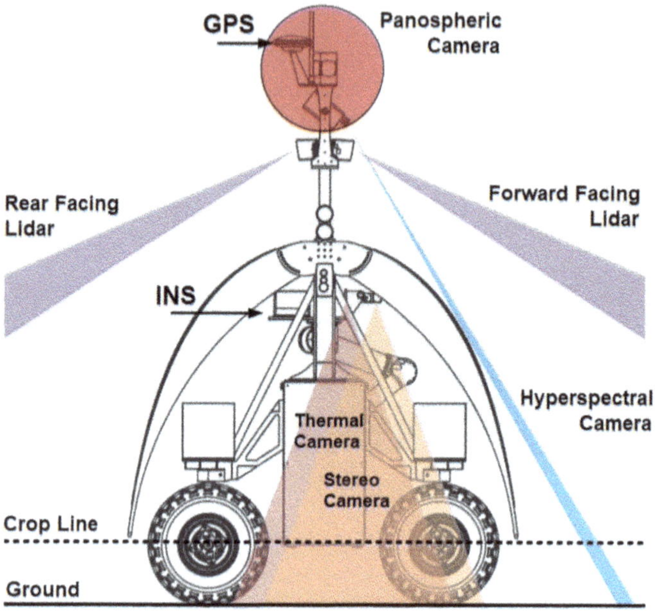

Fig. 17 Location of sensors on the robot [93]

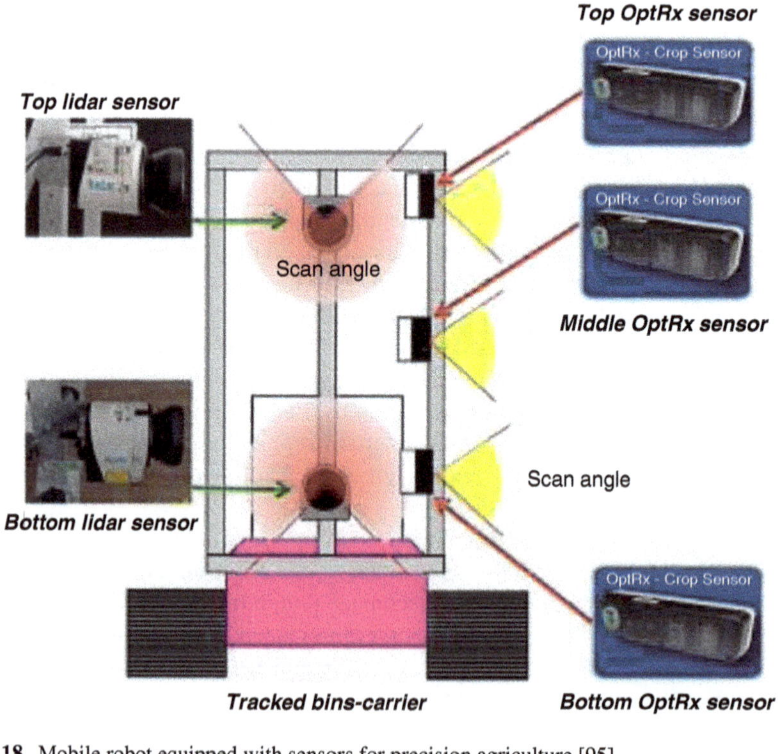

Fig. 18 Mobile robot equipped with sensors for precision agriculture [95]

Fig. 19 Robot BoniRob. (Photo by the authors from the Agritechnica exhibition 2019)

safety for the people around it and the environment as a whole, as safety for humans and the environment becomes increasingly important when considering a fully autonomous system that does not require any human intervention. This robot used a crawler chassis, which provides a wide range of versatility for this robot, excellent contact with the soil, which ensures high grip and low soil pressure. The remotely operated compact (length × width: 1.14 × 1.12 m) robot provides high lifting capacity (500 kg), although its own mass is small (250 kg) [95].

The main problems faced by crop monitoring robots are lighting, background separation, and data manipulation [92]. Robotic technologies for plant monitoring and phenotyping are very similar, using a variety of sensors and vision systems to gather as much information as possible about the plants being monitored. One of the first robots to emerge was the autonomous phenotyping robot BoniRob [96], using multi-sensor data synthesis to measure crop density, uniformity, height, stem thickness, and other parameters. An improved platform, BoniRob (Fig. 19), was later developed that can be used to implement precision spraying and soil hardness determination programs [97].

Accurate weed detection and identification is one of the key factors determining the effectiveness of a robotic weed control system. Using the *Drop on Demand* system, which accurately detects weeds in the plant row and selectively sprays herbicide droplets on the weed leaves, 100% efficiency has been achieved [98].

The ability to precisely separate weeds from crop plants is a very important task for the *ecoRobotix* robot by adapting it to a precision spraying system for plant protection products. According to the manufacturers, this robot makes it possible to detect and destroy weeds with up to 95% accuracy. Using a video camera, RTK-GPS systems, sensors and artificial intelligence to analyze the images, this weed

Fig. 20 Robot Dino. (Photo by the authors from the Agritechnica exhibition 2019)

control robot can detect and locate weeds in and between rows as it moves through the crop field. Thereafter, two robotic arms capable of performing up to 4000 movements per minute can spray each detected weed with a very low dose of herbicide [99].

The self-propelled electric robot *Dino* (Fig. 20) of the agricultural robot developers Naïo Technologies can mechanically kill weeds in the rows and rows of vegetables. The weed control solution is based on deep learning technology that allows the machine to recognize key plants and weeds regardless of their color, variety, and growth stage [100].

With video cameras and RTK-GPS systems, the robot *Dino* is able to accurately and in real-time determine the location of plants in rows. Although Dino's first purpose was precision mechanical weed control, it could be called an autonomous crop care robot when it installed additional intelligent systems. In addition to precision weeding, the robot has extensive capabilities in data collection, as it can count plants, estimate their size, calculate the date of harvest, future thinning operations, or fertilize [100].

French manufacturer Carre crop care robot *Anatis* (Fig. 21), using different sensors, a built-in camera and a GPS system, analyzes each individual soil area, collects data on soil and air humidity, temperature, plant density, growth stage, weeds or other undetected objects. All this collected data can be monitored by the farmer in real time on a smartphone or tablet, and then analyzed while the appropriate decisions are made [101].

The decision that a robot should be as versatile as possible and capable of performing several different agricultural tasks is followed by most research centers and commercial enterprises [102, 103]. The Danish-developed robot *Robotti* (Fig. 22) can independently perform various agricultural tasks, such as crop monitoring, sowing, planting, fertilizing, weed control, or harvesting. Video camera systems (RGB,

Fig. 21 Carre Anatis. (Photo by the authors from the Agritechnica exhibition 2019)

Fig. 22 Agrointelli Robotti. (Photo by the authors from the Agritechnica exhibition 2019)

NIR, LiDar, Thermal, etc.) and sensors can determine various crop characteristics: leaf area index, vegetation index, plant germination, height, color, weed density, etc. However, the ability to perform each task properly and accurately results in more complex hardware and software and the need for a variety of sensors [92, 103].

The ability of robots to move as close to plants as possible is a particularly important step in precision farming as the accuracy of the data obtained increases. With much smaller dimensions, lighter and manageable sensor and data analysis, robots are better suited to individual soil characteristics research and plant care compared to traditional tractors. Robotics in agriculture enables a return to small-scale farming practices. Collecting data with smart sensors can make it easier for farmers to monitor relevant soil and plant characteristics, maintain their good condition and productivity, while adapting to changing environmental conditions.

6 Estonian Use Case of N-Fertilization with GreenSeeker Handheld Crop Sensor

A fertilization experiment was conducted at the Rõhu Experimental Station (58°21.3´N, 26°31.3´E) of Estonian University of Life Sciences on spring wheat (Fig. 23). A mean annual precipitation of this region is 680 mm and mean annual temperature 5.8 °C (Estonian Weather Service, average of 1981–2010). The size of a trial field was 1 ha and the size of each treatment plot was 20 m^2 (2 m × 10 m). Mineral nitrogen treatments were 0, 40, 80, 120, 160 and 200 kg N ha^{-1}. The soil of the field site is Glossic Retisol with sandy loam texture.

The aim of this experiment was to examine the relationship between NDVI and the fertilization rate set with Greenseeker, and how fertilization rate affects yield in years with different weather conditions. GreenSeeker Handheld Crop Sensor was used to assess NDVI (Fig. 24)

To calculate the fertilizer application rates, fertilizer tables have been prepared for the most cultivated crops (Fig. 25). To calculate the optimal fertilizer rate, a reference area is used in the field that has received the maximum amount of nitrogen fertilizer during or after sowing. Based on the NDVI of the reference area receiving the maximum nitrogen fertilizer and the NDVI of the area waiting to be fertilized, a

Fig. 23 Fertilization experiment trial fields at Rõhu Experimental Station. (Photo by Toomas Tõrra)

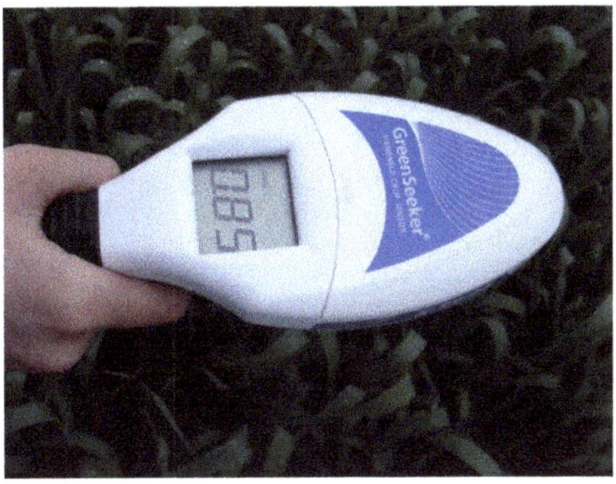

Fig. 24 GreenSeeker Handheld Crop Sensor. (Photo by Toomas Tõrra)

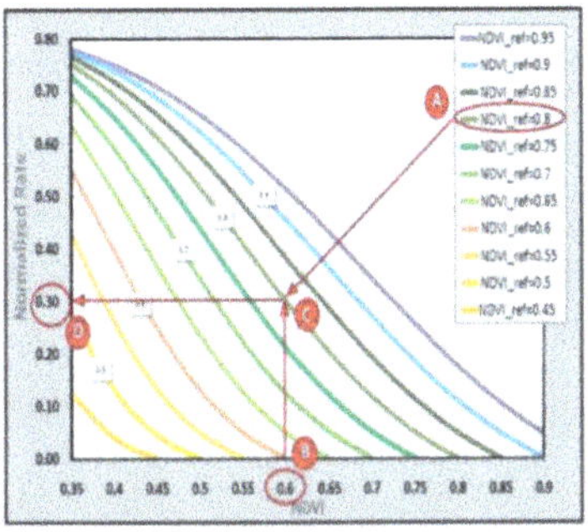

Normalized rate graph

Crop	%N	lb/bu	Maximum yield (bu/ac)											
			15	25	50	75	100	125	150	175	200	225	250	275
Spring wheat	2.45	60		66.8	134	200	267	334	401	468				
Winter wheat	2.30	60	37.6	62.7	125	188	251	314	376	439				
Dryland corn	1.30	56				99.3	132	165	199	232	265	298	331	
Irrigated corn	1.25	56					127	159	191	223	255	286	318	350
Barley	1.70	48	22.3	37.1	74.2	111	148	185	223	260				
Triticale	2.10	54	30.9	51.5	103	155	206	258	309	361				
Sorghum	1.34	56				102	136	171	205	239	273	307		
Canola	3.10	50	42.3	70.5	141	211	282							

Crop table

Fig. 25 Calculation of fertilizer rate based on NDVI measured with GreenSeeker

coefficient is obtained by multiplying the amount of fertilizer required to obtain the maximum yield. This gives the fertilization rate for the area to be fertilized.

The experiment was conducted in 2 years (2017 and 2018). Summer 2018 was unusually warm, and a severe drought occurred. Long-term (1981–2010) average precipitation for July is 72 mm in this region. Precipitations in July 2017 were 57 mm but in July 2018 only 23 mm. Precipitations in May 2017 were 28 mm but in May 2018 only 10 mm (long-term average for May is 56 mm). In June, precipitation was 65 mm in both 2017 and 2018 (long-term average 84 mm in June).

The results of these 2 years clearly show the great impact of the weather. If there is no precipitation during the vegetation period, as it was in 2018, then the plant will not be able to absorb the nutrients given to the field and the development of plants will be inhibited, and this is also reflected in the NDVI reading (Fig. 26). At the same time, in 2017 there was enough rainfall during the vegetation period and the plants absorbed well the nutrients given by fertilizers, which is also reflected in the stable growth of NDVI, as the fertilization rate increased. Similar to the increase in NDVI as the fertilization rate increases (Fig. 26), there is also an increase in yield according to the increase in fertilization rate (Fig. 27). A comparison of both figures

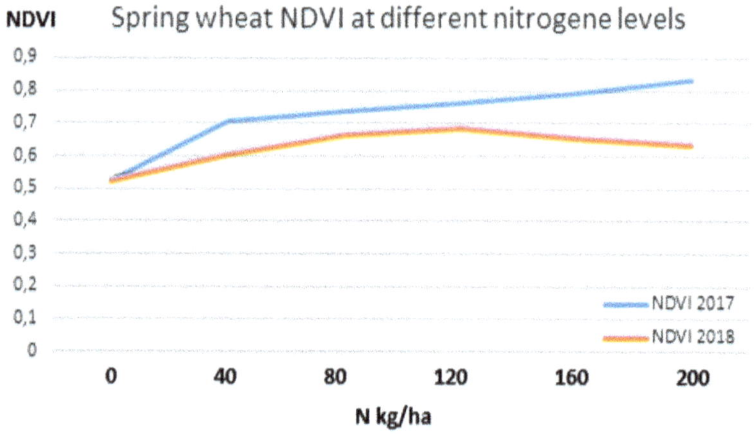

Fig. 26 Spring wheat NDVI measured with GreenSeeker at different fertilization levels in 2 years 2017 and 2018

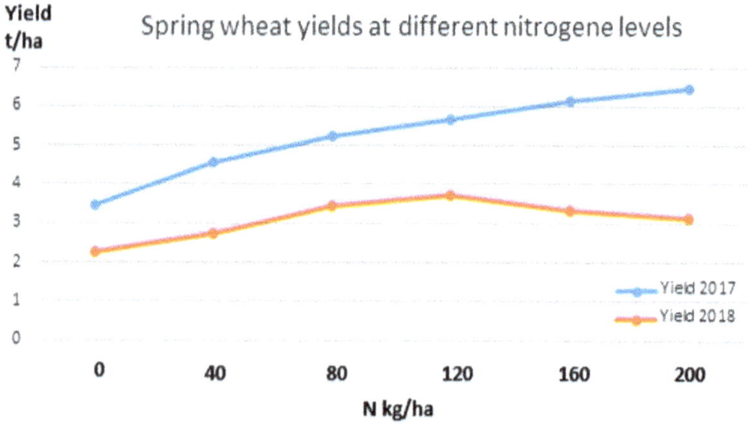

Fig. 27 Spring wheat yield at different fertilization levels in 2017 and 2018

shows similar behavior for both NDVI and yield with a similar increase with increasing nitrogen rate. Measurements show that higher yields can be expected as NDVI increases. Although the yield depends largely on weather conditions, by measuring NDVI we can predict the yield both in drought year and during year with increased precipitation. In the drought year, the NDVI reading on the same N background as well as the yield was lower than the results on the same N level in the rainier year. For example, in 2017, at the fertilization level N 80 kg ha^{-1} NDVI was 0.74 and a yield was 5.3 t ha^{-1}, while in 2018, N 80 kg ha^{-1} treatment had a NDVI of 0.67 and a yield of 3.5 t ha^{-1}.

7 Experience of Precision Farming in Lithuania – Use Cases

Modern precision farming systems have become more widespread in the last decade. Precision spraying and fertilization technologies are the most widely used. In assessing the soil properties, the application of phosphorus, potassium and other fertilizers necessary for the growth of plants is performed using a variable fertilization rate according to the individual field areas and the amounts of nutrients determined in them. According to the soil properties, the main fertilization with nitrogen fertilizers is also performed, only later when the plants grow, additional nitrogen fertilization is performed by scanning the crop. This subsection presents the relevant results of precision fertilization research carried out in Lithuania in 2014–2019 in different locations, including descriptions of the equipment used.

7.1 Measurement of Soil Electrical Conductivity

Electrical conductivity is the ability of a material to transmit an electric current and is usually expressed in millis Siemens per meter (mS m^{-1}) [104]. Soil electrical conductivity is a measure that correlates with soil properties that affect crop productivity, including soil structure, cation exchange capacity, drainage conditions, organic matter levels, salinity, and subsoil properties.

The electrical conductivity of the soil varies depending on the amount of moisture that is retained by the soil particles. Thus, electrical conductivity strongly correlates with soil particle size and structure. Saline soils and clay have high conductivity, silt has medium conductivity, sand has low conductivity [12, 104, 105].

First, when driving an SUV around the field under study, the boundaries of the field are measured, and the exact size of the field is determined. Soil scanning was performed using an EM-38 MK-2 electrical conductivity scanner to determine soil differences in the field (Fig. 28). In Lithuania, measurements of soil electrical conductivity were performed by driving off-road Toyota Hilux technological tracks every 30 m and towing the device EM38-MK2 mounted on a plastic sled, which was placed in a plastic case to protect from adverse environmental conditions (Fig. 29).

After measuring the electrical conductivity of the soil in the field, all the collected information on the electrical conductivity of the field soil, from the computer Panasonic CF-19 (Fig. 30) in the SUV, was sent to the office computer via the Internet using a 4G connection. In the office, sent information, using the Convert EM38-MK2 program was converted to a CSV file. Further, using the QGIS program, 3-hectare polygons (zones) were formed, which were similar to soil structures. The generated polygons were transferred back to the Panasonic CF-19 in the SUV computer in digital shp format.

In 2012, measurements of the electrical conductivity of the selected study field were performed, and the distribution of the total field height was determined as well (Fig. 31). Electrical conductivity ranged from 57.9 to 64.1 mS m^{-1}. The largest part

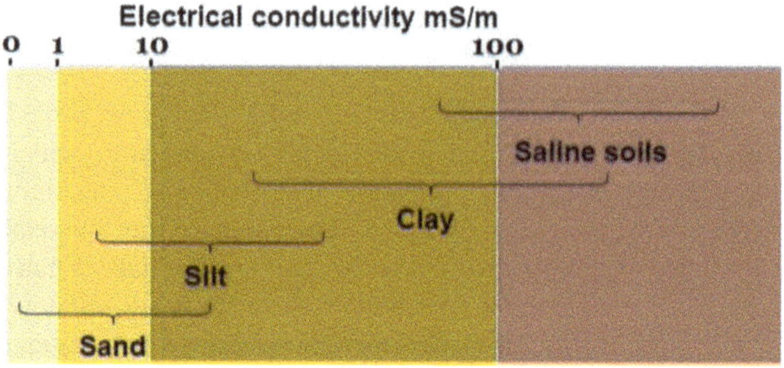

Fig. 28 Apparent electrical conductivity in different soils (Figure redesigned by the authors according to Stafford [12]

Fig. 29 Mobile determination of field boundaries and soil electrical conductivity: 1 – GPS antenna on the SUV roof; 2 – Trimble EZ-Guide 250 navigation mounted inside the SUV on the windshield; 3 – Electrical conductivity meter EM38-MK2 (Figure made by the authors)

(24%) of the field electrical conductivity was 64.1. These measurements can be easily linked to the readings of the GPS navigation system, which in this particular field showed that the altitude of the field varied from 45.6 to 51.6 m.

7.2 *Automated Soil Sampling*

Soil sampling was performed using Agricon SUV equipment (Fig. 32). The operator performs soil sampling according to the formed polygons of the same structure while driving through the field using the trajectory of the letter "Z" (Fig. 33). Samples are taken with a grooved needle mounted on the end of an automatically operated mechanical arm. From 12 to 20 subsamples are taken per soil sample. Sampling with automatic equipment is about 30% more efficient than sampling with stopping. Each time the needle is inserted into the soil to a depth of about 20–30 cm, the mechanical hand returns to its original position and the soil in the needle groove is pushed out into the box. The equipment has a total five boxes that

Fig. 30 Data acquisition equipment: 1 – GPS antenna for data collection from EM38-MK2 equipment and recording of sampling trajectories. GPS antenna Navilock 602U for transmission of position and altitude data to Panasonic CF-19 computer; 2 – Panasonic Toughbook CF-19 in the docking station (Figure made by the authors)

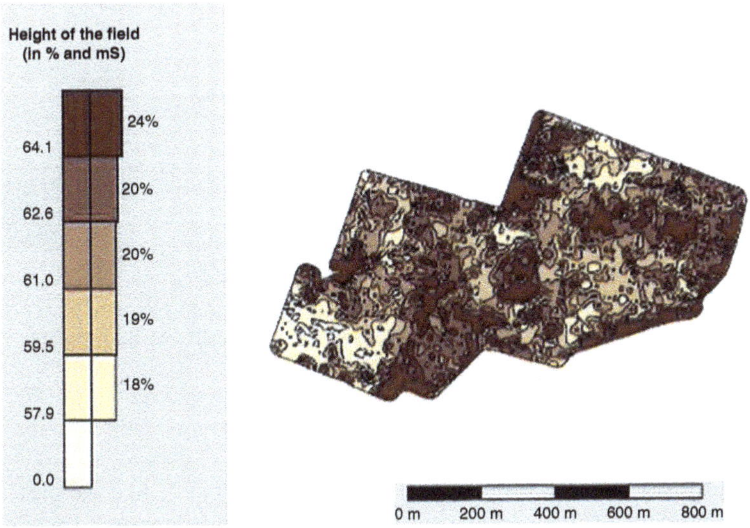

Fig. 31 Measured electrical conductivity of the test field (The figure was prepared by the authors)

Fig. 32 Automatic soil sampling equipment: 1 – Needle with internal notch for soil sampling; 2 – Automatically operated mechanical hand; 3 – Needle cleaning mechanism; 4 – Box for collecting soil samples (five boxes) (Figure made by the authors)

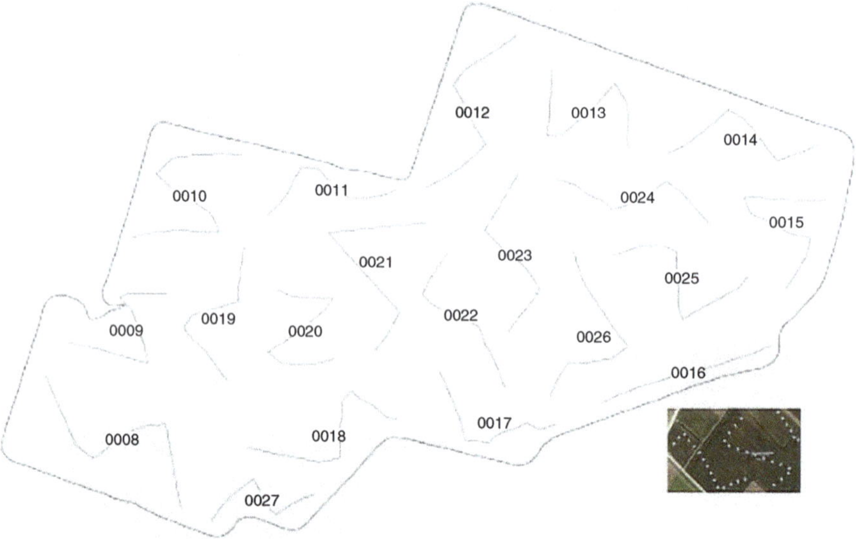

Fig. 33 Field sampling using the Z trajectory (The figure was prepared by the authors)

hold individual taken samples of different soil property zones. One box holds 300 to 500 g of soil.

One operator can take 60 to 100 subsamples non-stop, i.e., five samples. When all the boxes have been filled with soil, the operator removes the soil in each box and packs it into individual plastic bags with a bar code on them, which records all the necessary information about the sample: 1 – the coordinates of the sampling point; 2 – information about the researched field; 3 – operator who took the sample; 4 – date and time of sampling. For soil testing, all this collected information was sent to a laboratory accredited in Germany by Agrolab GmbH. One operator can take samples from approximately 200 ha per day.

Tests for pH, magnesium (Mg), potassium (K) and phosphorus (P) of the main elements are performed as standard. Mg and pH tests are performed by $CaCl_2$ method, K and P – CAL method. (Extraction of calcium acetate lactate (CAL) in Germany [106, 107] is a standard soil test method for determining the P and K value of available phosphorus and forms the basis for P and K fertilization recommendations [108].

The results of the tests obtained from the laboratory in digital format in PDF and CSV are sent to Agricon's head office in Germany, where the information obtained is uploaded to the Agriport online program to an account created for a specific farm. It takes about 30 working days from sampling to receipt of test results. Fertilization recommendations are also provided along with the results of soil analyses. Based on the crop rotation planned in a specific field and the expected yield of each crop, a 4-year fertilization plan for pH, P, K, Mg and fertilization maps for the spreader computer in pfb, shp and iso formats is created. The Agriport program provides opportunities to independently plan crop rotation, create a fertilization plan and prepare digital fertilization maps for the spreader computer. The program shows the results of specific field soil studies and maps of the distribution of pH, Mg, K and P elements. It is also possible to see the soil sampling trajectories and the information of the tests performed at each sampling point and the need for elements. The research results in the program are presented both in pure form and in the form of oxides.

7.3 *Changes of Mineral Fertilizer Elements in Soil Using VRF*

For the application of mineral fertilizers in a specific field at a variable rate, according to the maps, an agricultural unit consisting of a tractor, a mineral fertilizer spreader and a fertilization computer-terminal was used (Fig. 34).

The change of mineral fertilizers in the soil was observed in the field of 57.98 ha of farmer Majauskas R. farm in Antagyne village, Kaunas district, Lithuania (55°11′N, 23°86′ E). Initial soil surveys were conducted in 2012. It was found that according to the distribution of phosphorus in the soil in the whole field, the largest part (32 and 26%) was formed by B+ and B- soil groups, in which the phosphorus content in the soil varied from 3.2 to 5.6 mg 100 g^{-1} (Fig. 35).

After the implementation of fertilization at a variable rate in a 4-year crop rotation and repeated phosphorus studies in the soil, it was found that the whole field became significantly more uniform in terms of phosphorus distribution (Fig. 35). In 2016, in the same field, the largest part (60%) was formed by one C-group soil, in which the phosphorus content ranged from 5.1 to 5.9 mg 100 g^{-1}.

Fig. 34 Variable rate of fertilizer application according to fertilization maps equipment: 1 – Rauch spreader terminal CCI 100 (Muller); 2 – Mineral fertilizer spreader Rauch Axis-H; 3 – John Deer 6630 tractor (The figure was prepared by the authors)

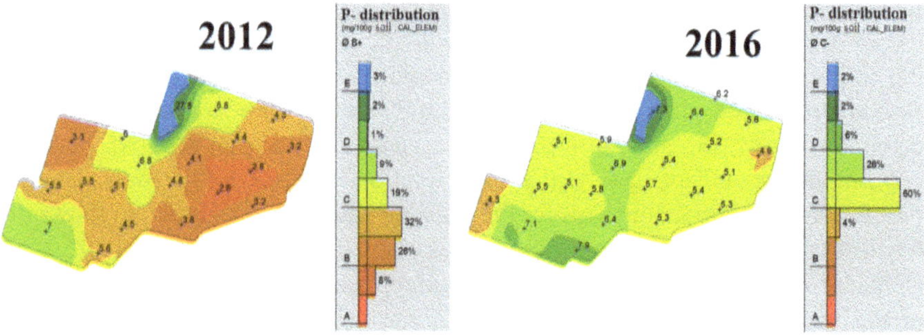

Fig. 35 Distribution of phosphorus in the soil in 2012 and 2016 using VRF (The figure was prepared by the authors)

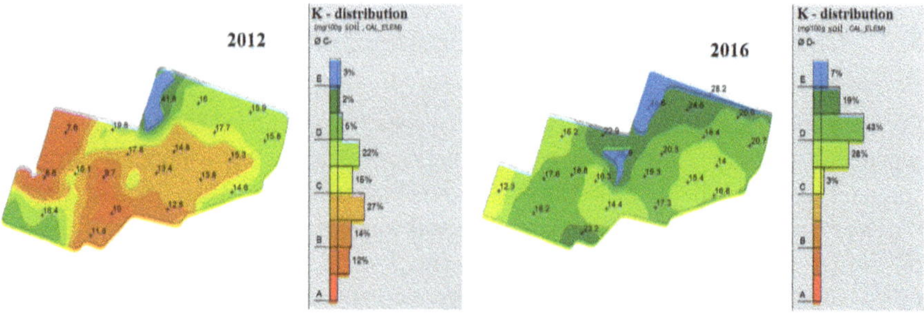

Fig. 36 Potassium distribution in the soil in 2012 and 2016 using VRF (The figure was prepared by the authors)

In the same field, similar studies were performed with the distribution of potassium in the soil (Fig. 36). If in 2012 the distribution of potassium in the field was quite uneven, it mostly belonged to groups B and C, 41% and 37%, respectively. As much as 12% of the soil in this field belonged to the lowest group A, according to

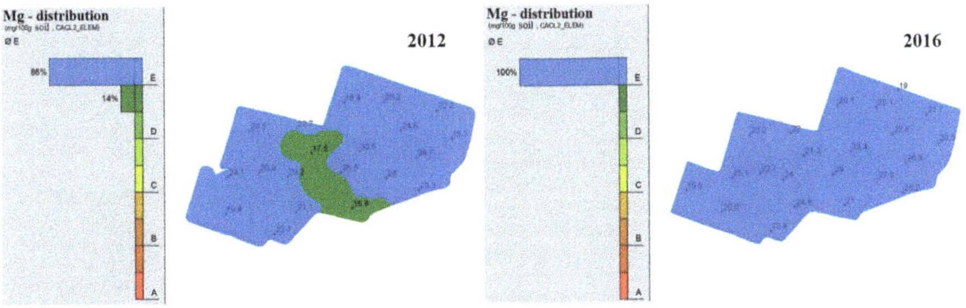

Fig. 37 Magnesium distribution in soil in 2012 and 2016 without any application (The figure was pre-pared by the authors)

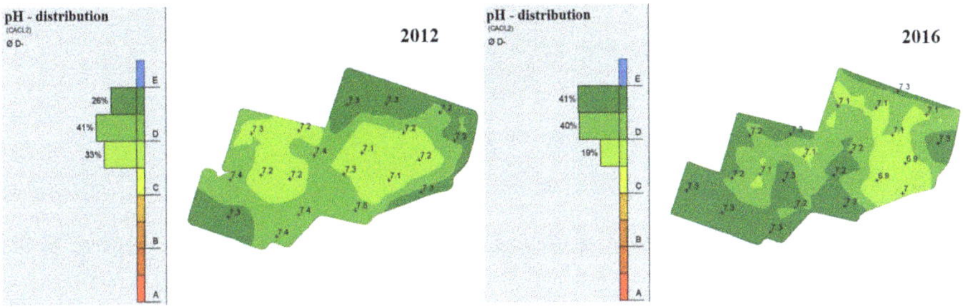

Fig. 38 Soil pH distribution in 2012 and 2016 using VRF (The figure was prepared by the authors)

which the potassium content was less than 10 mg 100 g^{-1}. After 4 years of precision fertilization, this field became much more uniform in terms of potassium distribution in the soil, at most 62% belonged to class D, according to which the potassium content in the soil varied from 14.3 to 20.7 mg 100 g^{-1}.

Magnesium distribution maps demonstrate exceeded element amount in all field and no fertilizer application have been made (Fig. 37), then after four years the whole field (100%) according to the magnesium distribution was one E soil group. It was calculated that will be enough for 4-year crop rotation yield.

When assessing the influence of the variable rate of precision fertilization on the soil pH distribution maps, no major differences were observed. Both in 2012 and after 4 years, the soil belonged to the same D- group on average (Fig. 38).

7.4 Nitrogen Fertilization at a Variable Rate Using Yara N-Sensors

In 2019, 40 operating Yara sensors were used in Lithuania, which worked on an area of about 50,000 ha. This Yara N-Sensor system is most widely used in central Lithuania (Fig. 39), where soils are the most fertile and farmers receive the highest

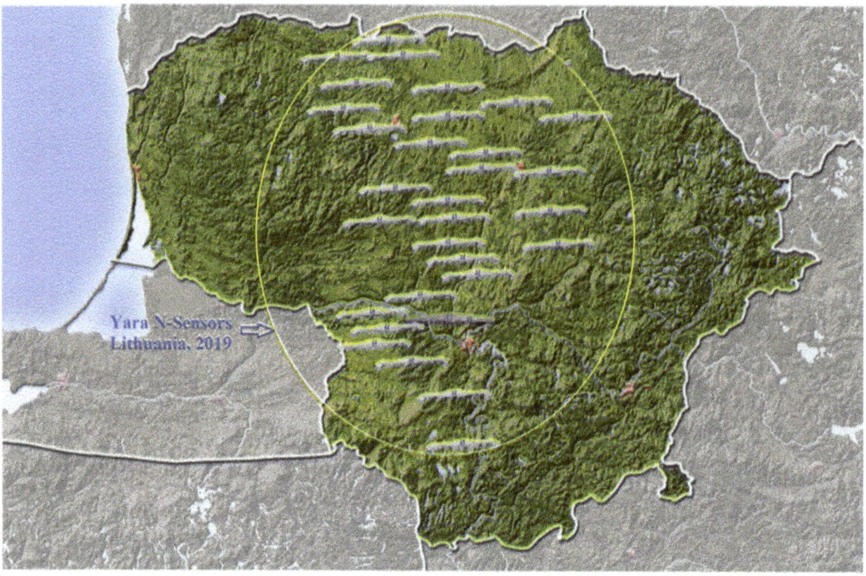

Fig. 39 Distribution of Yara N-Sensor use in Lithuania in 2019 (The figure was prepared by the authors)

income from crop production. Over 250 trials have been conducted on farms over the past 10 years, comparing fertilization at a constant rate with fertilization at a variable rate using a Yara N-Sensor.

In most agricultural fields, nitrogen uptake is not uniform across the field, as it is highly dependent on soil water regime, particle size distribution, nitrogen leaching, nitrogen uptake by other key elements (e.g., phosphorus, potassium, pH) quantity. Agricultural plants mirror the real situation in the soil. Yara N-Sensor equipment is used to reduce nitrogen uptake differences in individual field locations. With this equipment it is possible to see up to 1000 times better than with the human eye, so even for a uniform-looking field, thanks to this sensor, fertilizer differences between different parts of the field can be as much as tens of kilograms of active substance. The Yara N-Sensor scans the crop and determines how much N (kg ha^{-1}) is absorbed by the plants, after the N-fertilization program performs the calculation and sends the calculated fertilizer variable rate to the spreader computer. It is one of the modern precision farming systems that determines the exact location in the field and can decide in real time where and how much N fertilizer should be spread.

Experimental research was carried out in different Lithuanian farms in 2018–2020. The nitrogen sensor was mounted on the roof of the tractor (Fig. 40). The computer terminal with N-fertilization program is mounted in the tractor cab. The nitrogen sensor measures in both directions at a 45-degree angle (Fig. 41), so it only sees plants. In this way, the influence of soil and smaller weeds in the rows is avoided than when measured directly from the top, especially in the early stages of

Fig. 40 Equipment used for precision spreading of nitrogen fertilizers: 1 – Yara N-Sensor ALS; 2 – Agricon TOUCH800 terminal (Muller computer); 3 – John Deer 6630 tractor; 4 – Yara N-fertilization software (The figure was prepared by the authors)

Fig. 41 Nitrogen fertilizer application using a 45-degree Yara N-Sensor ALS [84]

plant development. The accuracy of the sensor is about 93%, which does not depend on the plant variety, growth phase or plant crop density. The program used for nitrogen fertilization is based on agronomic algorithms, which, depending on the plant and growth stage, decide what rate of N fertilizer should be given to the plant according to the condition of the crop at a specific location.

The principle of N-sensor measurement is based on the measurement of the far-red wavelength. Band-pass filters with 730, 760, 900 and 970 nm center wavelength are used. These wavebands have been identified as optimal for determining the nutritional status of the crop [109]. The xenon light sent the light to the crop area and if the reflected into the optics wavelength is short, then photosynthesis is active in that field location, which means that the plants have sufficient nitrogen supply

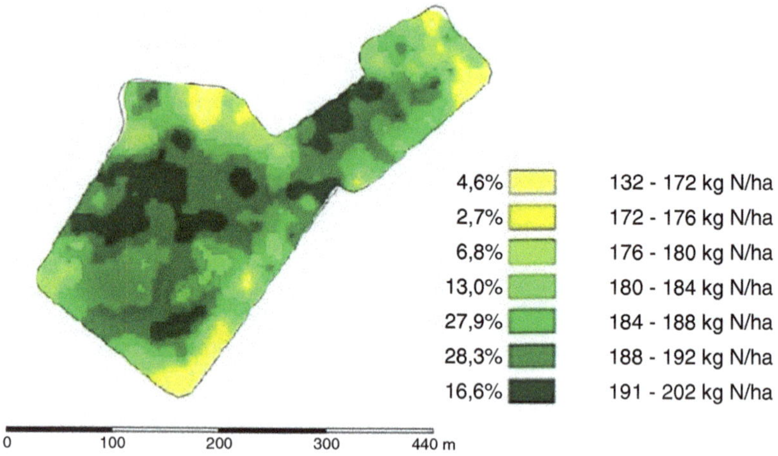

Fig. 42 N-uptake map in winter wheat crop in 2019 May 30 (The figure was prepared by the authors)

and vice versa. The N-sensor simultaneously measures not only the intensity of photosynthesis but also the biomass of the plant crop at a specific field location. According to the N-sensor, the terminal display shows how much nitrogen (kg ha^{-1}) has been uptaken in the plants at a specific field location.

The Yara N-sensor is used during plant vegetation for additional two fertilizations and for qualitative 3rd or 4th fertilization in cereals. In oil seed rape this equipment is used for first and second fertilization. If the rapeseed biomass is no more than 60% dead in the spring compared to the autumn biomass, can be used N-sensor, and if more, we use N-application map made according a biomass map of the rapeseed field scanned from the autumn. In addition to conventional nitro-gen fertilization, the N-sensor can also be used for: N (liquid, granular) fertilization, cereals, oilseed rape, corn, potatoes, as well as for spraying growth regulators, fungicides, desiccants.

7.5 *VRF Maps for N-Fertilization Using Proximal Sensors*

7.5.1 N-Fertilization Maps for Winter Wheat

Experimental research was carried out in 2019 in Joniskis district in the field (56°18′ N, 23°64′ E) of Kurmaiciai agricultural company, where winter wheat was grown. The total field area is 10.72 ha. Studies of nitrogen uptake by whole crop using Yara N-Sensor ALS were performed and mapped (Fig. 42). N-uptake data showed that the worst plants absorbed nitrogen in areas at the edges of the field. The main reasons why these areas had a harder time absorbing nitrogen may

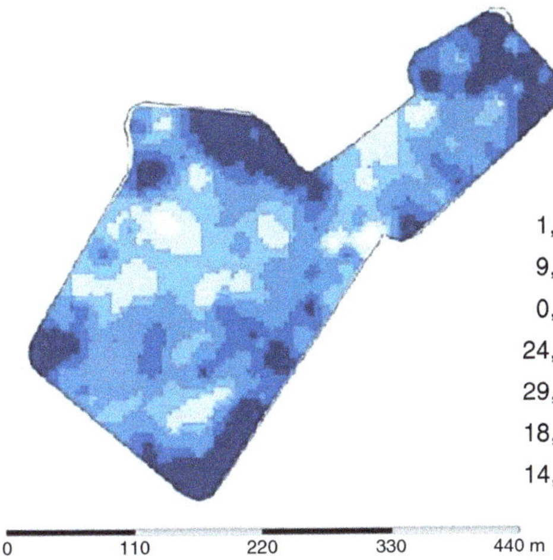

Fig. 43 Recommended N-fertilization map of winter wheat in 2019 May 30 (The figure was prepared by the authors)

have been less favorable physical properties of the soil for plants to grow, higher soil hardness.

Based on the data obtained for N-uptake, a recommended winter wheat nitrogen fertilization map (Fig. 43) for variable fertilization was created. The average recommended fertilization rate was 41 kg N ha^{-1}, the standard deviation was 7.2 kg N ha^{-1}. In the central part of the field there were zones where the recommended rate of additional fertilization was up to two times lower than the average and up to three times lower than at the edge of the field.

The winter wheat crop was harvested on July 24, 2019. Yields of winter wheat ranged from 4.9 to 7.54 t ha^{-1} in the field (Fig. 44). The average yield of this field was 6.30 t ha^{-1}.

The yield results obtained showed that the highest winter wheat yields were found in the central part of the field, where N-uptake was highest and recommended additional N-fertilization rates were lowest. Due to more frequent machine movements and stronger soil compaction, the field margins gave the lowest yield, in some cases less than 4.9 t ha^{-1}, although the recommended rates of additional N fertilization were higher than average. In total, such a low yield of winter wheat was 7% of the total field area.

No direct relationship has been established between the yield of cereals and the amount of N fertilizer applied. One year we can spread 150 N kg ha^{-1} and get 8 t ha^{-1} of winter wheat, but the next year we can spread 180 N kg ha^{-1}, but we will get a lower yield of 6.5 t ha^{-1} of winter wheat. Precipitation and temperature usually determine the availability of nitrogen in the soil to plants and its uptake during vegetation.

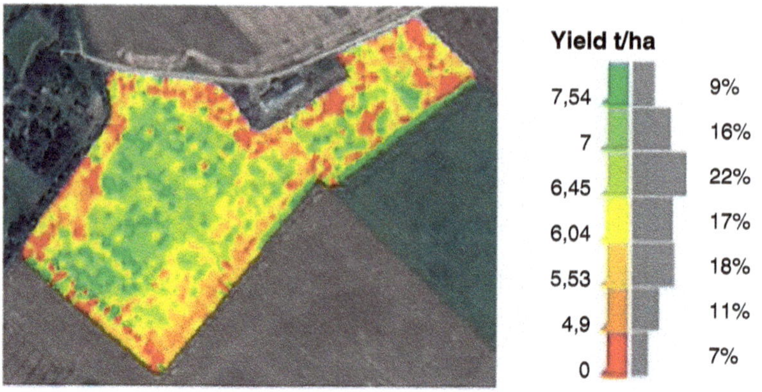

Fig. 44 Map of the winter wheat harvest July 2019 (The figure was prepared by the authors)

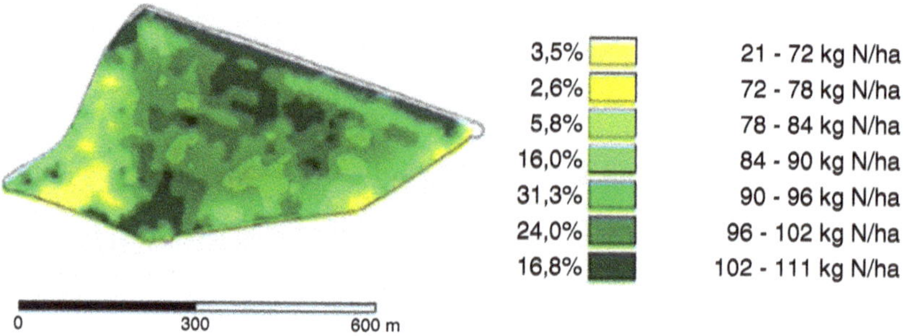

Fig. 45 N-uptake map in winter oilseed rape crop in 2019 October 8 (The figure was prepared by the authors)

7.5.2 N-Fertlization Maps for Winter Rapeseed

The latest research was carried out in 2019–2020 in the field (55°92′ N, 23°91′ E) of Navickas A. farm in Pakruojis district, where winter oilseed rape was grown. The field area is 17.46 ha. After sowing and winter rape germination, in 2019 October 8 the N-uptake was evaluated using Yara N-Sensor ALS equipment (Fig. 45). The average N-uptake in the field was found to be 93 kg N ha^{-1}. The minimum and maximum N-uptake varied from 21 to 111 kg N ha^{-1}. The standard deviation was 10.0 kg N ha^{-1}.

In the spring of 2020 (March 15), fertilization was performed according to the N-fertilization map (Fig. 46). The main statistics of the N-fertilization map were as follows: mean 86 kg N ha^{-1}, standard deviation 7.6 kg N ha^{-1}.

After 20 days (April 5), a repeat study of N-uptake was performed, and a map was drawn (Fig. 47). The average N-uptake of winter wheat crop was found to be 52 kg N ha^{-1}, with a standard deviation of 8.8 kg N ha^{-1}. Based on the obtained data,

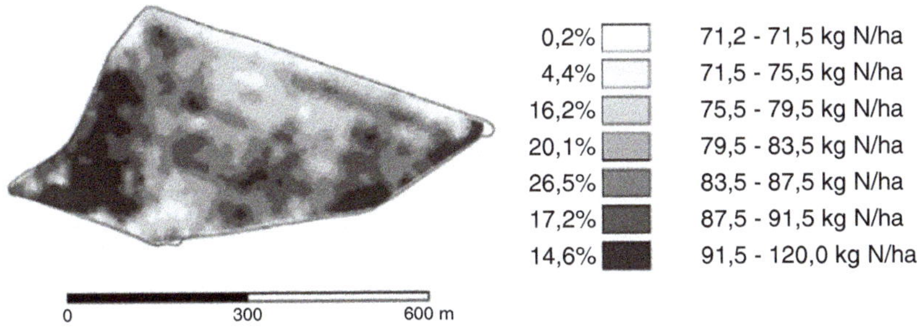

Fig. 46 Additional fertilization of winter oilseed rape according to the N-fertilization map in 2020 March 15 (The figure was prepared by the authors)

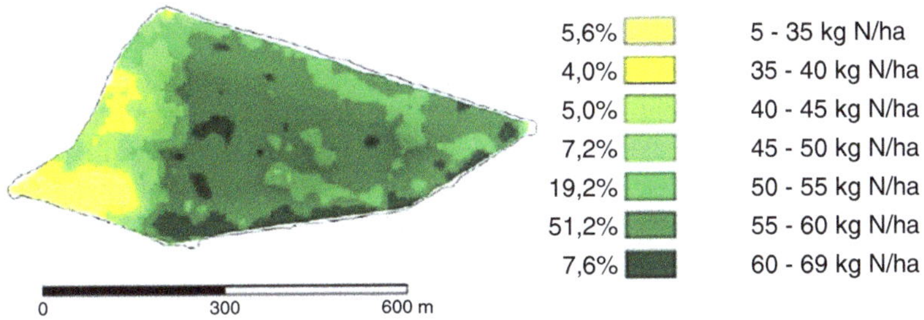

Fig. 47 N-uptake map in winter oilseed rape crop in 2020 April 5 (The figure was prepared by the authors)

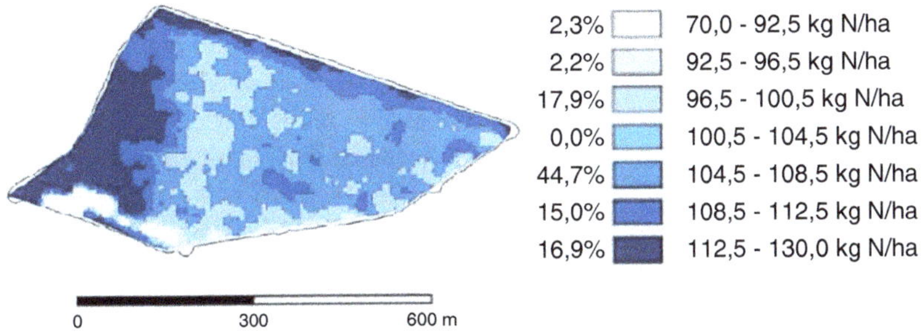

Fig. 48 Recommended N-fertilization map of winter oilseed rape in 2020 April 5 (The figure was prepared by the authors)

a recommended nitrogen fertilization map (Fig. 48) was created for the whole field fertilization at a variable rate. According to the map, it is recommended to fertilize in different places from 70 to 130 kg N ha^{-1} of N-fertilizer. The average recommended fertilization rate was 107 kg N ha^{-1}, the standard deviation was 7.5 kg N ha^{-1}.

Spring N-uptake map showed that 70% of the field was in the limits of 10 kg N ha^{-1} uptake. It became more even after the N-application based on autumn rape scanning map.

As with winter wheat in the previous study, a similar trend was also observed in this field, with the N-fertilization program recommending a higher fertilizer rate in areas with lower winter rapeseed nitrogen uptake. However, an inverse relationship has been found that in areas with particularly low nitrogen uptake, additional nitrogen fertilization is no longer recommended, as no higher yield increase is expected.

Many years of experience have shown that, from an economic point of view, variable rate fertilization with a variable rate of Yara N-Sensor can increase significant economic benefits, e.g. for cereals this benefit starts from 60 Eur ha^{-1}, for rapeseed it ranges from 60 to 90 Eur ha^{-1}. If this equipment is also used for precise application of growth regulators and fungicides, then the total economic effect of using the N-Sensor can increase up to 100–120 Eur ha^{-1}. The main reasons for the economic benefits are the following advantages: 80–100% lower crop lodging, 12–20% higher combine productivity, 10% fuel savings at harvest, more even crop, 40% more uniform protein, up to 30 kg ha^{-1} better nitrogen balance (less need for additional nitrogen than taken by crop), 0–14% savings in nitrogen fertilizers, 3–7% higher yields, reduced disease prevalence, lower drying costs, 5 days earlier harvesting due to harvest more even mature [110–116].

7.6 *N-Fertilization with Yara N-Tester*

It is recommended to apply the fertilization rate for the main fertilization in early spring from 30 to 80 kg N ha^{-1}, depending on the condition of the crop, pre-sowing crop, nitrogen content in the soil after winter. Yara N-tester readings for plant N-monitoring can be used to determine the timing and rate of additional fertilization for winter wheat. This method makes it possible to reduce nitrogen losses and perform fertilization according to the precise needs of the plants, depending on how much N is needed at a given moment, rather than as much as is written in the textbooks or as much as appears from the eye.

In order to use the fertilizer efficiently and optimally, taking into account the price of the fertilizer and the potential damage of fertilization to the environment, it is very important to know the right time and the exact rate before applying N fertilization. Fertilizing plants when they are not needed, loses the effectiveness of N fertilizers. In addition, unused nitrogen from plants can leach into groundwater or evaporate, leading to severe pollution. In case of using too much N fertilizer, the plants may lodge spread the disease in the crop.

The exact time and rate of additional fertilization can be determined with a Yara N-tester (Fig. 49), which is used to measure the intensity of photosynthesis and the amount of chlorophyll present in the leaves of plants. Before the measurement, the plant species, variety and growth phase are selected. After selecting a location corresponding to the average of the field and measuring 30 fully

Fig. 49 Yara N-tester for additional fertilization to determine nitrogen demand and time (The figure was prepared by the authors)

developed leaves of different plants, the nitrogen demand per kg N ha^{-1} of active substance for a specific cereal variety is immediately displayed on the screen of the device. There are about a thousand different plant variety corrections in this device. The correction shows how one plant variety differs from another, as an example it can be given that with one N-tester unit value one N demand can be close to zero and another variety with the same unit value nitrogen demand can be as much as 45 kg N ha^{-1}.

The exact timing of additional fertilization can be determined from observations of photosynthesis in plants, termed N-monitoring. In the spring, during the plant phases (BBCH-30-51), an average field-specific place is selected for N-monitoring. A flag or other clearly visible observation point shall then be inserted in the area reflecting the average of the crop in that field. From that day onwards, measurements shall be made every 3 days with an N-tester and an N-monitoring table shall be completed, recording regularly every 3 days the growth phase of the plants, the N-tester's readings in units (e.g., 650), average precipitation and temperature, and the amount of fertilized nitrogen in the active substance.

The principle of measuring the device itself is not complicated. With this device, the reflected red and infrared wavelengths in the leaves of a plant are measured, which determines the intensity of photosynthesis and the amount of chlorophyll in the leaf of the plant. The device automatically calculates the wavelength and immediately displays the nitrogen demand in kg N ha^{-1}.

In Lithuania, in the spring of 2019, the farmer's winter wheat field (56°12´ N, 22°91´ E) was monitored with an N-tester and additional nitrogen fertilization was performed accordingly. Figure 50 presents the results of N-monitoring, which show how the adjusted values of the N-tester over time and when next additional nitrogen fertilization was applied. In this figure, it is also possible to see on what day and how much precipitation fell and at which N-tester values N is considered to be optimal.

According to the N-monitoring schedule, the most suitable fertilization time for the plant can be predicted. To avoid higher measurement errors of the device, it is recommended to fertilize with sulfur as the main spring fertilizer, as this is the second element, after nitrogen, that has the greatest effect on plant photosynthesis.

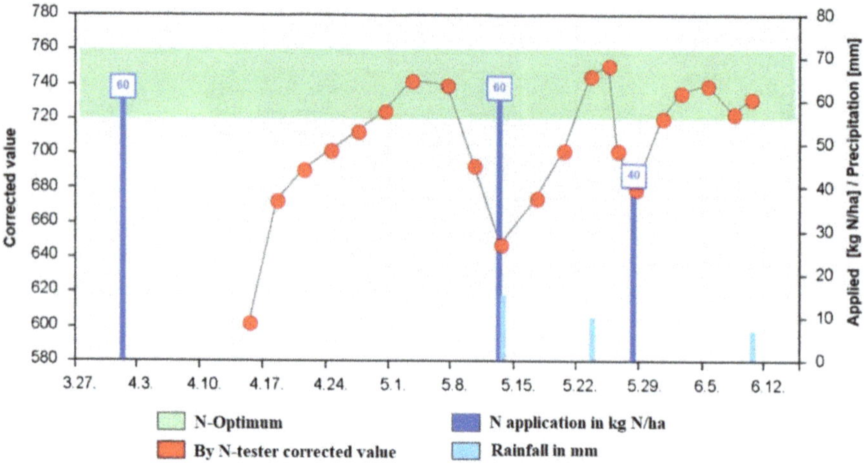

Fig. 50 Additional winter wheat fertilization according to N-tester N-monitoring (The figure was prepared by the authors)

In addition, measurements are not recommended if the plants are sprayed with pesticides, as the plants are under stress and the measurements may not reflect the true situation of photosynthesis. In this case, one should wait a few days before measuring.

8 Adoption of Precision Agricultural Technologies

The application of precision farming is closely related to new agronomic and agroengineering knowledge, knowledge of new information technologies, farm characteristics, available agricultural machinery, farmer and farm worker education, machine service, agronomic consulting, national agricultural policy and economic incentive to purchase state-of-the-art agricultural machinery, farm opportunities for additional investment and learning.

One of the most important factors in introducing elements of precision farming technology is farm size [117, 118]. Say et al. [118] argue that countries with larger farms tend to use precision agricultural technologies much faster. Based on the results of an analysis of research conducted in Germany, it was found that the sociodemographic and farm characteristics have a significant influence on the adoption of precision agriculture in crop farms [119]. The results of the analysis showed that the probability of precision farming adoption in crop farms increases if the farmer's experience in crop production is more than 16 years or less than 5 years [119]. These are two groups of farmers, one of whom is educated and experienced farmers, the other group is made up of young transferees with knowledge of smart information technologies. Torbett et al. [120] published in their work that experienced farmers over the age of 50 tend to implement more precision farming technological operations on their farm. However, the results of the study analysis of other authors [121, 122] demonstrated that age has a negative impact on the application of precision farming technologies. Younger farmers with long-term insights are more likely to associate the modernization of their farms with smart precision technologies. Vecchio et al. [122] also found a positive interaction between the level of education and the adoption of precise agricultural technologies. The results of their study demonstrate that individuals adapting to new technologies have a high level of education, with almost two-thirds of such individuals having a master's degree.

Kernecker et al. [123] conducted a survey of farmers who do not apply precision farming technologies and found that the main reasons for non-application are high investment costs, too complex to use, technology not appropriate for farm context and size, added value is unclear and lack of access to live demonstrations of precision agriculture with neutral contact.

A survey was also conducted in Lithuania, the analysis of which showed that the majority of the respondents face problems in implementing precision farming systems due to inaccurate identification of equipment or inability of farmers to use it properly [124]. The adoption of precision technologies in agriculture is also influenced by the size of farms. Smaller farms find it more difficult to absorb more innovative technologies and have more problems trying to adapt them to farm activities. Large farms, meanwhile, are easy to innovate and face far fewer problems in managing them. This may be explained by the fact that large farms employ more skilled workers and are more focused on companies that disseminate innovation. In addition, it can be observed that farmers managing farms of more than 500 ha are much more likely to have problems due to incompatibility of engineering equipment. This can be explained by the fact that such farmers, when installing the most modern and innovative equipment, purchase equipment from different manufacturers, which is still complicated to communicate with each other [124].

9 Conclusions

Precision agriculture is a set of technologies consisting of: automatic steering, telematics and permanent tramline systems for agricultural machinery, precision tillage, variable rate seeding, crop spraying and harvesting machines, and also robots controlled by ground, aircraft or satellite sensor signals. In addition, precision agriculture includes methods and tools for collecting soil samples, analyzing soil and plant properties, and harvest mapping. The main goals of precision agriculture are the efficient and safe use of fuel, fertilizers, seeds, plant protection products and at the same time improving the quality of products, increasing the competitiveness of farms.

In precision agriculture, three components are most important: obtaining and accumulating comprehensive data, interpreting, and analyzing it and also making and implementing decisions at the right place and time. In a precision farming system, most processes are followed by continuous control sensors that provide accurate and relevant information throughout the process. Currently, proximal sensors are the most widely used to measure soil, plant condition, fertilization need, and yield. Using the obtained data, the technological parameters of agricultural machinery are adjusted during the work in order to use resources more efficiently: the consumption of fertilizers, pesticides, fuel and time are optimized.

In conventional agriculture, fertilizers and other chemicals are spread in the fields at the same rate. However, in most cases, the properties of the soil and plants in the sub-region of the field are significantly different, which results in too much or too little amounts of chemicals being applied in individual parts of the field. This problem can be solved by using one of the tools of precision agriculture, i.e., a variable rate technology based on the application of variable fertilization rates depending on the local needs of the soil and/or plants.

Balancing plant nutrient supply and uptake is necessary not only to optimize plant growth and economic returns, but also to reduce negative impacts on the environment and humans. It has been attempted to set recommended rates for variable rate fertilization to ensure crop demand, farm profitability, and environmental quality, and many studies have proposed various ways to optimize plant fertilization strategies. To this end, measuring equipment is being developed and tested, part of which is already commercialized and placed on the market by various manufacturers.

In recent years, non-invasive methods for the determination of nutrients (mainly nitrogen) in crops in the field have been developed and used quite rapidly. Most of these methods work depending on the optical properties of the plants, which are affected by water content, leaf aging, disease damage, and nutrients accumulated in the plants. The most commonly studied optical parameters of plants are: reflection of light rays from the leaf surface, transmission of rays through the leaves, and fluorescence of chlorophyll and polyphenols. In practice, multispectral proximal reflection sensors for optical properties of plants are commonly used, mounted on a tractor or agricultural machinery, that scans a relatively narrow field band. Remote field scanning with manned or unmanned aerial vehicles (drones) is promising. By scanning the entire field area, more accurate maps of plant properties can be obtained. Hyperspectral sensors can also be used in an aircraft. Most of the remote sensing sensors used in practice are multispectral, i.e., capturing radiation in broad spectral bands-wider than 40 nm with a center of sensitivity coinciding with the blue, green, red and near-infrared zones. Remote sensing based on hyperspectral scanning is designed to collect reflection data in very narrow spectral bands (about 10 nm) in various spectral ranges. The use of hyperspectral cameras is also very promising in soil properties studies.

There are currently a number of measuring instruments on the market based on leaf permeability (SPAD Konica Minolta), chlorophyll fluorescence in leaves (Dualex Dynamax Inc., MiniVeg N Sensor Fritzmeier), light reflection from plants surface assessment (GreenSeeker NTech Industrines, Yara N sensor YARA International ASA, CropScan Cropscan Inc., FieldSpec ASD Inc., ISARIA Fritzmeier, CropSensor

Fritzmeier/Claas Agrosysytems, CropCircle Holland Scientific, CropSpec Topcon, OptRx AgLeader), satellite image data (QuickBird, etc.).

Recently, the development of a precision farming system has focused on creating of agricultural robots and research into their use. The robots can be equipped with various sensors that are used to monitor meteorological conditions, soil and plant condition (parameters). In addition, robots for sowing, planting, spraying, weed identification and eradication have been developed and are being further improved. It is safe to say that in the future, robots will play an increasingly important role in the cultivation technologies of various agricultural crops.

As the progress of precision agriculture is fundamentally dependent on the progress of the production of intelligent agricultural machinery, research and development of various sensors, "large data" processing programs, artificial intelligence and autonomous agricultural machines are establishing rapidly.

The elements of precision agriculture are still applied only in fragments not only in agricultural enterprises and farmers' farms in the Baltic States but also in other European countries. Maps of agrochemical properties of plants and soil are usually made and used to fertilize individual field areas with different fertilizer rates. Farmers often face problems with the interpretation and use of the obtained information: yield, fertilization, maps of harmful organisms and soil agrochemical properties, therefore it is necessary to conduct complex research of precise agricultural technological processes, implementation of their results and training of specialists in this field. Although precision farming, with its intelligent technologies and equipment, has great potential, but its implementation is too slow for the reasons listed above.

References

1. United Nations (2015) TRANSFORMING OUR WORLD: THE 2030 AGENDA FOR SUSTAINABLE DEVELOPMENT A/RES/70/1 https://sustainabledevelopment.un.org/content/documents/21252030 Agenda for Sustainable Development web.pdf
2. FAO (2019). Transforming the world through food and agriculture. http://www.fao.org/3/ca5299en/ca5299en.pdf
3. SmartAgriHubs (2019) Needs Assessment Summary, July 2019 https://smartagrihubs.eu/hubs/Content/SmartAgriHubs-Needs-Assessment-Summary-July-2019.pdf
4. Agriculture and Rural Development DG-AGRI (2019). A new tool to increase the sustainable use of nutrients across the EU. https://ec.europa.eu/info/news/new-tool-increase-sustainable-use-nutrients-across-eu-2019-feb-19_en
5. Thilakarathna, M. S., Raizada, M. N. (2018). Challenges in using precision agriculture to optimize symbiotic nitrogen fixation in legumes: Progress, limitations, and future improvements needed in diagnostic testing. Agronomy, 8(5), 78.
6. Diacono, M., Rubino, P., Montemurro, F. (2013). Precision nitrogen management of wheat. A review. Agronomy for Sustainable Development, 33(1), 219–241.
7. Fixen, P. E. (2007). Can we define a global framework within which fertilizer best management practices can be adapted to local conditions? Fertilizer Best Management Practices, 77–86.

8. Hunt Jr, E. R., Daughtry, C. S. (2018). What good are unmanned aircraft systems for agricultural remote sensing and precision agriculture? International Journal of Remote Sensing, 39(15–16), 5345–5376.
9. Roberts, T. L. (2007). Right product, right rate, right time and right place... the foundation of best management practices for fertilizer. Fertilizer Best Management Practices, 29–32.
10. Germanas L. (2008). Precision agriculture. My Farm, 7, 62–64. (in Lithuanian)
11. Bogužas V., Arvasas J., Šniauka P. (2013) Agriculture. Handbook. Aleksandras Stulginskis university, 176 p.
12. Stafford, J. (2019). Precision agriculture for sustainability. Burleigh Dodds Science Publishing Limited, 494 p.
13. United Nations, Department of Economic and Social Affairs, Population Division (2019). World Population Prospects 2019: Highlights (ST/ESA/SER.A/423).
14. Colaço, A. F., Povh, F. P., Molin, J. P., Romanelli, T. L. (2012). Energy assessment for variable rate nitrogen application. Agric Eng Int: CIGR Journal, 14(3), 85–90.
15. Savci, S. (2012). An agricultural pollutant: chemical fertilizer. International Journal of Environmental Science and Development, 3(1), 77–80.
16. Stewart, W. M., Dibb, D. W., Johnston, A. E., Smyth, T. J. (2005). The contribution of commercial fertilizer nutrients to food production. Agronomy Journal, 97(1), 1–6.
17. Stewart, W. M., Roberts, T. L. (2012). Food security and the role of fertilizer in supporting it. Procedia Engineering, 46, 76–82.
18. FAO, (2020). Mineral and Chemical Fertilizers: 2002–2017. Food and Agriculture Organization of the United Nations. Available online: http://www.fao.org/economic/ess/environment/data/chemical-and-mineral-fertilizers/en/ (accessed on 15 April 2020).
19. Sadeghpour, A., Ketterings, Q. M., Godwin, G. S., Czymmek, K. J. (2017). Under-or over-application of nitrogen impact corn yield, quality, soil, and environment. Agronomy Journal, 109(1), 343–353.
20. Zhang, Q., Pierce, F.J. (2013). Agricultural Automation. Fundamentals and Practices. CRC Press, Taylor & Francis Group: Boca Raton, London & New York, 41–61.
21. Steponavičius, D., Kemzūraitė, A., Bauša, L., Zaleckas, E. (2019). Evaluation of the effectiveness of pod sealants in increasing pod shattering resistance in oilseed rape (Brassica napus L.). Energies, 12(12), 2256.
22. Mamo, M., Malzer, G.L., Mulla, D.J., Huggins, D.R., Strock, J. (2003). Spatial and temporal variation in economically optimum nitrogen rate for corn. Agronomy Journal. 95(4), 958–964.
23. Scharf, P.C., Kitchen, N.R., Sudduth, K.A, Davis, J.G., Hubbard, Y.C., Lory, J.A. (2005). Field-scale variability in economically-optimal N fertilizer rate for corn. Agronomy journal, 97(2), 452–461.
24. Mozgeris, G., Jonikavičius, D., Jovarauskas, D., Zinkevičius, R., Petkevičius, S., Steponavičius, D. (2018). Imaging from manned ultra-light and unmanned aerial vehicles for estimating properties of spring wheat. Precision Agriculture, 19(5), 876–894.
25. Mulla, D. J. (2013). Twenty five years of remote sensing in precision agriculture: Key advances and remaining knowledge gaps. Biosystems Engineering, 114(4), 358–371.
26. Zhang, J., Manske, G., Zhou, P. Q., Tischbein, B., Becker, M., & Li, Z. H. (2017). Factors influencing farmers' decisions on nitrogen fertilizer application in the Liangzihu Lake basin, Central China. Environment, Development and Sustainability, 19(3), 791–805.
27. Galloway, J. N., Townsend, A. R., Erisman, J. W., Bekunda, M., Cai, Z., Freney, J. R., Martinelli, L. A., Seitzinger, S. P., Sutton, M. A. (2008). Transformation of the nitrogen cycle: recent trends, questions, and potential solutions. Science, 320(5878), 889–892.
28. Hoffman, E., Cavigelli, M. A., Camargo, G., Ryan, M., Ackroyd, V. J., Richard, T. L., Mirsky, S. (2018). Energy use and greenhouse gas emissions in organic and conventional grain crop production: Accounting for nutrient inflows. Agricultural Systems, 162, 89–96.
29. Kongshaug, G. (1998). Energy consumption and greenhouse gas emissions in fertilizer production. In IFA Tech. Conf., Marrakech, Morocco, 1998. Int. Fertilizer Industry Assoc.

30. Swaminathan, B., Sukalac, K. E. (2004). Technology transfer and mitigation of climate change: The fertilizer industry perspective. In IPCC Expert Meeting on Industrial Technology Development, Transfer and Diffusion, Tokyo, Japan (Vol. 2123).
31. Rajaniemi, M., Mikkola, H., Ahokas, J. (2011). Greenhouse gas emissions from oats, barley, wheat and rye production. Agronomy Research, 9, 189–195.
32. Pierpaoli, E., Carli, G., Pignatti, E., Canavari, M. (2013). Drivers of precision agriculture technologies adoption: a literature review. Procedia Technology, 8, 61–69.
33. Tekin, A. B. (2010). Variable rate fertilizer application in Turkish wheat agriculture: Economic assessment. African Journal of Agricultural Research, 5(8), 647–652.
34. Shi, Y., Zhu, Y., Wang, X., Sun, X., Ding, Y., Cao, W., Hu, Z. (2020). Progress and development on biological information of crop phenotype research applied to real-time variable-rate fertilization. Plant Methods, 16(1), 11.
35. Yang, C. (2001). A variable rate applicator for controlling rates of two liquid fertilizers. Applied engineering in agriculture, 17(3), 409–417.
36. Nawar, S., Corstanje, R., Halcro, G., Mulla, D., Mouazen, A. M. (2017). Delineation of soil management zones for variable-rate fertilization: A review. Advances in Agronomy, 143, 175-245.
37. Biggar, S., Man, D., Moffroid, K., Pape, D., Riley-Gilbert, M., Steele, R., Thompson, V. (2013). Greenhouse gas mitigation options and costs for agricultural land and animal production within the United States. ICF International, Department of Agriculture Climate Change Program Office: Washington, DC, USA.
38. Bogue, R. (2017). Sensors key to advances in precision agriculture. Sensor Review. 37(1), 1–6.
39. Frasconi, C., Raffaelli, M., Emmi, L., Fontanelli, M., Martelloni, L., Peruzzi, A. (2017). An automatic machine able to perform variable rate application of flame weeding: design and assembly. Chemical Engineering Transaction, 58, 301–306.
40. Grisso, R. D., Alley, M. M., Thomason, W. E., Holshouser, D. L., Roberson, G. T. (2011). Precision farming tools: variable-rate application. Virginia Cooperative Extension. Publication 442–505, 1–16.
41. Khanal, S., Fulton, J., Shearer, S. (2017). An overview of current and potential applications of thermal remote sensing in precision agriculture. Computers and Electronics in Agriculture, 139, 22–32.
42. MacLeod, M., Eory, V., Gruère, G., Lankoski, J. (2015). Cost-effectiveness of greenhouse gas mitigation measures for agriculture: A Literature Review. OECD Food, Agriculture and Fisheries Papers, No. 89, OECD Publishing, Paris. https://doi.org/10.1787/5jrvvkq900vj-en
43. Schumann, A. W. (2010). Precise placement and variable rate fertilizer application technologies for horticultural crops. HortTechnology, 20(1), 34–40.
44. Adamchuk, V. I., Jasa, P. J. (2002). EC02-178 Precision Agriculture: On-the-Go Vehicle-Based Soil Sensors. Historical Materials from University of Nebraska-Lincoln Extension, 706.
45. Adamchuk, V. I., Hummel, J. W., Morgan, M. T., Upadhyaya, S. K. (2004). On-the-go soil sensors for precision agriculture. Computers and Electronics in Agriculture, 44(1), 71–91.
46. Corwin, D. L., Lesch, S. M. (2005). Apparent soil electrical conductivity measurements in agriculture. Computers and Electronics in Agriculture, 46, 11–43.
47. Lesch, S. M. (2005). Sensor-directed spatial response surface sampling designs for characterizing spatial variation in soil properties. Computers and Electronics in Agriculture, 46, 153–180.
48. Naderi-Boldaji, M., Sharifi, A., Hemmat, A., Alimardani, R., Keller, T. (2014). Feasibility study on the potential of electrical conductivity sensor Veris® 3100 for field mapping of topsoil strength. Biosystems Engineering, 126, 1–11.
49. Friedman, S. P. (2005). Soil properties influencing apparent electrical conductivity: a review. Computers and Electronics in Agriculture, 46, 45–70.
50. Romaneckas, K., Zinkevicius, R., Steponavicius, D., Maziliauskas, A., Butkus, V., Marcinkeviciene, A. (2015). Principles of precision agriculture in on-farm spring wheat fertilization experiment. Engineering for Rural Development, Proceedings, 14, 558–563.

51. Spogis L. (2019). Precision seeding maps for soils fertilized with ammonium sulphate enriched with biologically active substances. Master's thesis, 57 p.
52. Steponavičius, D., Kemzūraitė, A., Juknevičius, D., Katkauskas, A. (2017). The relationship between apparent soil electrical conductivity and particle size distribution of light-textured soils. In Proceedings of the 45th International Symposium on Agricultural Engineering, Actual Tasks on Agricultural Engineering, 21–24 February 2017, Opatija, Croatia, pp. 35–44.
53. Veris technologies. The sensors. Available online: https://www.veristech.com/the-sensors (accessed May 7, 2020).
54. Steponavičius, D., Kemzūraitė, A., Zinkevičius, R., Bartkus, T. Effect of soil pH on reasonable travel speed of mobile unit and lime application. Engineering for Rural Development: Proceedings of the 15th international scientific conference. (Ed. Aivars Aboltins). 25–27 May, 2016, Jelgava, Latvia. Vol. 15, p. 304–311.
55. Samborski, S. M., Tremblay, N., Fallon, E. (2009). Strategies to make use of plant sensors-based diagnostic information for nitrogen recommendations. Agronomy Journal, 101(4), 800–816.
56. GreenSeeker Crop Sensing Systems – Trimble Agriculture https://agriculture.trimble.com/product/greenseeker-system/ (accessed on 11 June 2020).
57. Yara N-Sensor – the perfect fertiliser strategy. (2020). Available online: http://www.cropservicesltd.co.uk/index.php?page=n-sensor (accessed May 5, 2020)
58. Aranguren, M., Castellón, A., Aizpurua, A. (2019). Crop Sensor-Based In-Season Nitrogen Management of Wheat with Manure Application. Remote Sensing, 11(9), 1094.
59. Padilla, F. M., Gallardo, M., Peña-Fleitas, M. T., De Souza, R., Thompson, R. B. (2018). Proximal optical sensors for nitrogen management of vegetable crops: A review. Sensors, 18(7), 2083.
60. Monje, O. A., Bugbee, B. (1992). Inherent limitations of nondestructive chlorophyll meters: a comparison of two types of meters. HortScience, 27(1), 69–71.
61. Cartelat, A., Cerovic, Z. G., Goulas, Y., Meyer, S., Lelarge, C., Prioul, J. L., Moya, I. (2005). Optically assessed contents of leaf polyphenolics and chlorophyll as indicators of nitrogen deficiency in wheat (Triticum aestivum L.). Field crops research, 91(1), 35–49.
62. Evans, J. R. (1989). Photosynthesis and nitrogen relationships in leaves of C 3 plants. Oecologia, 78(1), 9–19.
63. Schepers, J. S., Blackmer, T. M., Wilhelm, W. W., Resende, M. (1996). Transmittance and reflectance measurements of cornleaves from plants with different nitrogen and water supply. Journal of plant physiology, 148(5), 523–529.
64. Parry, C., Blonquist Jr, J. M., Bugbee, B. (2014). In situ measurement of leaf chlorophyll concentration: analysis of the optical/absolute relationship. Plant, Cell & Environment, 37(11), 2508-2520.
65. Cerovic, Z. G., Masdoumier, G., Ghozlen, N. B., Latouche, G. (2012). A new optical leaf-clip meter for simultaneous non-destructive assessment of leaf chlorophyll and epidermal flavonoids. Physiologia plantarum, 146(3), 251–260.
66. Daughtry, C. S. T., Walthall, C. L., Kim, M. S., De Colstoun, E. B., McMurtrey Iii, J. E. (2000). Estimating corn leaf chlorophyll concentration from leaf and canopy reflectance. Remote sensing of Environment, 74(2), 229–239.
67. Hu, J., He, D., Yang, P. (2010). Study on plant nutrition indicator using leaf spectral transmittance for nitrogen detection. In International Conference on Computer and Computing Technologies in Agriculture (pp. 504–513). Springer, Berlin, Heidelberg.
68. Buschmann, C. (2007). Variability and application of the chlorophyll fluorescence emission ratio red/far-red of leaves. Photosynthesis Research, 92(2), 261–271.
69. Tremblay, N., Wang, Z., Cerovic, Z. G. (2012). Sensing crop nitrogen status with fluorescence indicators. A review. Agronomy for sustainable development, 32(2), 451–464.
70. Zhang, Y., Tremblay, N., Zhu, J. (2012). A first comparison of Multiplex® for the assessment of corn nitrogen status. J. Food Agric. Environ, 10(1), 1008–1016.

71. Hatfield, J. L., Gitelson, A. A., Schepers, J. S., Walthall, C. L. (2008). Application of spectral remote sensing for agronomic decisions. Agronomy Journal, 100 (Supplement 3), S–117.
72. Ollinger, S. V. (2011). Sources of variability in canopy reflectance and the convergent properties of plants. New Phytologist, 189(2), 375–394.
73. Peñuelas, J., Gamon, J. A., Fredeen, A. L., Merino, J., Field, C. B. (1994). Physiological Changes in Nitrogen-and. Remote Sens Environ, 48, 135–146.
74. Knipling, E. B. (1970). Physical and physiological basis for the reflectance of visible and near-infrared radiation from vegetation. Remote sensing of environment, 1(3), 155–159.
75. Bragazza, L., Freeman, C. (2007). High nitrogen availability reduces polyphenol content in Sphagnum peat. Science of the Total Environment, 377(2–3), 439–443.
76. Liu, W., Zhu, D. W., Liu, D. H., Geng, M. J., Zhou, W. B., Mi, W. J., Hamilton, D. (2010). Influence of nitrogen on the primary and secondary metabolism and synthesis of flavonoids in *chrysanthemum morifolium* ramat. Journal of Plant Nutrition, 33(2), 240–254.
77. Agati, G., Pinelli, P., Cortés Ebner, S., Romani, A., Cartelat, A., Cerovic, Z. G. (2005). Nondestructive evaluation of anthocyanins in olive (Olea europaea) fruits by in situ chlorophyll fluorescence spectroscopy. Journal of agricultural and food chemistry, 53(5), 1354–1363.
78. Cerovic, Z. G., Ounis, A., Cartelat, A., Latouche, G., Goulas, Y., Meyer, S., Moya, I. (2002). The use of chlorophyll fluorescence excitation spectra for the non-destructive in situ assessment of UV-absorbing compounds in leaves. Plant, Cell & Environment, 25(12), 1663–1676.
79. Krause, G. H., Weis, E. (1991). Chlorophyll fluorescence and photosynthesis: the basics. Annual Review of Plant Biology, 42(1), 313–349.
80. Goulas, Y., Cerovic, Z. G., Cartelat, A., Moya, I. (2004). Dualex: a new instrument for field measurements of epidermal ultraviolet absorbance by chlorophyll fluorescence. Applied Optics, 43(23), 4488–4496.
81. Ghozlen, N. B., Cerovic, Z. G., Germain, C., Toutain, S., Latouche, G. (2010). Non-destructive optical monitoring of grape maturation by proximal sensing. Sensors, 10(11), 10040-10068.
82. Hallik, L., Kazantsev, T., Kuusk, A., Galmés, J., Tomás, M., Niinemets, Ü. (2017). Generality of relationships between leaf pigment contents and spectral vegetation indices in Mallorca (Spain). Regional Environmental Change, 17, 2097–2109.
83. Zebarth, B. J., Rees, H., Tremblay, N., Fournier, P., Leblon, B. (2002). Mapping spatial variation in potato nitrogen status using the N Sensor. In XXVI International Horticultural Congress: Toward Ecologically Sound Fertilization Strategies for Field Vegetable Production 627, 267–273.
84. Technical Data Sheet N-Sensor ALS. Research Centre Hanninghof. Brochure, 1–2.
85. OptRx crop sensors (2020). Available online: https://precisionagricultu.re/optrx-crop-sensors/ (accessed May 7, 2020)
86. Köksal, Ö., Tekinerdogan, B. (2019). Architecture design approach for IoT-based farm management information systems. Precision Agriculture, 20(5), 926–958.
87. Grieve, B. D., Duckett, T., Collison, M., Boyd, L., West, J., Yin, H., Arvin F., Pearson, S. (2019). The challenges posed by global broadacre crops in delivering smart agri-robotic solutions: A fundamental rethink is required. Global Food Security, 23, 116–124.
88. Bechar, A., Vigneault, C. (2017). Agricultural robots for field operations. Part 2: Operations and systems. Biosystems engineering, 153, 110–128.
89. Väljaots, E., Lehiste, H., Kiik, M., Leemet, T. (2018). Soil sampling automation using mobile robotic platform. Agronomy Research, 16(3), 917-922.
90. Krishna, K. R. (2016). Push button agriculture: Robotics, drones, satellite-guided soil and crop management. CRC Press, 450 p.
91. Sell, R., Väljaots, E., Pataraia, T., Malayjerdi, E. (2019). Modular smart control system architecture for the mobile robot platform. Proceedings of the Estonian Academy of Sciences, 68(4), 395–400.
92. Fountas, S., Mylonas, N., Malounas, I., Rodias, E., Hellmann Santos, C., Pekkeriet, E. (2020). Agricultural Robotics for Field Operations. Sensors, 20(9), 2672.

93. Underwood, J. P., Calleija, M., Taylor, Z., Hung, C., Nieto, J., Fitch, R., Sukkarieh, S. (2015). Real-time target detection and steerable spray for vegetable crops. In Proceedings of the International Conference on Robotics and Automation: Robotics in Agriculture Workshop, Seattle, WA, USA (pp. 26–30).
94. Vougioukas, S. G. (2019). Agricultural Robotics. Annual Review of Control, Robotics, and Autonomous Systems, 2, 365–392.
95. Vidoni, R., Gallo, R., Ristorto, G., Carabin, G., Mazzetto, F., Scalera, L., Gasparetto, A. (2017, November). ByeLab: an agricultural mobile robot prototype for proximal sensing and precision farming. In ASME 2017 International Mechanical Engineering Congress and Exposition. American Society of Mechanical Engineers Digital Collection, 1–7.
96. Ruckelshausen, A., Biber, P., Dorna, M., Gremmes, H., Klose, R., Linz, A., Weiss, U. (2009). BoniRob: an autonomous field robot platform for individual plant phenotyping. Precision agriculture, 9, 841–847.
97. Bangert, W., Kielhorn, A., Rahe, F., Albert, A., Biber, P., Grzonka, S., Kinski, D. (2013). Field-robot-based agriculture: "RemoteFarming. 1" and "BoniRob-Apps". VDI-Berichte, 2193, 439–446.
98. Utstumo, T., Urdal, F., Brevik, A., Dørum, J., Netland, J., Overskeid, Ø., Gravdahl, J. T. (2018). Robotic in-row weed control in vegetables. Computers and Electronics in Agriculture, 154, 36-45.
99. EcoRobotix. Available online: https://www.ecorobotix.com/en/ (accessed on 21 May 2020).
100. Dino. Available online: https://www.naio-technologies.com/en/agricultural-equipment/large-scale-vegetable-weeding-robot/ (accessed on 21 May 2020).
101. Anatis. Available online: https://www.carre.fr/entretien-des-cultures-et-prairies/anatis/?lang=en (accessed on 21 May 2020).
102. Grimstad, L., From, P. J. (2017). The Thorvald II agricultural robotic system. Robotics, 6(4), 24.
103. Robotti. Available online: http://agrointelli.com/robotti-diesel.html#rob.diesel (accessed on 21 May 2020).
104. Grisso, R. D., Alley, M. M., Holshouser, D. L., Thomason, W. E. (2005). Precision farming tools. Soil electrical conductivity. Virginia Cooperative Extension. Publication 442–508, 1–6 p.
105. Moral, F. J., Terrón, J. M., Da Silva, J. M. (2010). Delineation of management zones using mobile measurements of soil apparent electrical conductivity and multivariate geostatistical techniques. Soil and Tillage Research, 106(2), 335–343.
106. Njukeng, N. J., Nkeng, E. G., Ehabe, E. E., Schnug, E. (2013). A comparative study on the use of calcium acetate lactate, calcium chloride and acidic ammonium acetate-ethylene diaminetetra acetic acid (AAAc-EDTA) for the quantification of extractable, P, K and Mg from acidic soils. International Research Journal of Pure and Applied Chemistry, 22–31.
107. Schüller, H. (1969). Die CAL-Methode, eine neue Methode zur Bestimmung des pflanzenverfügbaren Phosphates in Böden. Zeitschrift für Pflanzenernährung und Bodenkunde, 123(1), 48–63.
108. AgroLab GmbH (2020). Anlage zur Akkreditierungsurkunde. Available online: https://www.agrolab.com/en/service/download/document-search/22-de-lufa-akkreditierung-dakks-anlage/file.html (accessed on 25 May 2020).
109. Reusch, S. (2006). N-Sensor ALS® – Basics, Application and Use. Landtechnik, 61(2), 76–77.
110. Agricon. (2020). Digital N-fertilization with sensors. Available online: https://www.agricon.de/en/n-fertilization (accessed on 25 May 2020).
111. Feiffer, A., Jasper, J., Leithold, P., Feiffer, P. (2007). Effects of N-Sensor based variable rate N fertilization on combine harvest. In Precision Agriculture'07". Proceedings of the 6th European Conference on Precision Agriculture. Wageningen Academic Publishers, Wageningen, The Netherlands (pp. 673–679).
112. Havránková, J. (2007). The evaluation of ground based remote sensing systems for canopy nitrogen management in winter wheat. Master's Thesis. Cranfield University, 146 p.

113. Havránková, J., Rataj, V., Godwin, R. J., Wood, G. A. (2007). The Evaluation of Ground Based Remote Sensing Systems for Canopy Nitrogen Management in Winter Wheat–Economic Efficiency. Agricultural Engineering International: CIGR Journal. IX, 1–9.
114. Mayfield, A. H., Trengove, S. P. (2009). Grain yield and protein responses in wheat using the N-Sensor for variable rate N application. Crop and Pasture Science, 60(9), 818–823.
115. MH, B. H. R. (2019). Profiling the Nitrogen Efficiency Using Agricultural Engineering Technique of YARA ALS Tractor Sensor. European Journal of Experimental Biology, 9(2), 1–10.
116. N-Sensor ALS – to variably apply nitrogen. Available online: https://www.yara.ie/crop-nutrition/farmers-toolbox/n-sensor/ (accessed on 23 May 2020).
117. Caffaro, F., Cavallo, E. (2019, September). Perceived Barriers to the Adoption of Smart Farming Technologies in Piedmont Region, Northwestern Italy: The Role of User and Farm Variables. In International Mid-Term Conference of the Italian Association of Agricultural Engineering (pp. 681–689). Springer, Cham.
118. Say, S. M., Keskin, M., Sehri, M., Sekerli, Y. E. (2018). Adoption of precision agriculture technologies in developed and developing countries. Online J. Sci. Technol, 8, 7–15.
119. Paustian, M., Theuvsen, L. (2017). Adoption of precision agriculture technologies by German crop farmers. Precision Agriculture, 18(5), 701–716.
120. Torbett, J. C., Roberts, R. K., Larson, J. A., English, B. C. (2007). Perceived importance of precision farming technologies in improving phosphorus and potassium efficiency in cotton production. Precision agriculture, 8(3), 127–137.
121. Kutter, T., Tiemann, S., Siebert, R., Fountas, S. (2011). The role of communication and co-operation in the adoption of precision farming. Precision Agriculture, 12(1), 2–17.
122. Vecchio, Y., Agnusdei, G. P., Miglietta, P. P., Capitanio, F. (2020). Adoption of Precision Farming Tools: The Case of Italian Farmers. International Journal of Environmental Research and Public Health, 17(3), 869.
123. Kernecker, M., Knierim, A., Wurbs, A., Kraus, T., Borges, F. (2020). Experience versus expectation: farmers' perceptions of smart farming technologies for cropping systems across Europe. Precision Agriculture, 21(1), 34–50.
124. Kazimierskis A. (2014). Innovation Management in UAB "Dotnuvos projektai". Master's Thesis. Aleksandras Stulginskis University, 62 p.

Part II
Wireless Network Systems Applications

Experimental Performance Evaluation Techniques of LoRa Radio Modules and Exploitation for Agricultural Use

Dimitrios Loukatos, Athanasios Fragkos, and Konstantinos G. Arvanitis

1 Introduction

In the twenty-first century agricultural productivity should be vastly increased to cover the nutrition demands of the earth's population under harsh and deteriorating environmental conditions which directly impact the quantity and quality of the crops [1]. Furthermore, the nature of the agricultural tasks is rapidly evolving and is very demanding, requiring network coverage, in wide rural areas with scarce or even inexistent communication infrastructures, and resilient signals that can travel through plants or warehouses. To tackle these issues, the agricultural practices have to become more productive and "climate-smart", by successfully exploiting a variety of existing and emerging technologies [2]. Thankfully, the progress in Information and Communication Technologies (ICT) allows for significant improvements in agri-production [3] and can have a considerable impact on efficiency, resilience and inclusion [4], as innovations such as the Internet of Things, Cloud Computing and Big Data are revolutionizing agriculture.

The modern wireless sensor networks (WSN) provide the capability of sensing and remotely controlling several environmental parameters and activities, including tasks related to agricultural farms, greenhouses, or production processes, and thus affecting almost every field of life. It must be noted though that, when working with real-world agricultural applications, many traditional network management techniques and protocols should be reconsidered. Indeed, the WSN should be able to meet the primary goals of optimizing the efficiency of the network and minimizing the energy use [5]. Into this concept, amongst the most promising network

D. Loukatos (✉) · A. Fragkos · K. G. Arvanitis
Department of Natural Resources Management & Agricultural Engineering, Agricultural University of Athens, Athens, Greece
e-mail: dlouka@aua.gr; athfrag@aua.gr; karvan@aua.gr

D. D. Bochtis et al. (eds.), *Information and Communication Technologies for Agriculture—Theme I: Sensors*, Springer Optimization and Its Applications 182,
https://doi.org/10.1007/978-3-030-84144-7_4

technologies are the high-end Low-Power Wide-Area Networks (LPWAN) [6] radios, using protocols like LoRa [7]. Indeed, the latter protocol meets the above-mentioned requirements as it is able to operate at long distances, consumes low energy and exhibits immunity to noisy environments.

These characteristics of the LoRa protocol made it suitable for many innovative applications, either in its pure version or in its LoRaWAN derived one [8]. Many of these applications have a strong impact on agriculture. The only tradeoff is the limited available bandwidth that the LoRa protocol is achieving, which cannot exceed a few kbps for longer distances [9], but even under this limitation the specific protocol is still valuable for the type of applications that have low data rate and long distance coverage requirements. The performance of LoRa is satisfactory especially when the 3 km distance limit is not exceeded [10].

The key factor for LoRa performance is the Chirp Spread-Spectrum (CSS PHY) modulation schema which is described in the IEEE Low-Rate Wireless Personal Area Networks (LRWPANs) standard 802.15.4 [11]. The chirp modulation is the method of transmitting symbols by encoding them into multiple signals of increasing or decreasing radio frequencies. Because of the changing frequencies, chirp-modulated signals are fairly robust to multi-path interference, fading, and Doppler shifts [12]. By increasing the bandwidth, and thus the number of "chirps" per symbol, signals will be transmitted over longer distances, as more information is transmitted per bit.

The important quantities according to the relevant specification [12] for the LoRa protocol can be described as:

- Spreading Factor (SF): The value of SF varies from seven to 12 and defines the number of chirps used to encode a bit. The higher SF values lead to more chirps used to represent each symbol. On the other hand, lower spreading factor values mean a higher data rate and thus more chirps are sent per second;
- Bandwidth (BW): The bandwidth is of dominant importance as a parameter of the LoRa modulation. Indeed, the BW is the range of frequencies that are available for transmission. Larger BW allows transmission at higher data rates, with shorter time "on-the-air" and lower sensitivity;
- Coding Rate (CR): The LoRa modulation schema also includes forward error correction (FEC) bits. This is achieved by encoding each "pure" 4-bit quantity with redundancies, actually into 5-bit, 6-bit, 7-bit, or even 8-bit transmission units. This bit redundancy allows the LoRa signal to tackle short interferences, with greater interference values to require greater CR ratios.

In order to successfully apply the communication benefits of the promising LoRa protocol, in real-world and large-scale applications, a meticulous testing and tuning of the LoRa modules is required. For this reason, many research works are focusing on the idiosyncrasies of the LoRa protocol and try to evaluate or optimize its performance [10, 13, 14].

Thankfully, this testing process does not always have to be demanding, in terms of the hardware and software components needed. Indeed, the appearance of low-cost credit card-sized computers, usually comprising open hardware implementations, ranging from arduino uno [15] units to raspberry pi [16] ones, can be used to orchestrate the scenarios for measuring the performance of various wireless radio modules. These systems can be combined with specific electronic components or even with conventional smart phone devices to form a simple, flexible and lightweight testbed for evaluating and/or modifying the behavior of the radio interfaces of interest. Apart from this, a wide variety of special software, either based on textual (e.g., Arduino IDE [17–19]) or visual programming methods (e.g., [20–22]), are addressing the programming of these credit card-sized systems. Even the behavior of the accompanying mobile phone/tablet devices is addressed in a similar manner, by tools like the MIT App Inventor [23]. These methods make even people with entry-level programming, able to create satisfactory software applications.

The experimental setup being presented involves components that are measuring the performance of LoRa interfaces during their activity, mainly in terms of radio coverage, energy consumption and packet losses. The whole methodology is based on and extends the techniques described in [24] and it is targeted at providing the means for an easy but yet effective way of collecting (and processing) radio-specific measurements of satisfactory accuracy. The measuring equipment and methods being presented are suitable for educational purposes or for use by not very experienced personnel. The impact of the LoRa radio status modifications (at short distances) is being verified by a suitable electromagnetic radiation meter instrument which is able to measure the surface power density at a given distance from the antenna. The radio behavior (e.g. transmit power, modulation characteristics, packet rate) can be controlled and monitored using simple commands, via a smart phone. Furthermore, the payload of the LoRa protocol packets, exchanged between the source and the destination node, can be modified to contain useful information, from signal level adjustment commands to time stamp and data loss ratio reports. This method has successfully been used in packet-switched network cases [25]. The performance evaluation techniques, described in this paper are tailored to be in pace with the LoRa idiosyncrasy but they can also be applied to other protocol testing cases, easily, because of the generic components and methods being used. For better comparison reasons, along with the LoRa specific results, indicative results involving Wi-Fi [26] radio modules are reported.

Apart from presenting performance evaluation techniques and results, this work also reports on the feasibility of the LoRa radios to carry out simple remote sensing and action tasks, by extending the basic measuring infrastructures, through typical examples in the area of agriculture. Following this introductory section, the rest of this chapter is organized as follows. Section 2 explains the design principles of the proposed measuring tools and their extensions. Section 3 provides characteristic implementation and measurement methodology details. Section 4 is dedicated to the evaluation of the proposed system, techniques and results. Finally, in Sect. 5, the description of this work ends with summarizing remarkable comments and presenting work plans for the future.

2 Design Overview

In order to evaluate the performance of the LoRa protocol two identical LoRa radio transceiver modules were used, in the form of arduino shields [27], to be fitted on two corresponding adruino uno units. The accompanying software [28] was also studied, in order to implement a simple client – server schema for communication among these radios. Methods for modifying the transmit power, the packet size, the inter-packet distance and for collecting parameters such as the Received Signal Strength Indicator (RSSI) [29], the packet error rate or the module's power consumption have been designed as well.

In order for things to become more efficient, the packets directed from the client (at the user's end) to the server (at the farm's end) were designed to contain, into their payload, additional commands for modifying the operating parameters (e.g., the transmit power) of the LoRa radio shield on the server. The replies sent back to the client also contain additional information reflecting the radio status of the server (e.g., the RSSI values). This technique was also extended to carry real world command requests (and the proper replies) for controlling farm-related actions, like watering the plants, or sensing actions and asking the status of environmental parameters (e.g., the soil humidity or its salinity level).

Although the user's commands can be provided through buttons and switches connected on the arduino unit, a raspberry pi unit was used to provide the necessary functionality to support user's intervention with the sensor nodes, via a conventional tablet or a smart phone device extending methods described in [24]. The hiring of a raspberry pi unit allowed for more sophisticated, and thus computationally demanding, processing operations such as database storage or web-based AI services [30] support (e.g. voice commands interpretation). The raspberry pi is a much more powerful unit than the arduino uno and their roles can be seen rather as complementary ones than competing ones.

On the other hand, the considerable gap between their potential, leaves plenty of space for other credit-card sized devices to flourish, like the ones based on the ESP8266 [31] or the ESP32 architecture. Indeed, instead of the raspberry pi, in the experimental and monitoring testbed being presented the WeMos board [32], an ESP8266-derived unit, can be used for quite simple tasks. The WeMos board is smaller than the raspberry pi, consumes less energy and also has a Wi-Fi built-in. In this regard, the latter board can be used to intercept the smart phone-derived messages from the user and to perform the bridging with the arduino/dragino system, via the serial interface, and other typical storage and processing tasks. Even beyond that, as the WeMos has a very similar pinout to the arduino uno, it looks and works, in many cases, like an arduino board and thus, it can replace this unit as well. A wide variety of sensor and actuator shields, manufactured for the arduino platform, can be exploited, with the LoRa dragino shield not to be an exception. More specifically, with minor jumper changes and by using the arduino IDE (Integrated Development Environment – IDE) environment for programming, at the human's end (client device), the system of a raspberry pi unit, plus an arduino uno, plus a LoRa shield, can be replaced by a WeMos unit equipped with a LoRa dragino shield, which is a more compact and cost effective arrangement, for a variety of comparatively simple measuring and processing scenarios.

A typical instance of the proposed architecture, allowing the realization of both LoRa protocol related performance evaluation actions and meaningful farming activities monitoring and controlling, is depicted in Fig. 1. The left part of this figure is dedicated to functions at the user's (i.e., the farmer's) end while the right part is dedicated to functions taking place at the farm's end.

3 Implementation Details and Measurement Methodology

In pace with the guidelines described in [24], the measurement process can be divided into three stages. The first stage is responsible for adjusting the experimental configuration settings. The second stage is dedicated to the recording of the activity during the LoRa radio modules interaction. For better results, during this stage, the fast-varying actions of interest are repeated many times (the number of repetitions is fixed or computed at the end of the process) before the recording process is completed. By inspecting the aggregate quantities, at the end of the process, interesting and quite accurate results are extracted, without the need for a short-term time granularity and its increased cost [33]. The third stage is responsible for collecting the results from the LoRa-equipped sensor nodes to the monitoring device

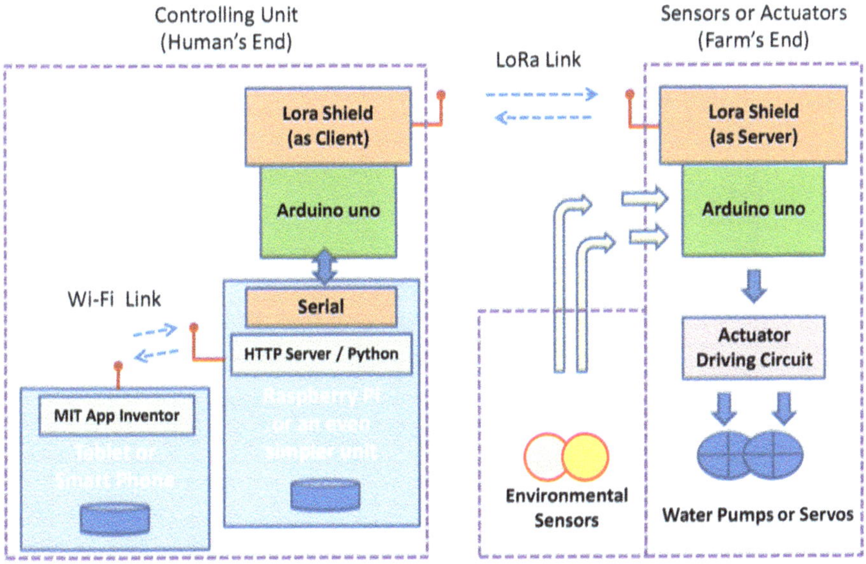

Fig. 1 A typical architecture allowing for both LoRa protocol performance evaluation actions and for farming activities monitoring and control

(e.g., the smart phone) and, possibly, for further data processing or delivery to a database or a web server.

Providing further details, educationally fruitful code was written to program the accompanying mobile phone/tablet devices, using tools like the MIT App Inventor, to tackle the communication tasks with the raspberry pi (or the WeMos) unit. This communication is based on a Hypertext Transfer Protocol (HTTP) message request-reply mechanism or on even faster counterparts that are implemented using the User Datagram Protocol (UDP) protocol. Bluetooth radio is also an option of more limited capabilities. The MIT App Inventor is an excellent programming environment that is based on visual blocks. By using this package, even quite inexperienced programmers can create fluent software applications. Despite the fact that similar environments have mostly been used on an educational basis involving primary and secondary school students [34], their potential has also been tested in more composite and demanding scenarios with satisfactory results [35, 36].

The use of a tablet or a smart phone device to monitor and control the whole experimental process can be seen as really valuable feature, if the harsh agricultural field measurement conditions are taken into account. The mobile application implemented by using the App Inventor environment has a main screen layout that allows for a flexible configuration and monitoring of the LoRa nodes, in a manner that is very familiar to the modern people. The initial design of this layout and the actual result on the smart phone or the tablet device, are shown on the left and the right part of Fig. 2.

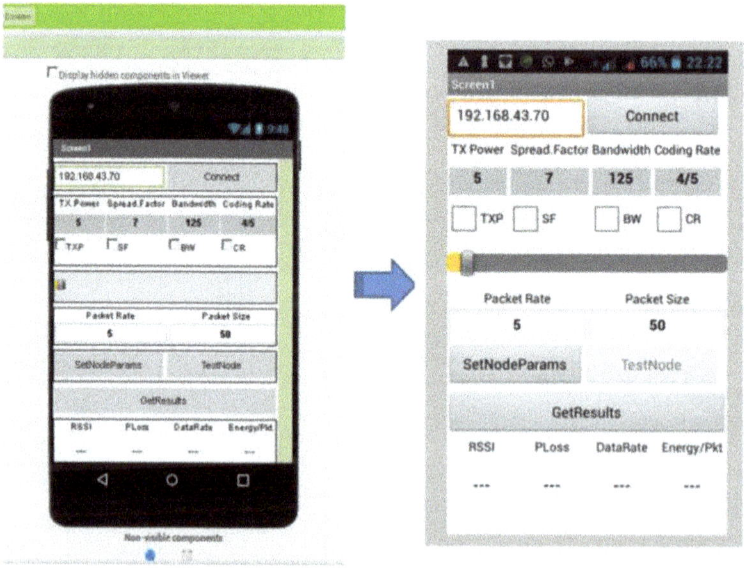

Fig. 2 The main screen layout of the mobile application allowing for a flexible configuration and monitoring of the LoRa nodes

The mobile application provides to the user fluent configuration options for the most critical operational parameters of the LoRa modules, namely: the transmit power (TXP), the spreading factor (SF), the bandwidth (BW) and the coding rate (CR). Furthermore, the traffic flow can also be adjusted via the packet rate (PktRate) and the packet size (PktSize) parameters. In terms of visual programming techniques buttons, edit boxes, check boxes, labels and scroll bar controls have been used to facilitate the interaction between the human and the LoRa radio modules. Invisible components like timers (for estimating the period of the testing) and Web objects (for providing HTTP access between mobile device and raspberry pi or WeMos equipped node) have been used as well.

More specifically, after establishing communication with the node, a typical measurement process requires setting and passing the proper values for the configuration parameters. These parameters are passed to the "client" node, via the Wi-Fi interface in the form of HTTP requests, (at human's end) and then are forwarded to the remote "server" node (at farm's) end, via the LoRa interface. It is the remote node that first alters its operational state, while the client node updates its configuration last. After this step is accomplished, the "TestNode" button is enabled and, if pressed, a countdown timer is triggered, providing information for the remaining duration of the experiment. During the experiment a large (and constant) number of packets, at the defined rate, are generated and directed towards the remote node. On completion of the experiment or if the "GetResults" button is pressed the proper results are shown on the screen of the mobile device. These results typically include the average RSSI record, the packet loss ratio indicator or even the achieved packet rate and the energy spent per each packet being transmitted. The RSSI and packet losses metrics are based on measurements taken at the remote node, while the data rate and the energy metrics are based on measurements at the "client" node. In general, the results are directed from the "server" node to the "client" node, via LoRa packets, while the "client" node delivers these data to the mobile device (i.e., the smart phone or the tablet) in the form of HTTP reply messages. Additional functionality tackling the issues of getting and storing the GPS (Global Positioning System) coordinates of each valid measurement along with the metrics of interest has been implemented as well, as part of an extended version of the mobile application. As the experimental portable equipment is intended for long distance measurements, under potentially harsh conditions in rural areas, the whole process has been kept as straightforward and simple as possible.

From the code aspect that runs on the testing nodes, in case of the raspberry pi based implementation, the necessary testing or monitoring commands, generated by the tablet or the smart phone device of the user and intercepted by the Wi-Fi interface of the raspberry pi, are forwarded through the USB connection to the arduino uno unit (via suitable code written in python) and finally to the LoRa modules through the RadioHead API (Application Programming Interface), and vice versa. Each valid command is served by a corresponding case handler routine that exploits the potential of the radio modules being interconnected. In case of the WeMos based implementation (programmed using the arduino IDE -Integrated Development Environment- environment and special additional environment and library components), the necessary testing or monitoring commands are circulating through the Wi-Fi interface of the WeMos device towards the pairing LoRa modules, through the RadioHead API, and vice versa. The code has been written so as to allow the Wi-Fi equipped node at the human's end (i.e., the raspberry pi or the WeMos node) to automatically establish a link with the mobile device of the user, with the latter device to act as an access point. This setup provides independence from fixed infrastructures. The LoRa radio modules at both ends were configured via the RadioHead library so as to send packets in a simple "unreliable" manner. This option means that the LoRa system will not to perform packet retransmissions, after unsuccessful packet delivery, and thus the process of estimating the energy spent for packet transmission actions can provide more accurate results. Additional software parts were programmed to provide detailed current consumption and packet transmission time readings to the end-user application.

The whole implementation (MIT App Inventor part) is following an event driven philosophy and comprises of several blocks. Figure 3 highlights exactly the role of these visual code blocks providing the mobile application functionality.

The corresponding arrangement for the experimental equipment for the client node (at the human's end), for the raspberry pi – arduino interconnection case, is shown in Fig. 4. This interconnection is feasible via a USB cable of type-A and type-B plugs. This solution is intercepted by the Linux OS based operating system of the raspberry pi unit (i.e., the raspbian distribution) as a serial connection/device, typically under the name "/dev/ttyACM0". The native serial port that both the raspberry pi and the arduino uno units have can be used as well, to reduce the energy being consumed by the arduino part, at the cost of increased wiring complexity. Alternatively, the hiring of a WeMos board to host the LoRa shield and to provide interaction with the operator (user's end), instead of the raspberry pi – arduino combination, resulted in reduced energy consumption and more compact layouts.

The nodes that participate in the measurement process combine raspberry pi with arduino boards (or WeMos based equivalents) and LoRa dragino shields and all these components are fixed on wooden bases at the top of wooden rods. These wooden parts are forming a capital "T"-shaped lightweight structure, which is very practical, increases portability and eliminates the electromagnetic interference that might lead to erroneous recordings. The wooden rods are long enough to provide an approximate 2.0 m elevation from the ground level for the LoRa transceiver modules, if necessary. The corresponding arrangement for the experimental equipment (for the WeMos – arduino case) is shown in Fig. 5, while Fig. 6 depicts these devices in action.

Furthermore, there are two additional components that are worth highlighting. The first one is a TM-195 RF Field Strength Meter [37]. This instrument intercepts any transmit power modification of the LoRa modules by measuring the surface power density (in $\mu W\ cm^{-2}$) at a specific distance from the radio module. Any external radio source should be eliminated or be subtracted from the readings. To ease this constraint, most measurements can refer to the LoRa module at the farm's end, that has no Wi-Fi radios nearby. The second component is an LTC4150 Coulomb counter [38], which is able to measure (aggregate) mAh of consumption as it

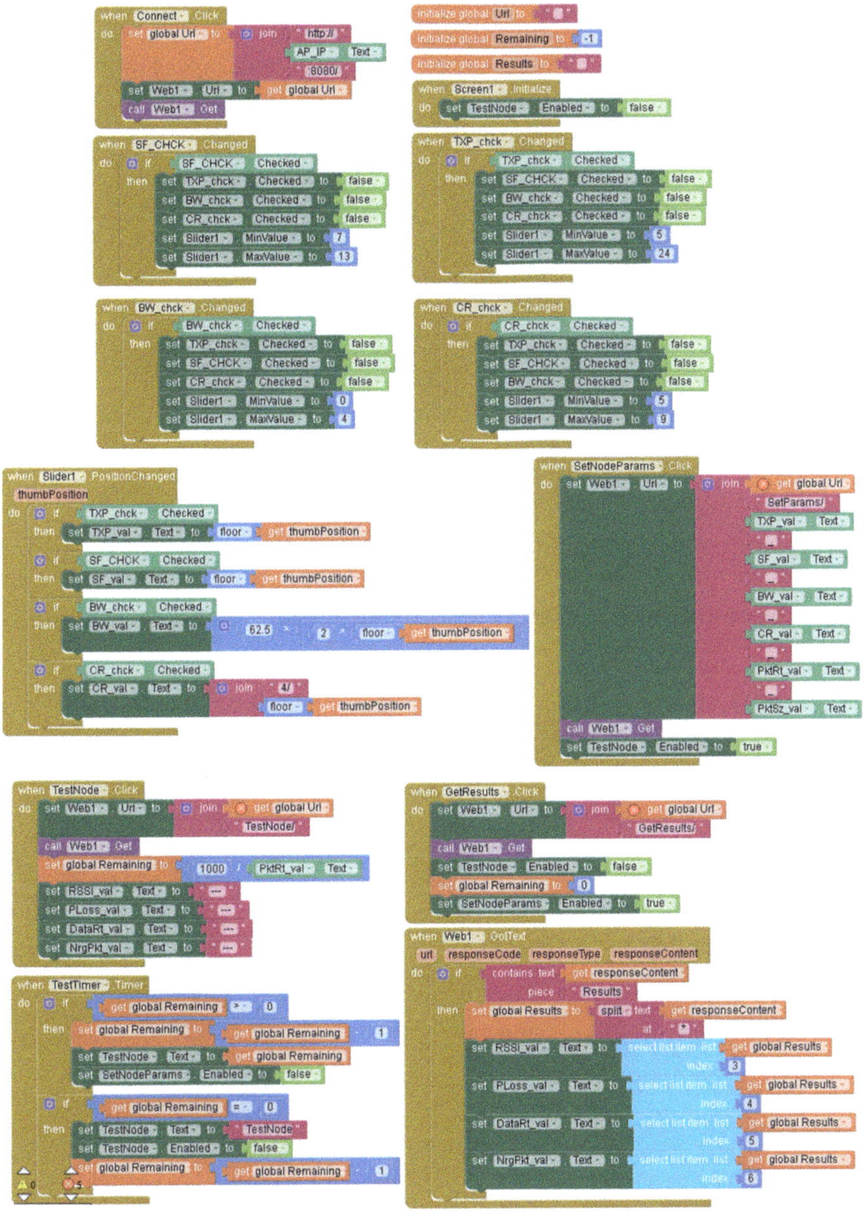

Fig. 3 Indicative visual blocks implementing the mobile application functionality

generates an interrupt every time that 0.0001707 Ah passes through it. The power is measured at electric charge level. This module is connected between the supply main and the arduino uno unit (or the WeMos unit) hosting the LoRa shield and reports to the arduino, its overall power consumption. Constant quantities (i.e., the

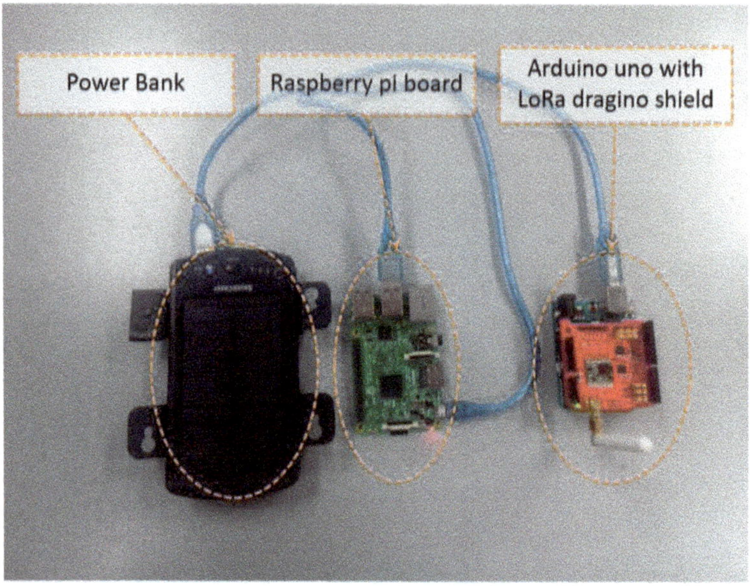

Fig. 4 The experimental arrangement for the client node (at human's end) equipment (raspberry pi – arduino case)

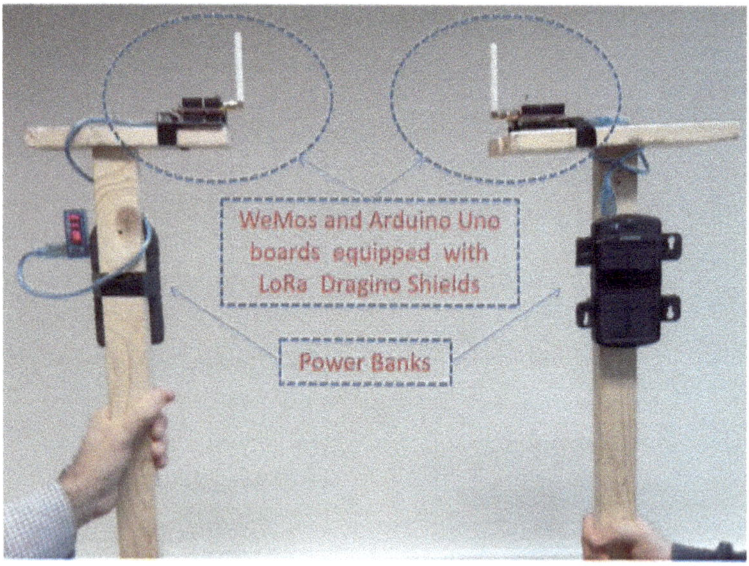

Fig. 5 Experimental arrangement for both client and server node equipment (WeMos – arduino case)

Fig. 6 The experimental measurement equipment, comprising of LoRa modules, in action

consumption of the LoRa module at idling) are subtracted to provide accurate estimations of the exact amount of energy spent during the activity phase.

As already mentioned, the measuring infrastructure can be slightly modified to provide remote control and monitoring of various sensor or actuator units. The main limitation is that the data rate to be supported has to be only a few kbps. Furthermore, only small packets are allowed (i.e., packets with less than 230 bytes payload, approximately or even with less than 50 bytes payload, for larger distance coverage specifications).

Provided that the network traffic is sparse enough, a wide variety of actuators can be supported; among these are ventilation motors, water pumps, electric (solenoid) valves and lamps. Typically, the commands arriving to the server node are interpreted in general purpose input/output [39] actions, which in turn are used to fire the necessary motor driving equipment. This driving equipment can be conventional relays or more reliable solid-state relays, for simple on-off operation. In case of variable operation, the driving circuits should be more complex involving pulse width modulation – PWM [40] and/or phase control – PFC [41], methods, with the latter to be applicable to alternating current – AC-powered systems.

Similarly, the farm-specific information can be acquired by a variety of sensors (e.g., measuring humidity, luminosity, distance, position or temperature) connected with the server node (at the farm's end) using its GPIO pins. A large variety of

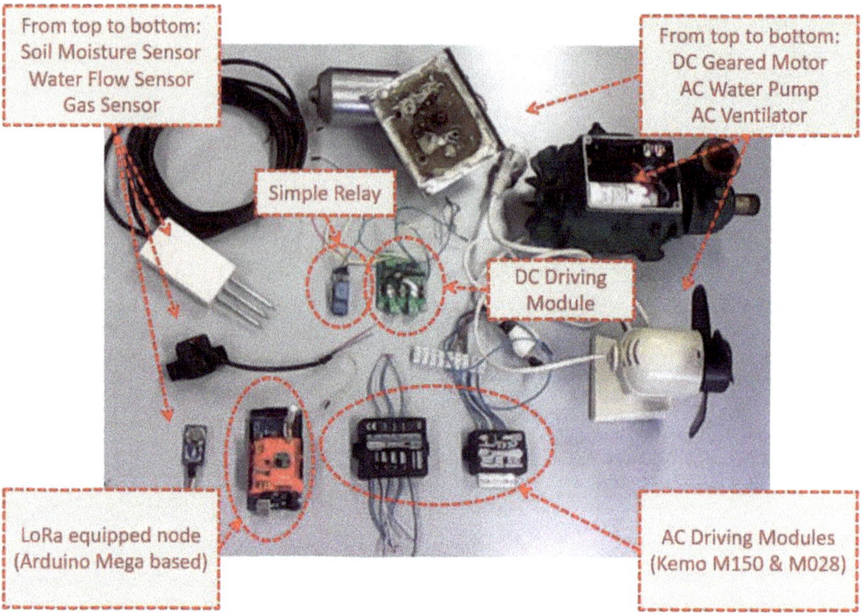

Fig. 7 Typical equipment used during the experiments, involving sensors and actuators that are meaningful for agricultural tasks, along with LoRa equipped nodes

interfacing protocols is supported by the system for communicating with the sensors (e.g., serial, I2C, SPI, single digital or analog inputs).

Figure 7 depicts typical equipment used during the experiments, involving sensors and actuators that are meaningful for agricultural tasks, along with LoRa equipped nodes. The arduino uno unit has been replaced with the larger and pin compatible arduino mega unit that offer more connectivity options.

4 Evaluation of Methods, Results and Discussion

This section presents characteristic measurements taken using the above-mentioned methodology, highlights strengths and weaknesses of the LoRa technology and reports on the feasibility of the proposed testbed to assist in typical agricultural monitoring tasks. The most interesting case is using the SF7 and BW125kHz configuration, which is the default setting for devices using the open-source LoRa implementation.

The first set of measurements is focusing on the Received Signal Strength Indicator – RSSI evaluation, for various distances between the LoRa radio modules participating in the experiments and for various transmit power level settings. The top left part of Fig. 8 depicts the RSSI values, intercepted by a LoRa module (blue points) as a function of the distance from the pairing LoRa module. Both LoRa

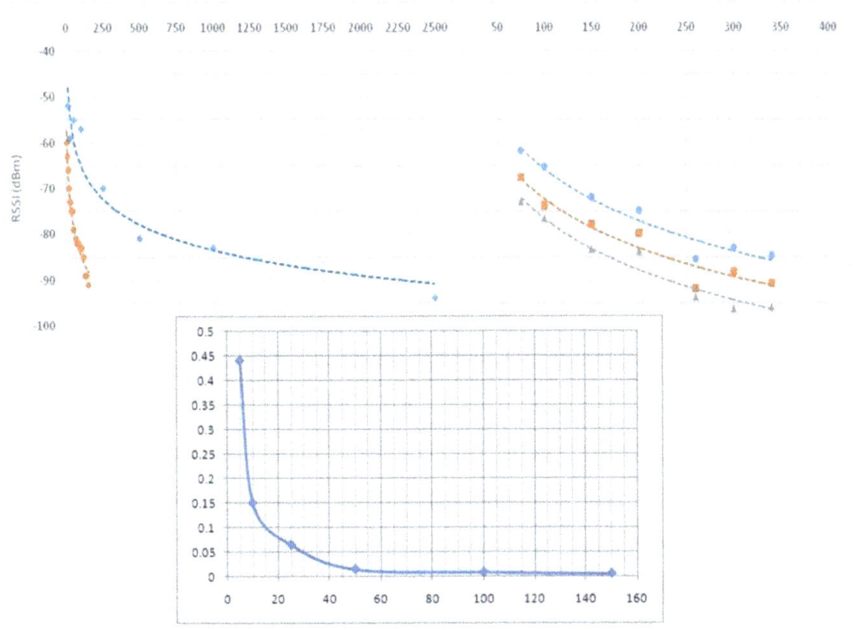

Fig. 8 Top left: The RSSI over the distance from the source, for LoRa and Wi-Fi transceivers. Top right: RSSI measurements over the distance, for different transmit power levels. Bottom: The surface power density as a function of distance from the source

radios had their transmit power adjusted to 10 dBm, while the spreading factor (SF) parameter was set to seven, the bandwidth (BW) was 125 kHz and the coding rate (CR) was 4/5. The RSSI measurement process was repeated using the Wi-Fi radio module of the raspberry pi board and an access point. The results for the Wi-Fi case are indicated by red points in the top left part of Fig. 8. The signal attenuation is less intense with the LoRa links, which are able for much longer distance communication as they can reach sensitivity levels of −140 dBm, while the Wi-Fi links usually stop before the −100 dBm border. The top right part of Fig. 8 depicts the RSSI measurements acquired using the LoRa radios, for distances from 50 m to 350 m and for transmit power levels of 5 dBm (in grey), 10 dBm (in red) and 15 dBm (in blue), respectively, at presence of some trees and buildings, in the university campus. By inspecting this graph, which shares the same vertical axis scale with the left graph, an almost 5 dBm offset is recorded, corresponding to these consecutive power settings, for the same distance between the radios. The bottom part of Fig. 8 depicts the surface power density, as intercepted by the TM-195 power meter, as a function of distance (in cm) from the LoRa radio module, while transmitting at 3 kbps and at 5 dBm. The vast decrease in the power density values being measured, over the distance from the transmitting LoRa equipped node (client), highlights the energy demanding nature of the wireless communications and the importance of having transceivers of high sensitivity.

The impact of typical LoRa-specific parameter combinations on the actual radio traffic characteristics was further evaluated. More specifically, the top left part of Fig. 9 shows the RSSI graph of the LoRa signal, as it is intercepted by the server node (at the farm's end), as a function of the distance from the client LoRa node (at the user's end), for two different cases: (a) with SF 7, BW 125 kHz, CR 4/8 and TXP 10 dBm and (b) with all parameters as in (a) except BW, which is now set at 250 kHz. As depicted in Fig. 9, the second case exhibits worse RSSI values. This deterioration of 3 dB, approximately, in the reception signal as the bandwidth increases, results in achieving shorter maximum communication distances. This behavior is one of the theoretically expected tradeoffs that the hunger for faster data rates (i.e., under increased bandwidth) has. The top left part of Fig. 9 also includes two shorter curves corresponding to the case where the above-mentioned settings (i.e., for 125 kHz and 250 kHz) were used for taking measurements inside the orchard at the university campus, between the trees. The case with the 125 kHz bandwidth selection keeps almost constant distance from its 250 kHz counterpart. In the orchard, the LoRa signals exhibit an approximate reception drop of 9 dBm, compared with the corresponding open space cases.

The bottom right part of Fig. 9 shows the time required for the transmission of each consecutive LoRa packet frame (time "on-the-air" or "airtime") as a function of the packet payload size (in Bytes), for different spreading factor (SF) values, with the bandwidth parameter at 125 kHz. More specifically, these times varied drastically from below 50 ms to above 2 s per packet, according to the selected

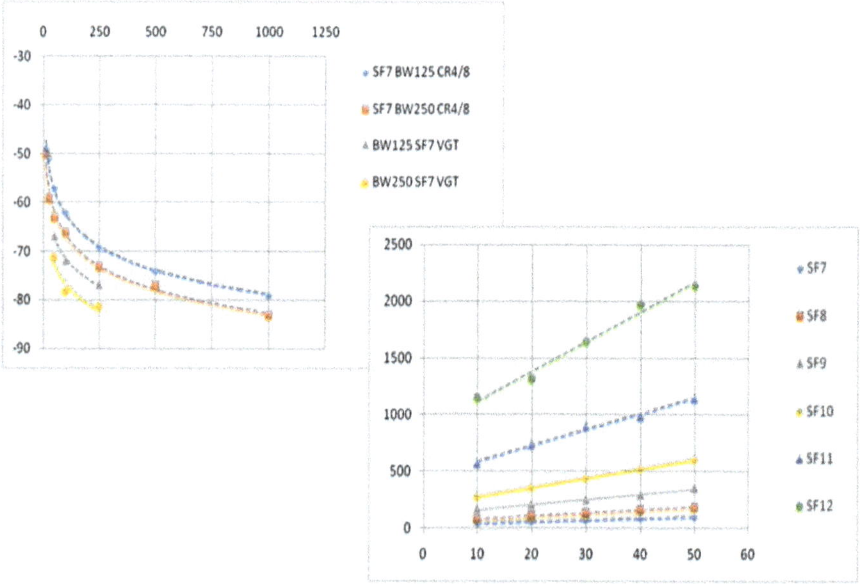

Fig. 9 Top left: The drop in RSSI values as the bandwidth increases, for LoRa transceivers, in open air and in orchard environment. Bottom right: The time "on-the-air" as a function of packet payload size, for different spreading factor values

configuration. The increase in the SF results in better noise immunity (i.e., the LoRa packets to be intercepted at weaker RSSI levels) at the cost of drastic reduction in the data rates being achieved. Indeed, higher SF implies result in fewer chirps per second and thus fewer data to be encoded per second. In contrast with transmissions at lower SF values, sending the same amount of data at higher SF requires more time.

The packet losses during the experiments using the LoRa radios were quite low. More specifically, at typical distances (e.g., 250 m – 350 m) and at presence of vegetation, the losses did not exceed 5% for packet delivery trials at low power (i.e., 5 dBm), they further dropped below 2%, for trials at higher power (i.e., 10 dBm), and even became negligible for transmissions near the maximum power (i.e., 20 dBm). At long distances (e.g., 5 km) and line-of-sight (LoS) conditions, at 10 dBm transmit power, the losses for spreading factor 7 were 10% approximately. By further increasing the signal immunity to noise, by a selection of spreading factor 11, the packet losses dropped to 3%, approximately. In both cases, at 5 km, the measured RSSI values slightly above −100 dBm, for the narrow bandwidth selection of 125 kHz which achieved better results. By increasing the power at the 20 dBm level, the losses dropped below 2%, even for the worst case of the SF7 setting selection.

Further power specific results are depicted in Fig. 10. These results involve the TM-195 power meter readings and the recordings provided by the LTC4150

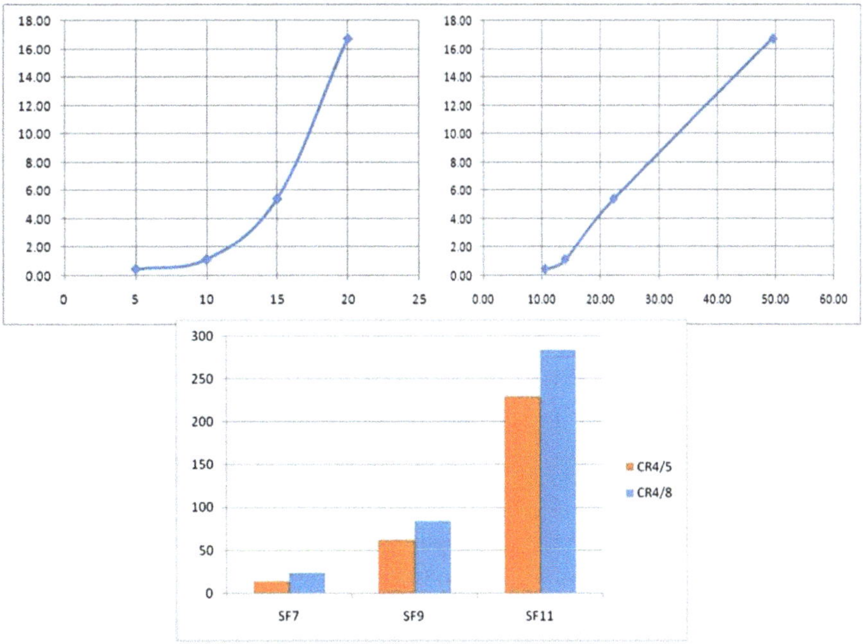

Fig. 10 Top left: The surface power density, at constant distance from the LoRa module, for various transmit power levels. Top right: The relation of the surface power density over the power consumption. Bottom: Energy required for transmission per packet, for different radio settings

Coulomb counter which has been properly attached on the arduino (or the WeMos) unit.

More specifically, the surface power density, in μW cm^{-2}, at constant and short distance (i.e., 5 cm, approximately) from the LoRa transceiver module, while transmitting at high constant rate (3 kbps, approximately), for various transmit power level arrangements (i.e., from 5 dBm to 20 dBm), is depicted at the top left part of Fig. 10. The corresponding chart of the surface power density over the average power consumption is depicted at the top right part of Fig. 10 and reveals the linear relation between the power being consumed and the power being emitted into the surrounding space. The average power consumption of the LoRa module, in mA, during the packet transmissions is calculated after subtracting the idle state values. More specifically, as indicated by processing of the LTC4150-generated measurements, the LoRa shield, along with the hosting arduino uno unit, consumed at about 60 mA, while the LoRa module itself, at idling state, at about 7 mA.

The bottom part of Fig. 10 depicts the energy amount, in mJ, required to send one packet of 15 B payload. These measurements were performed using a WeMos unit equipped with an LTC4150 counter module and a LoRa dragino shield, configured at the most frequently used radio setting of 125 kHz bandwidth and at a transmit power level of 10 dBm. The LTC4150, via providing interrupts, is exploited by the WeMos, which through the sum of electric charge being spent for the delivery of a specific and large number of packets, at constant supply voltage, calculates the total energy being spent. The energy per packet is calculated from this total energy mount, after subtracting the system's energy amount at idling for the same time period duration and dividing by the number of packets being transmitted during this period. The energy amount per packet drastically varies according to the SF parameter selection. Indeed, as already discussed, sending the same amount of data at higher SF requires more time, which means that the radio system should be active for a longer period and thus is consuming more energy. Furthermore, while transmitting with higher CR values (i.e., 4/8 instead of 4/5) this amount of energy becomes even greater, because more bits are added to the packet for redundancy purposes. The energy values being calculated are in accordance with other research works [13, 14] studying the same phenomenon. The increased energy amount per packet is attributed to the fact that, apart from the payload bytes, the header and the CRC bytes should be sent as well, and also to the extra processing that the preparation and delivery of each packet requires by the hosting board (i.e., by the WeMos unit).

The data rate values provided by the LoRa radios, compared with ones provided by the Wi-Fi radios, were considerably lower, especially with greater spreading factor values and narrow bandwidth. Even data rates even as low as 2–3 kbps, forming packets of 100 B to 200 B every half a second, are sometimes above the potential of the client – server communication needs. The delivery of much smaller packets is favored and suggested for the specific radio. The bandwidth utilization by the LoRa radios should follow additional fair policies like the 1% "on-the-air" time limitation [42] that applies to the total transmission time. However, most of the typical agricultural applications in a small-scale farm are lightweight enough to be satisfied by this low duty-cycle limitation. Indeed, the system of the two LoRa radios used during the experiments was adequate to carry typical commands for activating an electric pump or reporting on the soil humidity level, at a sparse time granularity, in packets containing a few bytes each. Such a system can be used on a permanent basis for farm activities monitoring, possibly assisted by a faster (but much more energy demanding) radio link (i.e., an IEEE802.11 link), if a high bandwidth is necessary to carry critical information (e.g., video streaming), in case of emergency triggers.

Finally, another interesting to mention issue, is the positive impression that the whole smart-phone assisted implementation and measurement process made on the students of agricultural engineering participating in the experiments. Indeed, the mobile devices are something that newer generations are very familiar with. Furthermore, the easy-to-use and portable nature of the equipment being used, along with its capability of long-range metrics gained good comments and assisted in better understanding the idiosyncrasies of the LoRa radio protocol. For these reasons, a more meticulous and well-documented evaluation of the potential benefits that the techniques being presented may have for the educational process will be necessary in the future.

5 Conclusions and Future Work

This paper presented a low-cost, generic and flexible methodology for evaluating and/or modifying the behavior of LoRa radio modules, forming easy-to-use and portable equipment, which is capable of getting fast results of satisfactory accuracy. The experimental testbed being described is exploiting innovative smart phone industry derived components and thus is providing the necessary user friendliness to monitor/modify the operational characteristics of LoRa-equipped sensor nodes, via simple commands. A pair of LoRa radio modules, hosted upon two arduino uno units (or similar counterparts), is the core of the measurement and control system. A satisfactory set of measurements were possible due to the embodiment of suitable additional electronic modules. Indeed, apart from the design and implementation details, characteristic measurement results, referring to radio coverage, packet losses and energy consumption are also reported. These results are in line with the underlying idiosyncrasies of the LoRa protocol and in accordance with similar research works. Many of the measurement series were performed in realistic, in terms of agriculture, conditions. The proposed testbed infrastructure was also further extended to be useful in real world scenario conditions, as it can easily be modified to serve the communication tasks of typical agricultural applications having sparse traffic requirements, like activating an electric pump or reporting on the soil humidity level.

The methodology being presented, due to its generic and modular design, has hardware agnostic features and thus, can be easily adapted, in the future, to provide a rich set of measurements for a large variety of LPWAN modules, able to serve a wider set of agricultural tasks. Furthermore, the discussed characteristics of the whole implementation, that allow for interaction via conventional smart phones, make the overall experimental setup ideal for use by not very experienced personnel and educationally beneficial for the students of agricultural engineering. The positive impact that the implementation and the measurement process had on the participants should be further evaluated and documented, for improving the educational process and students future careers.

Acknowledgments We would like to thank the personnel of the Farm Machinery Laboratory and the students (both the undergraduate and the postgraduate ones) of the Dept. of Natural Resources Management & Agricultural Engineering of the Agricultural University of Athens for their valuable assistance in the implementation, testing and measurement stages of this experimental platform.

References

1. FAO (2018) 'Climate-Smart Agriculture Sourcebook. Second Edition', online at: http://www.fao.org/climate-smart-agriculture-sourcebook/en/. Accessed in September 2019.
2. Symeonaki, E.G., Arvanitis, K.G. and Piromalis, D.D. (2019), 'Cloud computing for IoT applications in climate-smart agriculture: A review on the trends and challenges towards sustainability'. In Theodoridis, A., Ragkos, A. and Salampasis, M. (Eds.), Innovative Approaches and Applications for Sustainable Rural Development, HAICTA 2017, Springer, Cham, Earth System Sciences Series, Vol. 29, Chapter 9, pp. 147–167. https://doi.org/10.1007/978-3-030-02312-6_9.
3. O'Grady, M. and O'Hare, G., (2017) "Modelling the smart farm" in Information Processing in Agriculture, Vol. 4, Issue 3, May 2017, https://doi.org/10.1016/j.inpa.2017.05.001.
4. FAO (2017) 'Information and Communication Technology (ICT) in Agriculture', online at: http://www.fao.org/3/a-i7961e.pdf. Accessed in April 2019.
5. Welsh M. and Mainland G., (2004), 'Programming Sensor Networks Using Abstract Regions', Proceedings of USENIX NSDI Conf.
6. LPWAN (2019), Retrieved in April 2019 from the web site: https://en.wikipedia.org/wiki/LPWAN
7. LoRa (2019), LoRa protocol description on Wikipedia. Retrieved in April 2019 from: https://en.wikipedia.org/wiki/LoRa
8. Haxhibeqiri, J., De Poorter, E., Moerman, I., and Hoebeke, J. (2018). "A Survey of LoRaWAN for IoT: From Technology to Application. Sensors", Vol 18, https://doi.org/10.3390/s18113995.
9. Augustin, A., Yi, J., Clausen, T., and Townsley, W. M. (2016). A Study of LoRa: Long Range & Low Power Networks for the Internet of Things, Sensors, 16, 1466, https://doi.org/10.3390/s16091466.
10. Sanchez-Iborra, R., Sanchez-Gomez, J., Ballesta-Viñas, J., Cano, M.D. and Skarmeta, A.F. (2018), "Performance Evaluation of LoRa Considering Scenario Conditions.". MDPI Sensors, March 2018, Vol. 18, Issue3, doi:https://doi.org/10.3390/s18030772
11. IEEE (2019), IEEE Standard for Information Technology Telecommunications and Information Exchange Between Systems Local and Metropolitan Area Networks – Specific Requirements Part 15.4: Wireless Medium Access Control (MAC) and Physical Layer (PHY) Specifications for Low-Rate Wireless Personal Area Networks (WPANs). Technical report, IEEE, 2006.
12. Semtech (2019). LoRa FAQs. Retrieved in April 2019 from: https://docplayer.net/9473940-Lora-faqswww-semtech-com-1-of-4-semtech-semtech-corporation-lora-faq.html
13. Bor M. and Roedig, U. (2017). LoRa Transmission Parameter Selection. Proceedings of the "DCoSS 2017", Ottawa, Canada, June 2017, https://doi.org/10.1109/DCOSS.2017.10.

14. Faycal A., Magno M., Gautier M., Berder O. and Benini L. (2016). A Low Latency and Energy Efficient Communication Architecture for Heterogeneous Long-Short Range Communication. Proceedings of the Euromicro Conference on Digital System Design (DSD), August 2016, pp. 200–206, https://doi.org/10.1109/DSD.2016.97
15. Arduino (2019), Arduino Uno board description on the official Arduino site. Retrieved in April 2019 from the site: https://store.arduino.cc/arduino-uno-rev3
16. Raspberry (2019), Raspberry Pi 3 Model B board description on the official Raspberry site. Retrieved in April of 2019 from the site: https://www.raspberrypi.org/products/raspberry-pi-3-model-b/
17. C (2019). The C Programming Language. Retrieved in April 2019 from the site: https://en.wikipedia.org/wiki/C_(programming_language)
18. Python (2019). The Python Programming Language (Wikipedia). Retrieved in April 2019 from: https://en.wikipedia.org/wiki/Python
19. Arduino IDE, (2019). Description of the Arduino IDE software. Retrieved in April 2019 from the site: https://www.arduino.cc/en/software/.
20. Ardublock (2019). ArduBlock: A Graphical Programming Language for Arduino. Retrieved in April 2019 from the site: A Graphical Programming Language for Arduino
21. Node-RED (2019). Node-RED (Wikipedia): A flow-based development tool for visual programming. Retrieved in April 2019 from: https://www.google.com/search?q=node+red+wiki&ie=utf-8&oe=utf-8&client=firefox-b
22. Snap4Arduino (2019). Retrieved in May of 2019 from the site: http://snap4arduino.rocks/
23. App Inventor (2019), The MIT App Inventor programming environment. Retrieved in April 2019 from the site: http://appinventor.mit.edu/explore/
24. Loukatos D., Manolopoulos I., Arvaniti E., Arvanitis K., Sigrimis N., (2018a), 'Experimental Testbed for Monitoring the Energy Requirements of LPWAN Equipped Sensor Nodes', Proceedings of 6th IFAC Conference on Bio-Robotics (BIOROBOTICS 2018), Beijing, China, July 2018, https://doi.org/10.1016/j.ifacol.2018.08.196
25. Loukatos, D., Sarakis, L., Kontovasilis, K., Skianis, C. and Kormentzas, G., (2007). "Tools and Practices for Measurement-based Network Performance Evaluation", Proceedings of the 12th Computer Aided Modeling and Design of Communication Links and Networks (CAMAD '07) Part of PIMRC 2007, Athens, Greece pp. 1–5, https://doi.org/10.1109/PIMRC.2007.4394164.
26. Wi-Fi (2019). The Wi-Fi–IEEE 802.11 Standard. 2019. Available online: http://www.ieee802.org/11/ (accessed on 10 May 2019).
27. Dragino (2019), The LoRa Dragino shield for arduino. Retrieved in April 2019 from the site: http://www.dragino.com/products/module/item/102-lora-shield.html
28. RadioHead (2019), The RadioHead library to support LoRa modules. Available at the site: https://www.airspayce.com/mikem/arduino/RadioHead/. Accessed in April 2019
29. RSSI (2019), Received signal strength indicator – RSSI. Retrieved in April of 2019 from the site: https://en.wikipedia.org/wiki/Received_signal_strength_indication
30. Loukatos D., Arvanitis K.G., Armonis N. (2020) "Investigating Educationally Fruitful Speech-Based Methods to Assist People with Special Needs to Care Potted Plants", In: Human Interaction and Emerging Technologies. IHIET 2019. Advances in Intelligent Systems and Computing, vol 1018. Springer, https://doi.org/10.1007/978-3-030-25629-6_25.
31. ESP8266 (2019). The ESP8266 is a low-cost Wi-Fi microchip (Wikipedia). Retrieved in September 2019 from the site: https://en.wikipedia.org/wiki/ESP8266
32. WeMos (2019). The WeMos D1 R2 board. Retrieved in April 2019 from: https://wiki.wemos.cc/products:d1:d1
33. Wassie D., Loukatos D., Sarakis L., Kontovasilis K. and Skianis C., (2012). "On the energy requirements of vertical handover operations: Measurement-based results for the IEEE 802.21 framework", Proceedings of 17th IEEE International Workshop on Computer Aided Modeling and Design of Communication Links and Networks (CAMAD 2012), pp. 145–149, https://doi.org/10.1109/CAMAD.2012.6335316.

34. Kahn K., Winters, N. (2017), Child-Friendly Programming Interfaces to AI Cloud Services. In: Lavoué É., Drachsler H., Verbert K., Broisin J., Pérez-Sanagustín M. (eds) Data Driven Approaches in Digital Education. EC-TEL 2017. Lecture Notes in Computer Science, Vol. 10474. Springer, Cham
35. Loukatos D., Kahn K. and Alimisis D., (2018b) 'Flexible Techniques for Fast Developing and Remotely Controlling DIY Robots, with AI flavor', Proceedings of the 'Educational Robotics 2018 (EDUROBOTICS)', Rome, Italy, published by Springer, ISBN 978-3-030-18141-3, https://doi.org/10.1007/978-3-030-18141-3_14
36. Loukatos D., Arvanitis K.G., (2019) "Extending Smart Phone Based Techniques to Provide AI Flavored Interaction with DIY Robots, over Wi-Fi and LoRa interfaces", MDPI – Education Sciences, August 2019, Vol. 9, Issue 3, pp. 224-241, doi:https://doi.org/10.3390/educsci9030224.
37. TM-195 (2018), The TM-195 3-Axis RF Field Strength Meter. Retrieved in April 2018 from the site: http://www.tenmars.com/webls-en-us/TM-195.html
38. LTC4150 (2019), The LTC4150 coulomb meter, Retrieved in April 2019 from the web site: https://www.sparkfun.com/products/12052
39. GPIO (2019). General-purpose input/output pins. Retrieved in September 2019 from the site: https://en.wikipedia.org/wiki/General-purpose_input/output
40. PWM (2019). Pulse width modulation – PWM (Wikipedia). Retrieved in April 2019 from: https://en.wikipedia.org/wiki/Pulse-width_modulation
41. PFC (2019). Phase Control or Phase-fired control (PFC) (Wikipedia). Retrieved in April 2019 from: https://en.wikipedia.org/wiki/Phase-fired_controller
42. ETSI (2012) 'ETSI EN 300 220-1: Electromagnetic compatibility and radio spectrum matters (ERM); short range devices (SRD); radio equipment to be used in the 25 MHz to 1000 MHz frequency range with power levels ranging up to 500 mw; part 1 (2012)', online at: https://www.etsi.org/deliver/etsi_en/300200_300299/30022001/02.04.01_40/en_30022001v020401o.pdf. Accessed in September 2019.

Evaluating the Performance of a Simulated Softwarized Agricultural Wireless Sensor Network

José Olimpio R. Batista Jr, Gustavo M. Mostaço, Roberto F. Silva, Graça Bressan, Carlos E. Cugnasca, and Moacyr Martucci Jr

1 Introduction

Wireless Sensor Networks (WSNs) can be defined as a group of sensors, generally of a large number, spread over a specific area to collect and share information among them and with a gateway [1]. Different types of data can be collected, such as environmental variables (temperature, relative humidity, pressure, and vibration), presence, and movement of objects, among others.

Some of their main advantages over wired sensor networks are the reduced number of wires and cables, reducing infrastructure costs and problems, and increased flexibility of the networks used. In the specific case of agricultural production, they have been used on a variety of segments, such as vineyards [2], greenhouses [3], irrigation management [2], among others.

Of specific interest for this work are the uses of WSNs on animal production. There are several uses in the literature, such as monitoring animal feeding and animal welfare in confinement production [4]. Fewer works are describing their use on extensive pasture beef cattle production activities. WSNs can improve the gains of animal production farms due to better animal health and welfare management, better feeding, and productivity control.

Due to the Internet of Things (IoT) paradigm, an increasing number of WSNs have been adopted for farming activities [5]. This paradigm can be defined as how different technologies interact with each other in a specific environment [6], and

J. O. R. Batista Jr · G. M. Mostaço (✉) · R. F. Silva · G. Bressan · C. E. Cugnasca · M. Martucci Jr
Department of Computer Engineering and Digital Systems, Escola Politécnica da Universidade de São Paulo (USP), São Paulo, Brazil
e-mail: olimpio.rodrigues@usp.br; gmostaco@usp.br; graca.bressan@usp.br; carlos.cugnasca@usp.br; mmartucc@usp.br

D. D. Bochtis et al. (eds.), *Information and Communication Technologies for Agriculture—Theme I: Sensors*, Springer Optimization and Its Applications 182,
https://doi.org/10.1007/978-3-030-84144-7_5

includes devices, sensors, actuators, networks, routing protocols, databases as well as the processes to manage the data lifecycle using all those devices and processes. According to Dogra, Jha and Jain [7], 5G (Fifth Generation of Mobile Networks) and beyond will also increase the adoption of IoT technologies in many domains, which may include farm production.

Nevertheless, most papers on IoT implementation are on areas outside of the animal production domain, and most deal with theoretical aspects related to data collection, processing, storing, and access. Data routing, responsible for the protocols and processes needed for data transmission throughout the WSNs and IoT devices [8], is rarely studied in-depth in this domain.

For this reason, some routing protocols will be analyzed on the current work on an important scenario for animal production on extensive pastures: cattle weighing in the field. Better daily measurements of animals could help improve productivity and predict diseases. The daily measurements obtained in the field could be used, together with the more precise periodic measurements collected during vaccination procedures, on machine learning models to improve productivity.

The Collection Tree Protocol (CTP) and the IPv6 Routing Protocol for Low-Power and lossy networks (RPL), two of the most traditional routing protocols, will be evaluated, together with Software-Defined Networking (SDN). The last is an alternative, proposed for communication routing in the last years and has obtained considerably good results compared to the previous two [9].

The main objective of the present research is to evaluate the Packet Delivery Rate (PDR) and latency, two very important metrics for data routing, for cattle weighing in the field with WSN, considering the CTP and RPL protocols, as well as SDN. The main research question explored in this work is: "Is there a difference between routing protocols and methods for PDR and latency for cattle weighing in the field using WSN (Wireless Sensor Network)? If so, which is the best one?". This will be achieved through computer simulations of the routing protocols over the WSN.

This work is organized as follows: Section 2 contains a description of the main concepts used in this work, based on the state of the art literature; Sect. 3 contains the steps of the methodology used; Sect. 4 contains the main results obtained from the simulation and their analysis; Sect. 5 contains a discussion of the impacts of the results; and Sect. 6 contains the conclusions of the work, including its main limitations and future works.

2 Related Work

In this section, the main concepts that are important for conducting the simulation and the analyses are introduced: Sect. 2.1 contains a description of the main papers related to IoT implementation in the agricultural domain; Sect. 2.2 contains a description of the routing protocols used, including their main differences; and Sect. 2.3 contains a description of the SDN technology and its many uses.

2.1 IoT in the Agricultural Domain

The demand for animal-based food products is expected to increase by up to 70% by 2050 [10]. To meet such demand in a way that has a minimal impact on the environment and improves the quality and performance of livestock farming, the implementation of new technologies is required.

With the popularity of the IoT, WSNs have widely been researched. With the sensors' real-time monitoring and tracking functions, they can collect distributed data, analyze yielding, monitor agriculture environment conditions, decrease agricultural cost, and improve farm production. However, sensor nodes in agriculture may dynamically connect to the networks due to sensors powering off, unstable wireless signals affected by bad weather, distance, or movement, affecting network stability. Besides, with the adoption of precision agriculture with mobile autonomous vehicles carrying multi-sensors and precision livestock farming with monitored animals and systems, sensor nodes' connectivity may change frequently. Therefore, the stability and reliability of sensor network communication is an important topic nowadays [11].

Thus, IoT has a good potential to improve agribusiness, but connectivity is still a critical issue. 5G comes as a newly emerged communication and solution platform for rural areas. The authors Marchese, Moheddine, and Patrone [12] describe the integration of IoT devices, UAVs (Unmanned Aerial Vehicles), and satellites in the 5G environment. They state that solutions to extend the Internet connectivity to smart farming and smart villages, among other 5G application scenarios, need to go beyond the terrestrial infrastructure, and also that IoT and UAVs are promising in the agriculture field to monitor and control crop parameters and increase the quality and quantity of food.

The authors Fernandez, Vidal, and Valera [13] argue that, from the 5G concept, the agriculture IoT slice might be used to monitor and deliver relevant information in smart farming applications, e.g., values of temperature or humidity in different regions of a crop field. In this case, functions able to process high volumes of data may be required in nodes allocated in cloud computing infrastructures close to the core network.

IoT smart farming solutions may also integrate two services, one to control the humidity in some vineyards and another one to manage autonomous UAVs operations used for cattle surveillance. There are not important latency requirements in the first use case, and hence, everything can be deployed in the cloud infrastructure. In the second use case, there are significant latency requirements, and some parts of the solution may initially be deployed in the central cloud, while some others at the edge (e.g., in base stations or even in the UAVs themselves). By doing this, a cost-effective solution is built to offer the functionality of an IoT gateway in IoT scenarios where network infrastructures are insufficient (or simply unavailable) to support data communications with IoT devices over a delimited geographic area (e.g., a remote area in smart farming applications).

AlZu'bi et al. [14] propose an automatic and unsupervised smart irrigation system under a hardware implementation to accurately validate the proposed machine learning techniques and optimized equations. IoT sensors are attached in and around the proposed prototype to transfer the plants' environment status wirelessly. A smart irrigation system is implemented based on the analysis of the WSNs and the predefined environmental status for each plant under 5G.

When it comes to livestock production in general, one of the most critical aspects involves the management of feeding and drinking inside the farm, in a way that animals can stay healthy and may experience optimal weight gain, which is based on feed conversion ratio and energy expenditure. For that, the main requirement is related to the correct association of important production parameters with each animals' ID, such as weight, food, and water consumption over time (to obtain feed conversion ratio).

In free-range cattle-raising farms, the traditional methods involve weighing the animals once or a few times over the year, usually during vaccination or general herd health procedures, in which animals have to go through cattle crushes. If the animals could be weighed more often and without the need for human interaction, aspects such as farm management, animal welfare, and productivity would be benefited [15].

Works have been developed for the automated acquisition of individual animal mass by using depth images [16, 17]. Although this is very promising for confinements, its use for in-field weighing is still expensive due to many sensors and the use of high bandwidth of the deployed network.

The weighing of animals in grazing farms implies difficulties related to the large distances and the high number of animals that are usually observed. In this sense, the use of load cells next to the drinking troughs seems to be the best solution since it can detect both the animal's weight and water consumption. Concerning the granularity of measurements, it is capable of performing at least a few measurements a day, associated with the moments that the animals stay at the drinking troughs, and provides an automatic measurement, eliminating the need for human interaction. However, for that to work correctly, the network must provide a consistent and efficient connection between the nodes and the central point or gateway.

A critical aspect of IoT technologies and WSNs in the agricultural domain is the data routing protocol used. The next section contains a discussion of data routing protocols, focusing on the two main protocols analyzed in this work: CTP and RPL.

2.2 Routing Protocols

According to Simon Carbajo et al. [18], WSNs are limited in communication by the range of their radio modules. Thus, they need to form networks to transfer data from distant points. In this way, routing protocols are primary enablers of WSNs, and these require the implementation of reliable and energy-efficient mechanisms to maximize the reliability and availability of the network.

As most domains have specific requirements, it is vital to understand the adequate requirements and which routing protocol would be the most beneficial to use when studying the implementation of technologies in a specific domain. Section 4 contains a discussion of the main requirements for the use case analyzed in this work, while the next subsections contain discussions of the two routing protocols that will be simulated: Collection Tree Protocol or CTP and IPv6 Routing Protocol for Low power and Lossy Networks or RPL.

2.2.1 Collection Tree Protocol (CTP)

CTP is a tree-based routing protocol where a designated root, or set of roots, advertises itself and creates routing trees by spreading its gradients [19]. This protocol computes and maintains routes to one or more sinks, building and maintaining minimal cost trees with the sink as root, an adequate organization for converging exclusive traffic patterns [19].

CTP is responsible for broadcast communication and root discovery. It uses a component named link estimator to estimate the quality of a link with another neighbor node by sending and receiving acknowledgments for a series of packets [20].

CTP is a two-mechanism WSN routing protocol that offers high resilience to changes in topology: (i) route validation through data traffic to quickly detect inconsistencies in network topology; and (ii) the sending of control packages with adaptive periodicity to apply route repair mechanisms [21]. It is designed to quickly detect top-level cycling and recover from inconsistencies caused by high network dynamism [22].

According to Marques [22], CTP's route validation mechanism uses data packets to validate network topology and detect cycles. Cycle detection occurs by using data packets containing a metric to estimate the distance between the sending node and the receiving node measured in the number of hops. A cycle is detected when a node receives a packet whose distance from the sender to the recipient node is less than or equal to the distance from the node in question to the destination node. A topology repair mechanism is activated to correct this cycle [21].

Root nodes establish their gradient using a cost metric called Expected Transmission Count (ETX). It is a bidirectional single-hop link quality computation metric between two neighboring nodes that estimate the link quality based on the number of successfully delivered unicast packets between two given nodes from an acknowledgment of received packets [18].

For such computation, the PDR metric is used. It is calculated at the receiver node in terms of the division of Received Packets (RP) by Sent Packets (SP), as follows in Eq. 1 [23]:

$$\mathrm{PDR} = \frac{\mathrm{RP}}{\mathrm{SP}} \tag{1}$$

The value computed in Eq. 1 is also defined as in-quality, which is the quality from node Y to node Z measured by node Z by counting the successfully received packets from Y among all transmitted. It is also called PDRdown. For the actual ETX estimation, the out-quality is also needed. This is the in-quality estimated by node Y and is defined as PDRup at node Z. Then, node Z can calculate ETX, as shown in Eq. 2 [23]:

$$\text{ETX} = \frac{1}{\text{PDRdown} * \text{PDRup}} \quad (2)$$

Besides the CTP, the other protocol that will be analyzed in this work is the RPL. This will be described in more depth in the next subsection.

2.2.2 IPv6 Routing Protocol for Low Power and Lossy Networks (RPL)

RPL is a routing protocol that has been standardized by the Internet Engineering Task Force (IETF). It is widely used as a routing protocol for WSNs because it allows the usage of the Internet Protocol version 6 (IPv6) protocol on IoT devices. This enables a wide range of devices to be addressable over the Internet, considerably more than with the current IPv4 [22].

IETF specified the IPv6 over Low power Wireless Personal Area Networks (6LoWPAN) standard, which supports the idea of applying IPv6 even to the smallest machines. In this way, devices with limited hardware resources, typical in WSN, can participate in the IoT [23].

RPL uses a function known as Objective Function Zero (OF0) to assemble the topology and define the routing [20]. It uses the Destination-Oriented Directed Acyclic Graph (DODAG) concept to structure the topology formed by nodes in the WSNs.

There are two types of nodes in the WSNs that run RPL: (i) the first type is the sink node, which is responsible for initiating topology construction by disclosing its DODAG with DODAG Information Object (DIO) control messages. Sink nodes are usually pre-defined before the WSNs operation; (ii) the second type includes the nodes responsible for collecting and transmitting data (source nodes), hop-by-hop, towards the sink node. Source nodes also transmit DIO messages for topology maintenance [22].

RPL observes the node energy consumption to route packets by considering node neighbors' energy level before picking them as possible parents. Two units of information are used: (i) the type of the node, which indicates how it is supplied with power; and (ii) the Energy Estimation (EE). The RPL metric specification defines three possible states for the first: powered, on batteries, and scavenger. Such nodes report a maximum EE value and, in general, are preferable during parent selection.

If a network device is powered, it may be the root node connected to a personal computer or a special data collector (e.g. cluster-heads in hierarchical routing). If a node is on batteries, RPL computes the EE value using Eq. 3 [23]. The POWERnow value is the remaining energy, and POWERmax is the power estimation reported at boot up. However, if a node derives energy from external sources, it may report EE as a quantity value computed by dividing the amount of power that the node has acquired by the power consumed. This may be a poor estimation of how much load a node experiences in a certain period [23].

$$\mathrm{EE} = 100 * \frac{\mathrm{POWERnow}}{\mathrm{POWERmax}} \tag{3}$$

RPL also prevents routing loops by computing a node's position relative to other nodes concerning the DODAG root. This metric approximates the expected number of transmissions until a data packet reaches the gateway node. A node that is one hop away from the root, with excellent signal strength and little interference, may have an ETX equal to 1. Another node with a less reliable connection to the root will have a higher ETX [23].

Besides implementing CTP and RPL protocols on a simulated WSN in the agricultural domain, this work analyses the use of Software-Defined Networking (SDN) technology. The next section describes this technology, its importance, and its main uses.

2.3 Software-Defined Networking (SDN)

The issues of routing flexibility and provision of security services can potentially be addressed by the SDN, a technology approach for WSN and IoT, which can provide infrastructure sharing. This leads to lower operating costs, improved resource management, higher flexibility, and overall system sustainability.

The improvement in resource management is related to network resources (bandwidth, buffer allocation, among others), node management (battery level, tasks being performed, among others), and application management. The application manager knows which tasks are being performed by each node, what data is sent to which collector, what is the detection rate, which sensors are used by each task, and several other essential tasks [9].

Using SDN, a single controller or group of controllers may provide data routing control plan services for a larger number of nodes, thus allowing a system-wide view of network resources. The SDN enables data to be dynamically routed on a flow-by-stream basis, using source and destination information, then adapting to possible topology changes, providing better speeds and latency, eliminating potential bottleneck architectures and flexibility on network management [24].

According to Salman et al. [25], the SDN general architecture consists of four planes or layers: data, control, management, and application, separated from each

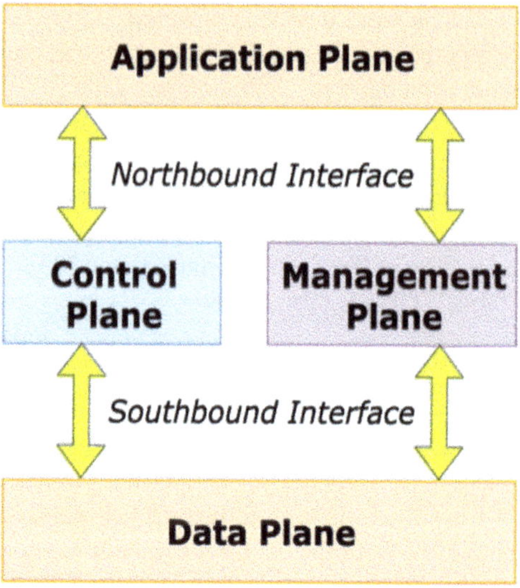

Fig. 1 SDN basic architecture

other, as illustrated in Fig. 1. This architecture provides network flexibility, dynamism, and management capabilities. SDN centralized control cannot be compared to a central telecommunications switch. Hence, SDN centralization is more logical than a physical concept [25].

The idea of a control plane and a data plane segregation involves a whole new way of networking when it becomes open, along with the intersection of open source and SDN. Building all this new virtualization technology, and bringing it to companies and the world, has created the need for a type of network programming.

Salman et al. [25] state that many technology leaders, governments, and researchers are making efforts to develop solutions that allow the broad deployment of IoT. Thus, the massive amount of data generated, the large scale of the network, concerns about security and privacy, the new requirements in terms of Quality of Service (QoS), and the heterogeneity in this ubiquitous network of networks, make its implementation a challenging task. SDN, a new network paradigm, has shown its usefulness in reducing management complexities in today's networks. Besides, the SDN presented effective security solutions by taking a global view of the network.

The control plane consists of the SDN controller, which has the role of network orchestration. Most of the calculations are done there, which gives it particular importance, being the "brain" of the network. The data plane consists of network devices (routers, switches, sensors, among others) responsible for simple matching operations to know how to forward packets. These simple devices forward all packets to the controller that they do not know how to act on. OpenFlow, the first standardized southbound interface, presented the central vision of the network's programming effectiveness. However, limiting the SDN to OpenFlow is an inappropriate limitation of the SDN horizons.

At the top of the control layer, there is the application layer. The communication between the control and application layers is carried out through the northbound interfaces. These interfaces, which provide the application with access to data collected from the network, offer most of the benefits of SDN. Control is fully distributed. Thus, reconfiguring the network and adding new features will be exhausting if done traditionally. Using SDN, this task becomes much more straightforward: centralized control provides the controller with a global view of the network, allowing it to hide management complexities and have more control over the entire network.

Margi et al. [9] also point out that a service orchestrator can use the SDN controller's centralized view to decide whether to admit new tasks or not. It can also use information related to the underlying communication technology to select the packet's path. For example, in heterogeneous IoT scenarios, devices may have two or more radios. The SDN controller will know the radios and use the different link layers to select the routes that meet the application requirements.

The next section contains a description of the main steps used in the methodology and the software used for the implementation of the simulations.

3 Methodology

The methodology adopted in this work consists of two main steps:

1. **Requirements gathering** for the implementation of WSN for cattle weighing on the farm;
2. **Simulation of the routing protocols CTP and RPL and the SDN** technology for a WSN defined by the requirements identified in step 1. Three scenarios were simulated: (i) using the CTP protocol; (ii) using the RPL protocol; and (iii) using the RPL protocol and the SDN technology.

The network simulations were conducted using the COOJA (Contiki OS Java) software. This is a WSNs simulator that focuses on simulating real hardware platforms and contains several network topologies models [26]. Due to compatibility and practicality reasons, the Instant Contiki virtual machine was chosen for the model implementation.

Table 1 contains the parameters of the simulation and the versions of the software used. These simulations were used to estimate the latency and PDR of the WSN on the three scenarios described above.

An essential definition of WSN simulation is the topology of the network. Fig. 2 illustrates the elementary topology used as a basis for the simulations [28]. It is important to note that only the behavior of Mesh 1 was simulated, as the problem addressed follows on the category of edge computing problems. The topologies used for the simulation on COOJA are illustrated in Figs. 3 and 4, and their main difference is related to the use of SDN technology for data routing.

Table 1 Parameters used in the simulations and software versions

Variable	Attribution	Variable	Attribution
Metrics	Latency and packet delivery rate	Radio-propagation model	UDGM distance loss
Time of each simulation	12 min	PHY layer	IEEE 802.15.4
Simulation speed	100%	COOJA	v1.8
Type of nodes	Tmote sky; Statics	Ubuntu	v18.04
Routing protocols	CTP, RPL	Qt Creator	v5.8
RDC/CCR	NullRDC/128 Hz	Instant Contiki	v3.0
MAC layer	CSMA	it-SDN [27]	v0.3

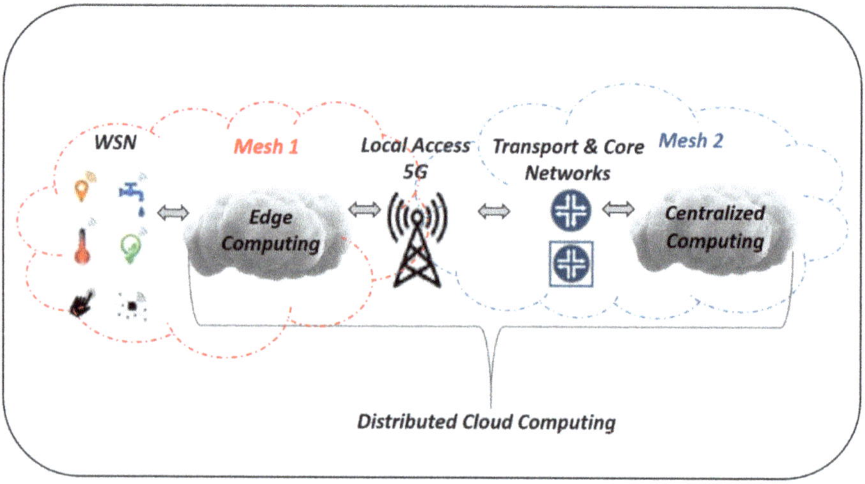

Fig. 2 The elementary topology for computational structure, based on WSN and 5G networks

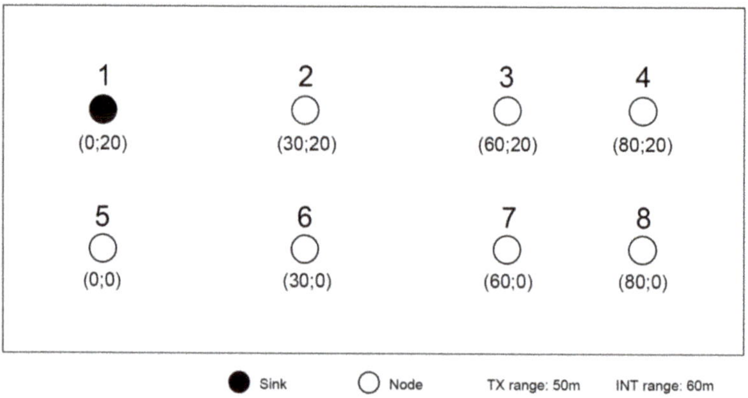

Fig. 3 Simulated topology without SDN and with node 1 as the sink node

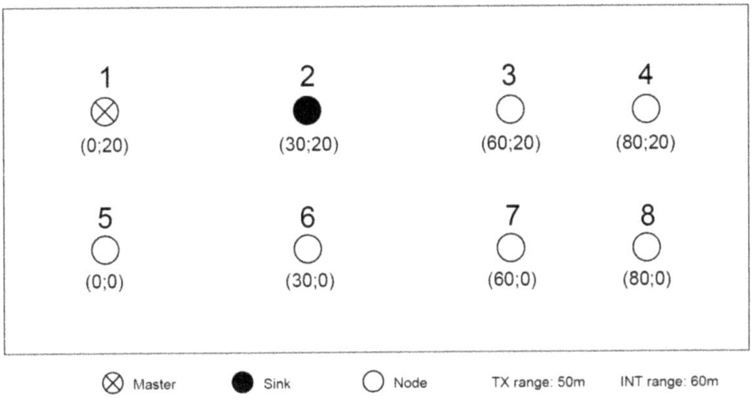

Fig. 4 Simulated topology with SDN and with node 1 as the master and node 2 as the sink

Table 2 Requirements gathered for using WSN for cattle weighing in the farm

Requirements
1. High network efficiency
2. High topology flexibility
3. Weekly or daily automatic data collection of animal weight and water consumption
4. Weekly or daily transmission of the data collected
5. Individual identification of the animal

In the next section, the results of both step 1 (requirements gathering for the use case) and step 2 (simulation of the use case, considering the different routing protocols and SDN technology) are analyzed.

4 Results

Table 2 summarizes the requirements gathered for the use case through an extensive literature review. They will be discussed in the following paragraphs.

Two essential requirements for the implementation of WSN for cattle weighing on the pastures are developed and discussed in the present work: (i) a high network efficiency, due to the difficulty of communication between nodes; and (ii) high topology flexibility, with redundancies, due to problems related to node failure.

The IoT technologies [29], such as sensors, actuators, and the network, should provide full coverage of all the beef cattle production processes. They should collect and transmit data of all the relevant processes along the entire supply chain. This would increase both processes' control, efficiency, and decision making.

Real-time monitoring of livestock indexes related to animal weight gain, disease detection, and control is of particular interest to the farmers. These can improve decision-making related to different processes involved in animal production, such

Table 3 Results of the simulations, considering PDR and latency on the different scenarios

Scenario	PDR (%)	Latency (ms)
1. CTP	98	132
2. RPL	99	107
3. RPL + SDN	99	56

as feeding, use of supplements, vaccination, among others. Nevertheless, it is considerably difficult to estimate those indexes on grass-fed cattle, as the animals are spread over a wide area.

In animal weighing, which is essential for defining activities related to providing feed and water for the animals and evaluating their body weight gain, this is even more problematic. The use of WSN on the farm, as specified in Sect. 2, would help to increase the number of data points, improving the velocity and effectiveness of decision making.

Therefore, two critical related requirements for improving decision making related to cattle weighing are: (i) automatic data collection of animal weight and water consumption throughout the animal's life, if possible, weekly or daily; and (ii) transmission of the data collected on a weekly or daily basis.

The research by Norton and Berckmans [30] and Condotta et al. [16] evaluated the use of automated camera-based systems to monitor animal behavior. Specific activities analyzed were related to diseases and individual animal mass. Nevertheless, even though this could be incorporated in the use case analyzed in this work, it would be costly to transmit images on the field and the investment needed on equipment.

Therefore, using an automatic weighing system specially designed to open field, grass-fed cattle by using a weighing scale (usually composed of load cells) placed next to cattle drinking troughs is proposed. This would allow for individual animal's weighing and identification tag collection daily. Additionally, it would improve monitoring water consumption by the herd.

The individual identification of the animal is essential to allow for better decision making. This is the last requirement that was identified in the present research. Throughout the next paragraphs, the computer simulation results of the proposed system for cattle weighing in the pasture, with weight plates on the water spots, are further described. The main focus of the present work is to evaluate the routing protocols and the SDN technology to transmit information between the nodes of the WSN. In this sense, further descriptions of the proposed system are out of the scope of the present chapter and will be published in a specific research paper.

Table 3 contains the results of the simulations implemented. As described in Sect. 3, two quality metrics were considered: PDR and latency. Two of the scenarios considered protocols without the use of SDN, and one considered the use of RPL associated with SDN. It is possible to observe that, although the PDR does not differ significantly among the scenarios, the latency shows a considerable decrease from scenarios 1 to 2 and from 2 to 3. The use of RPL together with SDN, in our specific use case, provided the lowest latency, by a considerable margin (48% lower than scenario 2 and 58% lower than scenario 1. One of the main reasons that explain these results is the centralized control and dynamic behavior of the WSN.

SDN can improve the results of the RPL due to the separation between the data plane (executed by sensor nodes) and the control plane (executed by the controller). Thus, a simple protocol defined to allow nodes to learn the shortest path towards the (closest) sink(s), to discover the neighboring nodes, and to periodically report local information to the controller (through the sink), could fulfill the requirements for this use case.

Therefore, it is possible to observe that the use of SDN, together with the RPL protocol, is the implementation that shows the best results for transmitting information on the use case of the implementation of WSN for cattle weighing in the farm. Also, through the use of this routing protocol, the system proposed fulfills all the requirements identified.

The next section contains a discussion of the main topics related to the implementation of the proposed system, the implementation of the SDN technology and the RPL protocol, and how this can impact other use cases in the agricultural domain, both on animal and vegetable production.

5 Discussion

WSN has emerged as a new type of distributed system, with applications in different areas such as target tracking, environment monitoring, traffic management, among others. Special mechanisms at the network layer are required to achieve reliable communication and guarantee a high delivery ratio with energy and memory efficiency.

As a result, the RPL routing protocol was specified and developed to overcome such requirements. The protocol is an end-to-end IP-based solution, which does not need the translation of gateways to address nodes within the network from the outside world, and it dynamically adapts the sending rate of routing control messages, which can be frequently generated only if the network is in unstable conditions [23]. In the simulations conducted in this work, the performance improvement was observed by using the RPL protocol instead of the CTP. This increase in performance is more pronounced with the use of SDN.

SDN has the potential to simplify network management. According to the Open Network Foundation [31], SDN addresses the fact that conventional networks' static architecture is inadequate for the dynamic computing and storage needs of today's data centers and operator environments. The benefits of improved management, greater flexibility, and sustainability tend to be the main reasons for choosing SDN over RPL. Additionally, situations in which the sensor nodes are attached to the animals, and therefore need dynamic allocation, should also benefit from adopting SDN technology.

The results presented in this work show that it is technically feasible to use SDN in cattle weighing and even other agricultural scenarios. The lower latencies obtained mean shorter response times, and consequently, the running applications become more reliable. This means that applications running in such network situations will experience better response times, thus enabling specific and high-demand uses, such as triggers or alarms for dangerous situations.

However, before a real deployment, a pilot test study is necessary to analyze and manage the challenges associated with network scalability, as increasing the number of nodes raises the challenges that SDN must face. Other aspects should also be explored, such as the type of hardware and software involved, as well as the evaluation of initial costs and improvement of operating costs, given that it demands the implementation of physical structure, such as exchanging routers and gateways for others that are capable of virtualization and software function control.

Another issue that should be researched in more depth is the use of mobile nodes on the WSN, a requirement for several use cases. One example would be if some animals carried sensors besides the static nodes on the farm's water spots. Concerning larger network sizes, one approach that can be taken for better performance is to organize the more extensive network into smaller clusters and to use distributed and hierarchical controllers.

In a more profound proposition, it is essential to mention that SDN can support the implementation of machine learning techniques and advanced analysis of the data collected [32]. This requires a coherent combination of intelligence with human control and supervision, software-controlled automated operational processes, and underlying programmable infrastructure. In this way, the networks will adapt, self-configure, monitor, repair, and optimize by constantly evaluating changes in their automatic routines and better reallocating resources according to the traffic patterns.

Centralized cloud computing has a global view of the network, but it is not suitable for applications that require low latency, real-time operation, and high-quality data for Artificial Intelligence services. On the other hand, the main goal of edge computing is to extend the functions of cloud computing to the edges of the network [24]. Because of its proximity to end-users and its decentralized deployment, edge computing can support quality applications and services that present requirements such as real-time execution, low latency, and high mobility with location recognition. This makes it more suitable for applications that generally do not have enough computing and storage resources, such as those on the farms. The distributed cloud computing allows for real-time data analysis on IoT networks with SDN.

The network slicing or segregation, an essential and vital aspect of 5G, is a solution for improving performance, reliability, energy efficiency, and economics towards implementing various 5G requirements and verticals. 5G radio access and its core networks are expected to be based on an SDN and NFV infrastructure that can orchestrate resources and control the network to provide efficient, flexible, and scalable services.

Internet Service Providers (ISPs) can implement slicing without 5G, but this will become much more prevalent with 5G and its emerging specifications, which require partitioning of data, control, and management plans to separate the environments that will be created. Thus, by serving individual customers or providing specific services, it gives ISP the opportunity to more easily support multi-tenancy, specific customers, and use cases, to meet each slice's unique SLA (Service Level Agreement) at different levels with guaranteed QoS. In this sense, the role of a horizontal and dynamic end-to-end 5G network orchestrator becomes critical to extend the concept of network softwarization, which must occur along the entire network (mesh 1 and mesh 2 in Fig. 2) to offer higher performance for the users (humans or machines) by using the best ISP available in each instant of communication according to the lower cost and consumption of network resources.

The 5G horizontal end-to-end orchestration may bring interesting and excellent results to the worldwide society:

- Improve sustainability, as all resources of 5G networks will be better harnessed, and smart applications can actually, effectively and simultaneously exist, minimizing the consumption of energy;
- Ensure the fulfillment of the requirements of each application and, therefore, better service qualities and user experience;
- Offer flexibility for new business models, which will imply new services with better competition, and consequently, better prices to users, among others.

The next section contains the final remarks of this work. It also contains its main limitations and suggestions for future works.

6 Conclusions

The implementation of WSNs in the agricultural domain, especially on animal production, has several specific requirements compared to other domains. In this work, the main requirements for the implementation of WSN on an important monitoring activity for beef cattle were identified. A system was proposed using WSN to weigh the animals, and the RPL and CTP routing protocols were evaluated through computer simulations. The use of SDN together with RPL was also evaluated, providing the best results in terms of latency and fulfilling all the requirements identified. Softwarization is a trend in the networks of the future.

The used methodology can be adapted for other use cases in the agricultural domain, both for animal and vegetable production. With the implementation of IoT technologies in this domain, evaluating and implementing the most resource-efficient routing protocols will be essential. In smart farming scenarios, this is an essential component of implementing autonomous processes and improving decision making. It will also be vital for correctly implementing 5G technologies.

The main limitations of this work are related to: (i) lack of open and available datasets for analyzing data routing on grass-fed beef cattle production; and (ii) the difficulty of implementing the WSN with the different routing protocols in field conditions due to the lack of interaction between academia and the agents of the beef cattle value chain. Future works are related to: (i) conducting experiments in the field to validate the results obtained; (ii) developing other important use cases on the agricultural domain, both for animal and vegetable production; and (iii) considering the possible impact of 5G technologies on data transmission on the farm.

Acknowledgment This work was supported by Itaú Unibanco S.A. through the Itaú Scholarship Program, at the Centro de Ciência de Dados (C^2D), Universidade de São Paulo, Brazil, by the National Council for Scientific and Technological Development (CNPq), and by the Coordination for the Improvement of Higher Education Personnel (CAPES) in Brazil.

References

1. Glisic S. and Lorenzo, B 'Advanced Wireless Networks', Second edition, John Wiley & Sons Publishing Company, United Kingdom, 2009.
2. Coates, R. W. et al. (2013) 'Wireless sensor network with irrigation valve control', Computers and Electronics in Agriculture. Elsevier B.V., 96, pp. 13–22. https://doi.org/10.1016/j.compag.2013.04.013.
3. Lea-Cox, J. D., Kantor, G. F. and Ristvey, A. G. (2008) 'Using wireless sensor technology to schedule irrigations and minimize water use in nursery and greenhouse production systems' In: Combined Proceedings of the International Plant Propagator's Society 58.
4. Karim, L. et al. (2014) 'An integrated framework for wireless sensor network management', Wireless Communications and Mobile Computing, 14(1), pp. 1143–1159. https://doi.org/10.1002/wcm.2260.
5. Keswani, B. et al. (2019) 'Adapting weather conditions based IoT enabled smart irrigation technique in precision agriculture mechanisms', Neural Computing and Applications, 31, pp. 277–292. https://doi.org/10.1007/s00521-018-3737-1.
6. Siow, E., Tiropanis, T. and Hall, W. (2018) 'Analytics for the Internet of Things: A Survey', ACM Computing Surveys, 51(74), pp. 1-33. https://doi.org/10.1145/3204947.
7. Dogra, A., Jha, R. K. and Jain, S. (2020) 'A Survey on beyond 5G network with the advent of 6G: Architecture and Emerging Technologies', IEEE Access, pp. 1–37. https://doi.org/10.1109/ACCESS.2020.3031234.
8. Ogawa, H. et al. (2016) 'Energy Consumption and Memory Footprint Evaluation of RPL and CTP in TinyOS', pp. 50–54. https://doi.org/10.14209/sbrt.2016.168.
9. Margi, C. B. et al. (2018) 'Software-Defined Wireless Sensor Networks Approach: Southbound Protocol and Its Performance Evaluation', Journal of Internet of Things (OJIOT), 4(1), pp. 99–108. Available at: http://www.ronpub.com/ojiot.
10. FAO-ONU (2017) The future of food and agriculture: trends and challenges, Food and Agriculture Organization of the United Nations. https://doi.org/10.4161/chan.4.6.12871.
11. Huang, T. et al. (2016) 'Multi-domain SDN survivability for agricultural wireless sensor networks', Sensors (Switzerland), 16(11), pp. 1–14. https://doi.org/10.3390/s16111861.
12. Marchese, M., Moheddine, A. and Patrone, F. (2019) 'IoT and UAV integration in 5G hybrid terrestrial-satellite networks', Sensors (Switzerland), 19(17). https://doi.org/10.3390/s19173704.
13. Fernandez, J. M., Vidal, I. and Valera, F. (2019) 'Enabling the orchestration of IoT slices through edge and cloud microservice platforms', Sensors (Switzerland), 19(13), pp. 1–29. https://doi.org/10.3390/s19132980.
14. AlZu'bi, S. et al. (2019) 'An efficient employment of internet of multimedia things in smart and future agriculture', Multimedia Tools and Applications. Multimedia Tools and Applications, 78(20), pp. 29581–29605. https://doi.org/10.1007/s11042-019-7367-0.
15. McAllister, T. A. et al. (2020) "Nutrition, feeding and management of beef cattle in intensive and extensive production systems', In: Animal Agriculture, pp. 75–98, Academic Press.
16. Condotta, I. C. F. S. et al. (2018) 'Using an artificial neural network to predict pig mass from depth images'. In: 10th International Livestock Environment Symposium (ILES X). American Society of Agricultural and Biological Engineers, pp. 1.

17. Sousa, R. V. et al. (2018) "Evaluating a LiDAR sensor and artificial neural network based-model to estimate cattle live weight', In 10th International Livestock Environment Symposium (ILES X), American Society of Agricultural and Biological Engineers, pp. 1.
18. Simon Carbajo, R. et al. (2017) 'Routing in wireless sensor networks for wind turbine monitoring', Pervasive and Mobile Computing. Elsevier B.V., 39, pp. 1–35. https://doi.org/10.1016/j.pmcj.2017.04.007.
19. Alves, R. C. et al. (2018) 'WS 3 N: Wireless Secure SDN-Based Communication for Sensor Networks', Security and Communication Networks. https://doi.org/10.1155/2018/8734389.
20. Oliveira, D. and Margi, C. (2017) 'Roteamento ciente de energia em redes de sensores sem fio definidas por software', pp. 811–815. https://doi.org/10.14209/sbrt.2017.108.
21. Gnawali, O. et al. (2009) 'Collection tree protocol', Proceedings of the 7th ACM Conference on Embedded Networked Sensor Systems, SenSys 2009, pp. 1–14. https://doi.org/10.1145/1644038.1644040.
22. Marques, V. F. (2019) 'Abordagem para suporte à mobilidade no protocolo RPL baseado na variabilidade da vizinhança' Dissertação submetida ao Programa de Pós-Graduação em Computação Aplicada do Centro de Ciências Tecnológicas da Universidade do Estado de Santa Catarina, Brasil.
23. Tsvetkov, T. (2011) 'RPL: IPv6 Routing Protocol for Low Power and Lossy Networks', Seminar SN SS2011. Network Architectures and Services, July. https://doi.org/10.2313/NET-2011-07-1_09.
24. Kaur, K. et al. (2018) 'Edge Computing in the Industrial Internet of Things Environment: Software-Defined-Networks-Based Edge-Cloud Interplay', Communications Magazine, 56(2), pp. 44-51. ISSN 1558-1896.
25. Salman, O. et al. (2018) 'IoT survey: An SDN and fog computing perspective', Computer Networks. Elsevier B.V., 143(2018), pp. 221–246. https://doi.org/10.1016/j.comnet.2018.07.020.
26. Mehmood, T. (2017) 'COOJA Network Simulator: Exploring the Infinite Possible Ways to Compute the Performance Metrics of IOT Based Smart Devices to Understand the Working of IOT Based Compression and Routing Protocols', Dept. of Electrical Engineering, SEECS, NUST Islamabad.
27. Alves, R. C. et al. (2017) 'IT-SDN: Installation Guide (for Linux 64 bits)' – March.
28. Parvez, I. et al. (2018) 'A Survey on Low Latency Towards 5G: RAN, Core Network and Caching Solutions', IEEE Communications Surveys & Tutorials, 20(4), pp. 3098-3130. https://doi.org/10.1109/COMST.2018.2841349.
29. ITU-T (2012) 'Y.2060 – Overview of the Internet of Things. Recommendation', ITU-T. https://doi.org/10.1021/ic00245a007.
30. Norton, T. and Berckmans, D. (2018) 'Precision Livestock Farming: the Future of Livestock Welfare Monitoring and Management?', Animal Welfare in a Changing World, pp. 130.
31. ONF – Open Networking Foundation (2020). Available at: https://www.opennetworking.org/.
32. Sandano, I. (2018) 'The Self-Driven Network, A trajetória rumo à automação e seus estágios´, São Paulo, Brasil.

Smart Agriculture: A Low-Cost Wireless Sensor Network Approach

Ioannis Angelis, Alexandros Zervopoulos, Aikaterini Georgia Alvanou, Spiridon Vergis, Asterios Papamichail, Konstantinos Bezas, Andreana Stylidou, Athanasios Tsipis, Vasileios Komianos, Georgios Tsoumanis, George Koufoudakis, and Konstantinos Oikonomou

1 Introduction

In the context of the increasing number of the world's population and ever-increasing rates of environmental pollution, issues regarding the ability to meet living and nutritional needs come to light [1]. Taking into account the modern technological and scientific tendency for the development of Internet of Things (IoT) environments and of Information Communication Technologies (ICT) their involvement in the field of agriculture is also imperative [2–4]. For a better understanding, the IoT consists of a set of things, such as sensors, mobile phones tags, cameras etc., which interact with each other and exchange data, while ICT enables users to handle, store, access, and transmit information [5, 6].

Agricultural production processes contribute significantly to the sustenance of the human population. Especially the olive trees cultivation is of utmost importance for the nutrition. It is worth noting that olive groves have various sensitivities, with temperature and humidity playing a leading role in their health [7–9]. Consequently, the need for real-time monitoring and dissemination of environmental parameters is emerging, especially considering their contribution in decision-making methods

I. Angelis (✉) · A. Zervopoulos · A. G. Alvanou · S. Vergis · A. Papamichail · K. Bezas · A. Stylidou · A. Tsipis · V. Komianos · G. Koufoudakis · K. Oikonomou
Ionian University, Corfu, Greece
e-mail: iangelis@ionio.gr; azervop@ionio.gr; akorina@ionio.gr; svergis@ionio.gr; aspapa@ionio.gr; kbezas@ionio.gr; astylidou@ionio.gr; atsipis@ionio.gr; vkomianos@ionio.gr; gkoufoud@ionio.gr; okon@ionio.gr

G. Tsoumanis
University of Ioannina, Arta, Greece
e-mail: gtsoum@uoi.gr

D. D. Bochtis et al. (eds.), *Information and Communication Technologies for Agriculture—Theme I: Sensors*, Springer Optimization and Its Applications 182,
https://doi.org/10.1007/978-3-030-84144-7_6

related to crop management, providing immediate and appropriate response, and meeting the requirements of precision agriculture [10].

The achievement of real-time monitoring and dissemination is facilitated by the involvement of Wireless Sensor Networks (WSNs) [11]. More specifically, the sensors in a WSN are small but powerful devices that can serve the needs of agricultural activities. The very same types of devices can be used as "things" on IoT or as "fog devices" on a cloud/fog computing environment.

In a system, which combines IoT environments, WSN technology and cloud computing enablers, it is necessary to take into consideration a plethora of issues. One of them is the selection of suitable, efficient, and low-cost equipment, as it is extremely important for the overall performance of the system because it determines the system's accountability and durability [12, 13].

Moreover, the selected equipment can affect the accuracy and synchronization of sensor measurements which is one more issue of major importance. In more detail, network synchronization is considered necessary, in order to offer time-correlated measurements from remote nodes. Therefore, synchronization enables the process of knowledge extraction, allowing the end user to interpret the results produced and act [14].

With the goal of achieving synchronized measurements, a simple synchronization scheme is described later in this chapter. This scheme provides synchronized measurements, using the clock of the network's sink as a reference point and enabling precision agriculture. It should be emphasized that this synchronization scheme uses as few transmissions as possible to save energy and elongate the network's lifetime, while also serving as a baseline for more complicated approaches. Despite the simplicity of the scheme, its evaluation through a previously conducted experimentation process confirms its suitability [15].

Additionally, it is very important to establish a cloud/fog computing environment, as it provides storage, networking and computing capabilities close to the end user [16, 17]. Thus, in large-scale deployment, the presence of cloud servers and fog devices is essential. In greater detail, cloud servers are characterized by increased computational capabilities and are located in large data centers [18]. Fog devices acquire similar, even though of less power, attributes and are located closer to the end user, facilitating data handling, and minimizing response time [18].

A cloud/fog network architecture that meets the above requirements and can be used in olive groves is described in Sect. 4. The selected latency adjustable architecture is capable of dynamically mapping the system behavior, allowing the identification of olive groves' needs. It is noteworthy, that based on the literature, latency can be considered as a critical factor for defining early environmental conditions [17]. The evaluation process of the selected architecture from a previous study is also presented, indicating its usefulness [19].

In addition to the need of monitoring environmental conditions, a major and unforeseen danger is the occurrence of a natural disaster, such as a flood, earthquake or wildfire, which is a complex phenomenon [20–23]. The need to deal with such disasters can be met by expanding the selected cloud/fog network architecture. This potential is analyzed in Sect. 4.

The rest of this chapter is organized as follows. An overview of wireless sensing technologies in smart agriculture is presented in Sect. 2. The synchronization scheme under consideration is described in Sect. 3, including its experimental evaluation, while Sect. 4 specifies the selected cloud/fog network architecture. Finally, conclusions and future work are drawn in Sect. 5.

2 Wireless Sensing Technologies in Smart Agriculture

This section presents a wide variety of WSN applications in agriculture as well as a basic overview of some fundamental technologies that have propelled the domain, transforming it into one of the most trending topics in ongoing, frontier research. Furthermore, a certain selection of technologies is thoroughly examined, as it is particularly prevalent in the literature, and is also utilized by the studies examined in the sequel.

2.1 Related Work

WSNs have been the subject of numerous scientific studies, and rightly so, since they are among the most promising technologies of the twenty-first century. Their use has been expanded to a variety of areas, particularly to IoT applications. More specifically, some of their properties, such as self-configuration, self-diagnosis, self-heal, and self-organization, make them ideal in food industry and smart agriculture [24]. WSNs, in conjunction with the rapidly evolving IoT paradigm, promise to provide a plethora of benefits to advance the domain of agriculture, even in developing countries, by enabling applications through a low-cost, energy-efficient, and wireless infrastructure [25, 26].

The technology behind WSNs has been preferred to meet a multitude of purposes. To start with, one study in this field uses the capacity of these networks to monitor humidity levels in real time, in order to enrich the irrigation process at a soccer stadium, in areas of the soil where the humidity is below a certain limit [27]. In addition, another study proposes a system for greenhouses, which aims to ensure ideal conditions, in order to avoid the development of pests and weeds that are favored in such artificial environments [28].

Yoo at al. deploy a WSN-based architecture in a greenhouse, which is capable of automatically detecting and altering relevant conditions, such as temperature and humidity, through sensor and actuator nodes [29]. This setup is experimentally evaluated in a melon greenhouse, proving to be effective, while the authors provide a variety of issues that were encountered during deployment, such as reduced communication range due to interference, premature battery exhaustion due to hardware inconsistencies, etc. Moreover, another WSN architecture is proposed, which is deployed in a pepper vegetable greenhouse, where temperature, humidity and lighting are monitored to provide optimal conditions for pepper cultivation [30]. The goal is for farmers to first receive data from the sensors and then take the appropriate measures for their crops.

WSNs are equally appropriate in greenhouses, as well as outdoors environments, where more parameters need to be considered, since the latter are exposed to uncontrolled environmental conditions. Díaz et al. (2011) provide a well-rounded overview of WSN deployment scenarios regarding outdoor agricultural monitoring, through which they identify all distinct phases that need to be considered in such applications [31]. The phases are compiled into a well laid-out methodology for WSN applications focusing on agriculture and they cover topics such as terrain considerations, architecture, implementation, and maintenance. Shinghal et al. (2017) design and deploy a WSN application to improve potato crop production by adjusting the irrigation system and taking into account various parameters, such as the depth of water, soil water tension and system capacity [32].

Aside from monitoring environmental parameters, additional considerations need to be made in natural environments. For instance, the deployment of a WSN application for crop protection by diverting animal intrusions is presented in [33]. The developed system attempts to spot and divert animal intrusions by employing four different types of nodes, each specialized in the tasks of detecting intrusions, diverting them using sounds and lights, as well as keeping track of their activity, occurrence, and frequency.

2.2 Equipment Overview

An important aspect to consider during the development stages of a system for the purposes of agricultural applications, is the utilized equipment [34]. The latter must be adjusted in order to meet the requirements of a system that is often exposed to harsh conditions (e.g., heavy rainfall, intense wind, bright sunlight, corrosive moisture). Also, the application-specific requirements, in terms of computational power, energy and wireless communication quality, are the main contributing factors to determining the most suitable per-application equipment. Finally, a system developed for agricultural applications must take into account the low-cost factor since on-field deployments usually require multiple devices in order efficiently monitor a large field [12, 13].

Along the years, the utilized equipment ranges in terms of price and capabilities [35]. Also, depending on the application's requirements, the hardware needs to be adjusted, thus the employed equipment among different applications comprises variant models of micro-controllers and antennas. Agriculture-specific applications, which require a system deployment on large fields, need to utilize hardware with low specifications (lowering the required cost and ensuring wider land coverage) for the collection of environmental measurements. In the case where further processing of the data is required, the information is then relayed back to a central node, hereafter referred to as sink, which usually includes higher cost and computational capabilities.

An essential part of agriculture oriented WSN systems is the information collected by the utilized sensors, which vary depending on the application [36, 37]. The most commonly measured information is soil moisture and temperature, air humidity, ultraviolet (UV) radiation, plant's leaves wetness and air CO_2 levels. This information can potentially aid in the evaluation of a field's health, which in turn might help prevent any future threats to plants or fruit production. Many different models of these sensors exist, all ranging in price, features and measurement accuracy [38].

Custom-designed hardware can potentially offer many additional benefits, including low power consumption and low development cost [37]. Such systems are called System on a Chip (SoC) design, which accommodate all necessary components (sensors, antennas and microcontrollers) in a single-board design. The complexity they introduce to the system, though, limits their widespread use. IoT applications usually include pre-designed hardware built specifically for such projects, thus allowing the designer to focus on properly collecting and distributing the data from each sensory device. Due to the low price, ease of configuration and variable models, a flexible alternative refers to the Arduino prototyping platform, which shows potential in this direction, offering capabilities that can be exploited to monitor variant properties of a field, like air humidity and soil moisture [39, 40].

A vital component for the proper operation of the system is the antenna, which is responsible for interconnecting all separate, remote nodes into one common network. Data in this network can be transferred at desired points of interest (e.g., the network's sink node). Each agricultural application design, and consequently the antenna of choice, focuses on the requirements each field might introduce. For example, large fields require low transmission radio frequencies in order to minimize weak signal points throughout the area [39]. Contrary, in cases where close range communication is preferred or required for data collection, close proximity higher frequency communication antennas are utilized accordingly [40]. Literature contains a multitude of different antennas each employing variable characteristics for WSN applications [41–43]. A few examples include transceivers like CC2420 and MC1319x, Bluetooth-enabled antennas like BlueCore and, also, Zigbee-based XBee antennas.

The literature, especially the last decade's, shows that many research activities lean towards the utilization of the Arduino prototyping platform and the Zigbee antennas, which can be configured to operate seamlessly together [44]. Such applications include, but are not limited to, soil and air parameters' monitoring for precision agriculture, autonomous irrigation systems based on plant's requirements, intrusion detection, scheduling of farming activities (e.g., harvesting, irrigation, fertilization, etc.), and farm cost-based maintenance of the collected information [39, 45–49].

2.3 Selected Low-Cost Equipment

The systems described in this study (Sects. 3 and 4) utilize various low-cost electronic components. These systems are based on the Arduino prototyping platform[1], which is responsible for the measurement sensing procedure. For a WSN to exist, wireless communication must be enabled, which is accomplished by using the XBee-PRO S2C module[2]. The XBee modules cannot be placed on the Arduino prototyping platform directly, therefore an Arduino Wireless SD Shield[3] is used. These components are selected due to their simplicity, ease of use, extensive documentation, and large community of followers, all of which can significantly expedite and assist the development phase. A comparison between the costs of the various components and the characteristic of antennas is conducted in the literature, further showcasing why they stand out [50, 51]. Ergo, this equipment is considered quite versatile and suitable for fast prototyping, making it ideal for research and industry agents alike.

2.3.1 Arduino

The base component of the systems considered in the sequel is the Arduino prototyping platform, which consists of a wide variety of available boards, each defined by a wide range of specifications in terms of hardware and software. The presented experiments make use of two variations of this platform: (i) the Arduino Uno Rev. 3[4] board; and (ii) the Arduino Mega 2560[5] board.

While the Arduino Uno is the most well-known Arduino board for beginners in the world of electronics, the Arduino Mega 2560 exceeds the capabilities of Arduino Uno. The main reasons for a researcher to choose the Arduino Mega instead of the Arduino Uno are the amount of SRAM and the number of Input/Output pins. The complete set of specifications for the particular Arduino boards, are encased in Table 1.

2.3.2 Arduino Wireless SD Shield and XBee

The Arduino Wireless SD Shield[6] is used as an intermediary between the Arduino board and the wireless antenna, which also enhances the capabilities of a node by giving it the capacity to store the produced data into a micro Secure Digital (SD)

[1] https://arduino.cc

[2] https://tinyurl.com/y5posdyh

[3] https://store.arduino.cc/arduino-wirelss-sd-shield

[4] https://store.arduino.cc/arduino-uno-rev3

[5] https://store.arduino.cc/mega-2560-r3

[6] https://store.arduino.cc/arduino-wirelss-sd-shield

Table 1 The specifications of the Arduino Uno and the Arduino Mega

Component	Arduino Uno Rev. 3	Arduino Mega 2560
Microcontroller	ATmega328P	ATmega2560
Operating Voltage	5 V	5 V
Input Voltage (recommended)	7–12 V	7–12 V
Input Voltage (limit)	6–20 V	6–20 V
Digital I/O Pins	14 (6 PWM output)	54 (14 PWM output)
Analog Input Pins	6	16
DC Current per I/O Pin	20 mA	20 mA
DC Current for 3.3V Pin	50 mA	50 mA
Flash Memory	32 KB	256 KB
Boot Loader	0.5 KB	8 KB
SRAM	2 KB	8 KB
EEPROM	1 KB	4 KB
Clock Speed	16 MHz	16 MHz
LED_BUILTIN	13	13
Length	68.6 mm	101.52 mm
Width	53.4 mm	53.3 mm
Weight	25 g	37 g

card. The wireless communication of the WSN's nodes is handled by the Digi XBee-PRO S2C[7] module. In order for the wireless module to communicate with the Arduino board, Serial Communication in various baud rates ranging from 9600 to 115,200 bps is utilized. The Digi XBee-PRO S2C module operates in a license free Industrial, Scientific and Medical Band of 2.4–2.5 GHz and its maximum transmission rate is 250 Kbps. Moreover, it facilitates three different protocols, namely: 802.15.4, DigiMesh and Zigbee. These protocols are in turn based on the IEEE 802.15.4 protocol, which is built on top of Carrier Sense Multiple Access with Collision Avoidance (CSMA/CA) protocol. The modules fill in the role of transceivers since their energy consumption is low. In particular, in transmit mode the energy consumption[8] is 120 mA, whereas in idle/receive mode the consumption is 31 mA respectively.

Zigbee Protocol

The considered systems make use of the Zigbee[9] protocol, which provides high reliability, low power consumption and high security. The particular protocol is built for mesh networks, augmenting them with automated routing, ad-hoc network

[7] https://www.digi.com/products/models/xbp24cz7wit-004

[8] https://tinyurl.com/y5posdyh

[9] https://tinyurl.com/y5posdyh

creation, and self-healing. With that said, it comprises three (3) different types of devices, these being: (i) Coordinator; (ii) Routers; and (iii) End Devices.

Coordinator

The Coordinator is responsible for the initialization of the network. While in this phase, the Coordinator establishes the operating channel and the Personal Area Network (PAN) ID for the entire network. Afterwards, it allows Routers and End Devices to join the network, at which point it assigns to them a unique 16-bit identifier, called Network Address. It is noteworthy that the Zigbee protocol allows only one Coordinator per PAN ID. During network operation, the Coordinator's functionality becomes similar to a Router.

Routers

In order for a Router to participate in a Zigbee network, it must first discover and join a valid network. After the join process is complete, the Router takes up responsibilities of routing data-packets from and to other nodes in the network. A Router is also responsible for allowing new devices to join their associated network, e.g., other Routers and End Devices.

End Devices

When a network node (either the Coordinator or a Router) enables an End Device to join the network, the former becomes the parent of the latter. The parent of the End Device then, acts as a buffer for the data-packets that are addressed to the specific End Device. This renders the End Devices capable of going into low power modes ("sleep" state) to minimize their energy consumption, a factor that is essential for applications where the available energy is limited.

2.3.3 Raspberry Pi 3

The Raspberry Pi 3 Model B[10] is a low-cost credit-card sized computer with low energy consumption. Its board comes with a Quad-Core 64-bit processor running at 1.2GHz[11] and 1GB of RAM. Hence, Raspberry Pi is a fully assembled computer, that supports a wide variety of operating systems. The operating system itself is installed on a microSD card. A noteworthy attribute of a Raspberry Pi is the

[10] https://www.raspberrypi.org/products/raspberry-pi-3-model-b/

[11] https://www.raspberrypi.org/products/raspberry-pi-3-model-b/

connectivity it provides via USB ports, which makes it a suitable candidate for hosting gateway properties in a WSN-based environment [52].

2.3.4 Sensors and Others

In the research area of WSNs, a variety of applications are developed in order to tackle daily problems in various agricultural domains [36]. In order for the applications to be functional, sensing of various climate phenomena is necessary, these being namely temperature, relative humidity, soil moisture and UV radiation. Further, the utilization of an accelerometer is common practice in such applications, in order to detect unauthorized node movements, caused by human intervention, animal intrusion or misplacement due to extreme weather conditions (e.g., strong wind).

With that said, to obtain these measurements a set of commonly used sensors are presented here. The air temperature and relative humidity degrees can be monitored by the RHT-03[12] sensor, whereas the soil moisture levels can be obtained using the YL-69[13] sensor. The VEML-6070[14] is popular for measuring the UV radiation, while the MPU-6050[15] accelerometer is used for detecting node movements. In Fig. 1, a fully assembled node, using the above modules, is depicted.

Moreover, to prolong a WSN's lifetime, a power bank is often employed to provide power to a node in a reliable fashion. In fact, it is considered a good practice for the power bank to feature a solar panel (or an alternative power source), in order to further elongate the WSN's lifespan. Another welcoming attribute refers to its resistance to extreme weather conditions in order to avoid malfunctions caused by environmental agents, e.g., atmospheric corrosion. A power bank, that satisfies all the aforementioned attributes, is the Sandberg Outdoor Solar Powerbank[16] with a capacity of 16,000 mAh. Note that, besides the power bank, all the other parts of the node need also protection against the weather elements, which is commonly accomplished by enclosing all the electronic parts in an appropriate case.

3 Synchronized Monitoring

In this section, the need for synchronized monitoring is highlighted and the methods used to achieve it are discussed. Additionally, a simple synchronization scheme is investigated as an efficient, network-based means of acquiring time-correlated

[12] https://www.sparkfun.com/products/10167

[13] https://www.oddwires.com/yl-69-soil-hygrometer-humidity-soil-moisture-detection-sensor/

[14] https://www.vishay.com/ppg?84277

[15] https://www.invensense.com/products/motion-tracking/6-axis/mpu-6050/

[16] https://sandberg.it/en-mt/product/Outdoor-Solar-Powerbank-16000

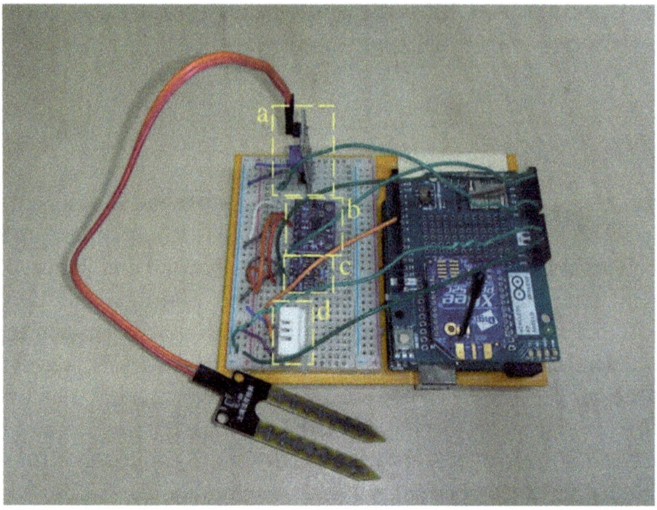

Fig. 1 A fully assembled node, in right side of the figure, an Arduino Uno Rev. 3 is depicted with all its components that allow wireless communication (i.e. Wireless SD Shield and Digi XBee PRO S2C module). In the left side of the figure, the aforementioned sensors are also depicted namely, (**a**) the first part of the YL-69 sensor, (**b**) the MPU-6050 accelerometer, (**c**) the VEML-6070 sensor and (**d**) the RHT-03 sensor

measurements. Finally, a set of experiments is presented to evaluate the effectiveness of this simple synchronization scheme in a realistic setting.

3.1 Related Work

Zhang et al. (2002) provide a detailed overview of technologies related to precision agriculture [53]. One of the thematic fields covered by this study concerns temporal variability, which determines the effectiveness of the monitoring and decision-making process. Consequently, there is a need for knowledge of the physical time point, during which a measurement is generated. One option to find the relevant timestamp—which is associated with each measurement and comes from the device that carries it out—is to use the Internet. However, Internet connection cannot be achieved on WSNs because of their limitations in size, cost, and power consumption.

Each WSN node owns a local clock to track time, which is characterized by a degree of accuracy. Accuracy is largely determined by the manufacturing parameters of the device and is commensurate with its cost and energy consumption rates. More precisely, greater accuracy equates to higher costs and higher levels of energy consumption [54]. As for the Arduino microcontroller, its clock depends entirely on how long the device has been active.

Since each device is equipped with its own clock, it is necessary to associate the timestamps with a point in physical time so that human interpretation can be possible. However, it is worth noting that the timestamps are likely to be of no use, if the acquired timestamp indications, associated with a common point in time, are completely different. In the case of WSNs, the nodes that make up a network and are scattered over an area are almost impossible to start functioning at the same time. As a result, their clocks are desynchronized and so supplementation of measurements by these nodes is considered pointless.

As for achieving correlation with physical time, Global Positioning System (GPS) modules are commonly used, which provide satellite-related measurements over time [55]. However, supplying the nodes of a WSN with GPS modules is a costly process and perhaps problematic, in terms of signal loss, when deployed in suburban areas. Correlation can also be achieved with the help of an external measuring device. Such a device could be a remote server or even a laptop.

As already stated, most WSNs include a sink node, responsible for collecting the measured data from the entire network and subsequently relaying/delivering them to the outside world. Given this role, it is reasonable to specify the sink's clock as the reference point for the correlation of the other nodes' clock, since the sink's clock can be associated with a point in physical time through an external device, such as a laptop or a fog device [56]. In this way, there is no need to associate the clock of each node separately with physical time.

The synchronization process is certainly demanding. In greater detail, the need for communication-based synchronization is reinforced in the case of time-keeping clock devices, as these devices' indications are known to differ gradually, especially when exposed to extreme environmental conditions [57, 58]. Regarding the equipment used and presented in Sect. 2.3, the Arduino microcontroller resonator operating at 16 MHz, loses tens of seconds per day [59]. This performance is quite poor for time-critical applications. In fact, the clock discrepancy, even though small, indicates that it is not enough to perform the synchronization process only once, but on the contrary, it needs to take place frequently to account for accumulated inaccuracy.

Whether the need for clock synchronization arises for the purpose of time stamping or process synchronization, various approaches focusing on WSN synchronization have been presented over the years. One of the most well-known is Flooding Time Synchronization Protocol, which exploits timestamping at the Medium Access Control (MAC) layer, which is then forwarded to the rest of the network, with clock inaccuracies being corrected on a per-node basis along the way [60].

A similar recent approach, found in Skiadopoulos et al. (2019), validates the effectiveness of the MAC layer timestamping solutions [61]. However, one major challenge, that arises, is that most applications do not have access to the lower levels of the networking protocol, deeming these approaches hard to exploit. One interesting approach, that addresses this challenge, achieves high accuracy by utilizing both external hardware (in this case Real Time Clocks) and signals sent by the radio module to time transmissions [62]. There are also other approaches, which require multiple rounds of communication in order to increase precision [63]. Yet, these are particularly costly in terms of energy consumption, leading to a reduction in network lifetime.

3.2 A Simple Synchronization Scheme

A synchronization scheme for smart agriculture applications should be designed with minimum energy cost as a priority. This is especially true, if one considers the pinpoint accurate synchronization that is achievable through more complex approaches and the fact that such a high degree of accuracy is redundant in the vast majority of applications. As such, a very simple scheme is considered in this chapter, requiring only a single broadcast transmission to achieve adequate synchronization accuracy across the network in order to obtain time-correlated measurements. Its simplicity it the key enabler which makes this scheme interesting to study, as it serves as a baseline for just about any other synchronization approach.

In the considered network, consisting of nodes as presented in Sect. 2.3, each node maintains its own local clock, which represents the node's uptime. The sink node, corresponding to the Coordinator device type, is responsible for collecting the entire network's measurements, which are then transmitted to an external device via serial connection, at which point they are correlated to physical time. Thus, it is sufficient to provide the Coordinator's clock to all remote nodes as a reference point, in order for the latter to correlate their clocks according to the former's time. All nodes, aside from the one executing an action, are referred to as remote.

Let t be a point in physical time, when the Coordinator's and remote node*i's* clock is t_c and t_i, respectively. The simplest way to propagate the Coordinator's clock t_c to the network is through a broadcast transmission, which is a common operation in the Zigbee protocol. Once a remote node i has acquired the Coordinator's clock t_c, it is possible to calculate the difference Δt_i between the acquired and its own clock t_i, i.e.,

$$\Delta t_i = t_c - t_i$$

Do note that Δt_i may be negative, if the remote node was powered on before the Coordinator, which does not impact subsequent calculations, but should be taken into account during implementation by using signed integers. The propagation of the Coordinator's clock is depicted in Fig. 2a.

Under normal conditions, the difference Δt_i is assumed to remain constant for each node i in the passage of time and can be used to estimate the Coordinator's clock. Consider a later point in time $t' = t + r$, where $r > 0$ is the elapsed time since t. Remote node i is capable of estimating the Coordinator's clock t'_c by adding the previously computed difference Δt_i to its current clock t'_i, that is

$$t'_c = t'_i + \Delta t_i$$

To illustrate this, consider the case where a remote node i starts operating five seconds (5 s) before the Coordinator. Clearly at any point in time, the remote node's clock will be ahead of the Coordinator's by five seconds. Therefore, the node simply

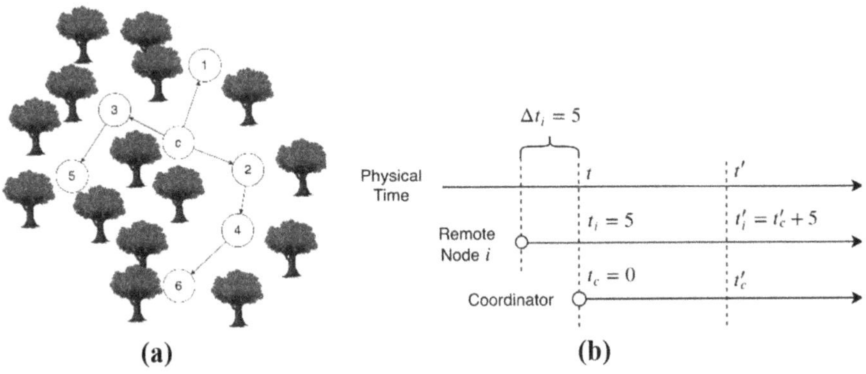

Fig. 2 A showcase of the considered synchronization scheme: (**a**) The network's Coordinator *c* broadcasts its clock to all other nodes. (**b**) Example of remote node *i* powering on five seconds (5 s) before the Coordinator

needs to have a way of finding out the initial difference, which is accomplished through the described scheme. This example is visually depicted in Fig. 2b.

However, there is a major issue in one of the assumptions made, that complicates this process. The node difference Δt_i for each node i does not remain constant as time goes on. Indeed, jitter, which is inherent in time-keeping devices due to manufacturing parameters, that influence clock frequency, results in actual time difference gradually deviating from Δt_i. In order to tackle this issue, it is possible to repeat the synchronization process, so that Δt_i can be recomputed, taking this deviation into account. The period between synchronizations T_{sync} is determined by the utilized hardware's accuracy and the application's requirements. The issue of gradual time difference deviation, due to clock jitter and its resolution through recurring synchronization events, is graphically illustrated in Fig. 3a.

To further clarify how T_{sync} may vary across different situations, some specific applications are considered next. The most important factors to acknowledge are the parameters being monitored, their variability across time and the cost of inaccurately measuring them. A secondary concern, which is hardware specific, is the clock's accuracy; the more accurate the clock, the less frequent the need for synchronization. For instance, consider the application of olive grove irrigation and health monitoring.

The parameters that are most commonly monitored (temperature, relative humidity, etc.) are relatively predictable across short time spans, while the risk of identifying a disease or the cost of initiating irrigation a few milliseconds later is quite low. Trends across longer time spans are more relevant, making synchronization requirements low, and so T_{sync} can be unusually high in these cases. On the other hand, in wildfire monitoring, similar parameters are monitored, but their effects on how the fire spreads can be dramatic, while the cost of not measuring these in a timely manner can incur resounding costs in crops and potentially human lives; thus, T_{sync} would have to be exceedingly low in such circumstances.

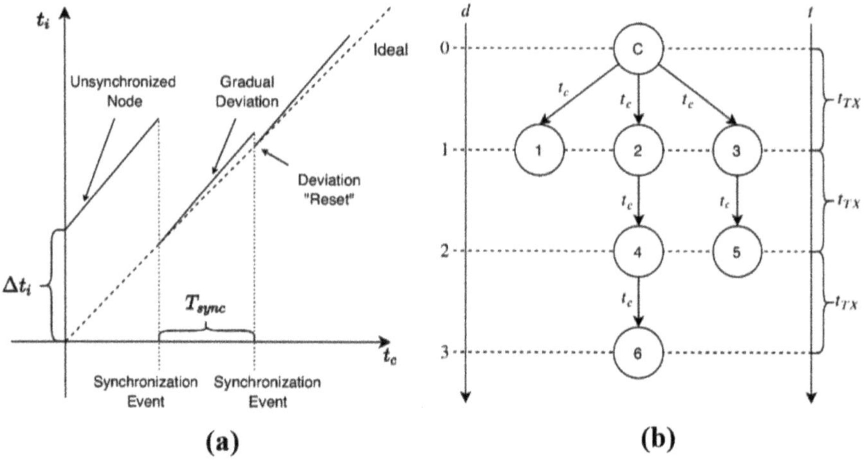

Fig. 3 Issues associated with the simple synchronization scheme: (**a**) Gradual deviation of time difference between two nodes due to clock jitter. (**b**) Nodes further away from the Coordinator are guaranteed to have a larger synchronization error than those that are closer in large networks

Another problem that requires attention, as it results in a certain degree of inaccuracy and has not yet been addressed, relates to the fact that providing remote nodes with the Coordinator's clock, through a transmission, requires time, which is not accounted for in the considered scheme. In particular, the Coordinator's clock t_c is not being updated, while it is being transmitted and forwarded to the rest of the network. Transmission delays are both hardware and software specific. For instance, MAC protocols often employ random delays to avoid collisions, the various hardware components need time to pass around data at certain transfer rates, etc.; all contributing to the accumulation of transmission delays. In conjunction with the fact that these delays are often not deterministic, the task of accurately identifying them is intricate.

Synchronization inaccuracy caused by transmission delays is also proportional to the distance between a remote node and the Coordinator. For simplicity, suppose that the overall time required for the successful transmission and reception of a message between two adjacent nodes (always) lasts one second (1 s), which is certainly an exaggeration. The Coordinator initiates the synchronization process at a certain point in time by broadcasting a message containing its clock t_c. Each remote node $t'=t+r$ within the Coordinator's transmission range, i.e., each 1-hop neighbor, receives the clock and calculates the difference Δt_i, which is already off by one second (1 s). These same nodes forward the message identically, so it still contains t_c, to their adjacent nodes; the Coordinator's 2-hop neighbors. Their difference Δt_i will now be off by two seconds (2 s), when compared to the Coordinator's actual clock. Thus, intuitively, the synchronization inaccuracy of nodes d hops away from the Coordinator will be $d \times t_{TX}$, where t_{TX} denotes the transmission delay between two adjacent nodes. This behavior is showcased in Fig. 3b.

Fundamentally, most communication-based clock synchronization approaches attempt to deal with the presented issues, which can be summarized as: (i) accurately computing the initial clock offset, or Δt_i, while minimizing propagation error; and (ii) accounting for the gradual synchronization error caused by different clock frequencies. The considered synchronization scheme attempts to deal with the latter in a rather simplistic manner but does not incorporate any mechanisms to address the former. However, the point of the particular synchronization scheme is to evaluate its effectiveness in practical applications, as it is the most basic of available approaches and it can serve as a baseline for more advanced ones, while maintaining a low energy cost requirement.

3.3 Experimental Evaluation in Olive Groves

In the sequel, an experimental evaluation of the considered synchronization scheme is described, which was carried out in [76]. This evaluation aims to showcase the scheme's effectiveness in providing measurements that are accurately correlated to physical time. To evaluate the synchronization scheme's effectiveness, when operating under realistic conditions, a WSN consisting of 30 nodes was deployed on an olive grove situated in Chalidiata, Corfu, Greece. In particular, the network comprised 29 Routers hosted on Arduino Uno boards and a single Coordinator hosted on an Arduino Mega. An installation site inside the olive grove is showcased in Fig. 4. This is almost an ideal use case for the considered synchronization scheme, due to the lenient requirements for synchronization accuracy.

Regarding the deployment, nodes were set up in a star topology, with the Coordinator being a single hop away from all nodes. This minimized the accumulation of synchronization error per hop, which is important in real applications, but it doesn't enable further studying of the scheme's performance. Nonetheless, the results were collected from an experiment that lasted roughly 17 min. The synchronization process was repeated every T_{sync}=100s, while remote nodes forwarded an indication of their clocks at intervals of 10 s, starting from the time they were powered on. When the Coordinator received a message from a remote node, which included the node's estimated timestamp and a handful of environmental measurements, the message was immediately stored on a micro-SD card along with the Coordinator's actual timestamp.

The difference between the Coordinator's and the remote nodes' clocks Δt is depicted in Fig. 5. Incoming indications are grouped into windows of 10 s, so that the time difference Δt, for all 29 remote nodes' indications, is included in each window. The mean Δt is relatively stable, fluctuating between approximately 170 ms and 250 ms, even during the last stages of the experiment. Maximum values of Δt are also upper bounded at approximately 500 ms. This indicates the overall stability and effectiveness of the synchronization scheme: a simple broadcast transmission is capable of accomplishing timestamped measurements, achieving high precision compared to a lot of external hardware, such as Real Time Clocks.

Fig. 4 An installation site on the olive grove, featuring multiple Arduino Uno boards within their protective cases scattered around an olive tree

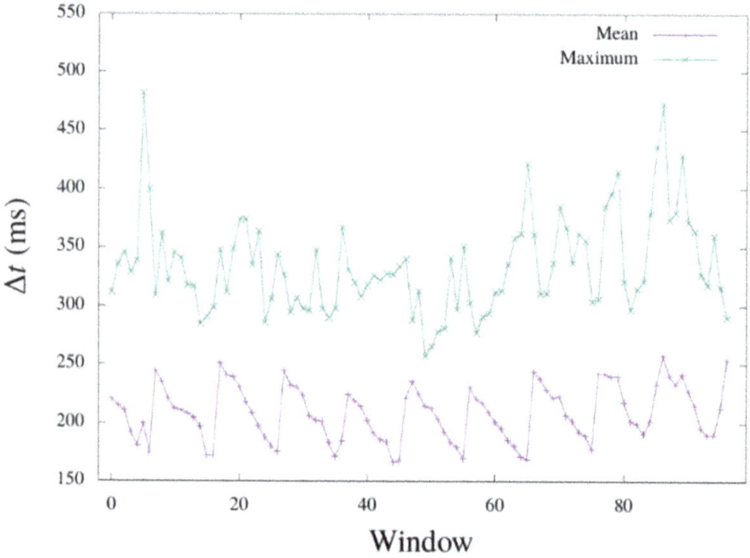

Fig. 5 The mean and maximum values of time difference Δt between remote nodes and the Coordinator. Each data point represents the mean and maximum values of a window, containing a single Δt measurement from each of the 29 remote nodes

Interestingly, the slope of Δt follows a downward trend, which seems to spike every 10 measurements, then gradually decrease again. These resets are caused by the synchronization event being repeated every 100 s. The time difference Δt at the reset points is partly due to the transmission delay t_{TX}, which is not accounted for. The downwards trend, after the synchronization event takes place, indicates that remote nodes' clocks tick slower than that of the Coordinator's. The consistency of this trend is peculiar, and it is unclear why it occurs. Nevertheless, two factors to consider here regarding this behavior include: (i) the fact that the Coordinator is hosted on an Arduino Mega, featuring different processing characteristics than the Arduino Uno boards; and (ii) that it is also burdened with an increased workload, due to receiving more messages, having to store received measurements on an SD card etc.

As indicated by Fig. 6, where the distribution of the time difference Δt between all remote nodes' and the Coordinator's clock is depicted, Δt values are mostly clustered between 150 ms and 300 ms. A few further descriptive statistics are provided to further illustrate the stability of the synchronization scheme. The Δt values have a mean of $\overline{\Delta T} = 208.295\text{ms}$, a standard deviation of 76.906 ms, a median of 230 ms, as well as a 25th and 75th percentile of 167 ms and 251 ms, respectively. Overall, it should be evident that despite its simplicity, the considered synchronization scheme is capable of providing adequately accurate timestamps fairly consistently.

Besides, it should be noted that the study, in which these experiments were conducted, utilized a serial baud rate of 9600 bps [15]. This parameter affects the rate at which symbols are transmitted over a serial connection, i.e., the rate of

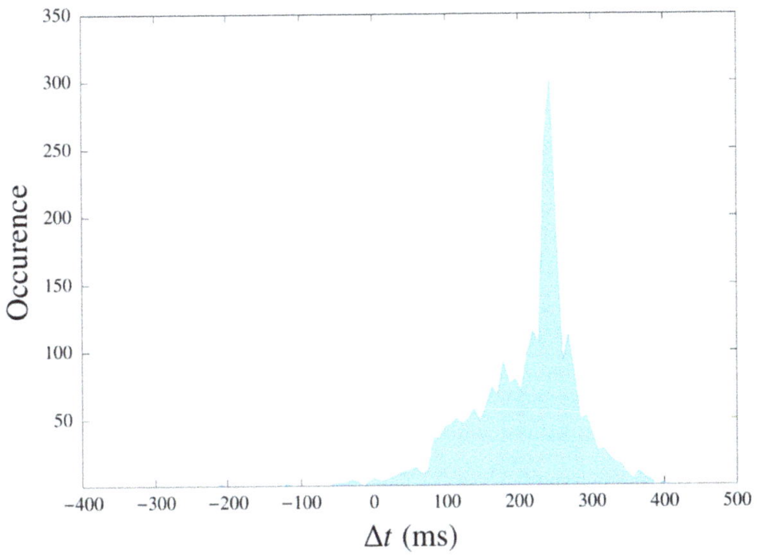

Fig. 6 Distribution of Δt values received from all remote nodes

communication between the Arduino microcontroller and the XBee module, leading to the creation of major bottlenecks in the transmission and reception of messages. With that said, the baud rate can go as high as 115,200 bps, which was estimated to reduce transmission delay by more than 90%. Thus, it should be apparent that the timestamping precision of even the simplest synchronization schemes is capable of surpassing the performance of most other hardware-based approaches, enabling even extremely time-critical applications, such as wildfire monitoring, hydroponic cultivation and olive grove irrigation scheduling [64–66].

4 Advanced Monitoring Architecture

The synchronization of WSN-based systems for agricultural applications is an important step towards the collection of reliable data. Equally important is also the system's monitoring architecture which can potentially enable the advanced operations capable of processing and cross-referencing information collected from multiple WSNs and finally delivering the result to the farmer's personal device. This section goes through related work on advanced monitoring architectures. Additionally, it details a cloud/fog architecture, which aims at minimizing the system's response time. The efficacy of the presented architecture is evaluated based on olive grove deployment-specific requirements. Finally, directions for future work are drawn in the final part, exploring the special case of wildfires, which by nature demand stringent monitoring and robust data handling.

4.1 Related Work

In most cases, the volume of data produced in the field is enormous, making storage and preprocessing a difficult task for the end user. Advances in cloudification offer a promising solution in this regard, as cloud systems provide storage, networking, and processing power, while being more flexible than a personal computer. Further, the requirements for low latency and high mobility have pushed the functionality of the cloud to the edges of the network, augmenting its overall performance and leading to reduced latency [67].

To this end, the wide proliferation and adoption of IoT and smart devices have paved the way for greater elasticity, by offloading cloud computational demanding processes and resources in closer proximity to the end user, alleviating the overall system's capacity utilization, while increasing accessibility. Ergo, the cloud/fog computing paradigm has come into light, increasing data-handling efficiency, and minimizing response time [17, 18, 68]. In the near future, due to the expected increase in the number of sensors on the field and the generated data, the need for instant decisions with low power consumption will be even more prevalent, thus, making cloud/fog approaches an excellent candidate for realizing such applications in various domains, precision agriculture included [69].

With that in mind, Guardo et al. (2018) have proposed a framework for agriculture, which showcases that the cloud/fog architecture can be used effectively in time-sensitive agricultural devices [17]. Evaluation is carried out with respect to the performance gain this framework brings about. Additionally, some information processing occurs at the fog device as a form of load balancing, which further reduces response times, when compared to cloud-only systems.

The number of WSNs has also substantially increased with the rise of IoT, as they constitute a fundamental structural component for these kinds of implementations. Several platforms have been developed around the cloud architecture and the use of WSNs for monitoring the field [70]. One example is a multidisciplinary system, named "ArgoCloud", that utilizes cloud and mobile computing for transferring, storing and preprocessing data acquired from sensors, with the goal of assisting in crop fertilization [71].

Despite the aforementioned benefits, it is evident that most farmlands do not have unlimited network information resources for monitoring. Given this restriction, another study proposes a creative service process, based on the cloud computing platform of the IoT, which is able to improve the current cloud-to-physical networking, along with the computing speed of the IoT [72]. Experimental results highlight that, under the considered IoT service platform, the cost of network transmission can be effectively reduced, allowing large areas with limited network resources to use agricultural monitoring automation.

The WSNs are useful beyond their role in agriculture and crop monitoring. More specifically, WSNs can be employed for other purposes, like decision making and predictions related to wildfires in forests and rural regions. One study in this field has shown, through a number of conducted simulations, that WSNs can better detect and predict a fire, in comparison to conventional methods, such as satellite imagery [73].

At this point, especially for systems responsible for extreme phenomena, like wildfire monitoring, an important WSN-related challenge, which has been studied by a plethora of researchers (e.g., [74]), needs to be acknowledged. This refers to the minimization of energy consumption due to limited battery life. However, a regular grid with homogeneous deployment can minimize the power consumption, allowing the detection of forest fire outbreaks early on, while ensuring the system's sustainability [75].

4.2 Cloud/Fog Architecture

In Sect. 3.2 a simple synchronization scheme, utilized by a WSN for smart agriculture applications, is analyzed. Yet, when it comes to large-scale deployments over vast areas, it is quite common to separate the nodes into many WSNs instead of deploying just one. This is practical because it is easier to monitor smaller sections of an enormous area, i.e., in case of a fire hazard it is clearer to spot the point of origin. On other occasions it is simply the result of having a significant distance between two groups of sensors that cannot communicate with each other, creating two independent WSNs. Therefore, the crucial issue of how the data of all these WSNs can be collected, stored, and processed effectively, as a whole, arises.

A simple cloud/fog architecture is adopted in many similar installations, like those presented in the previous subsection. This architecture has been broadly utilized, even in other fields (vehicular applications, health, and fitness applications, etc.), for its simplicity and ability to adapt to the application's specific requirements. It is for these reasons that the aforementioned architecture is proper for environmental overseeing, such as olive grove monitoring. Noteworthy is the fact that this architecture has already been used in many agricultural applications, where its utilization targets time-sensitive needs of olive groves [19]. The considered cloud/fog architecture for agricultural monitoring consists of three layers, as graphically presented in Fig. 7.

At the top layer (Cloud Layer), a central infrastructure is formed by utilizing cloud computing services of the deployed servers, that in turn form a complete data center infrastructure. These servers can be used both as storage and computational units and are responsible for the overall system administration. Since the integration and maintenance of the servers is quite expensive and their location (in network distance) is generally not close to the end users, the fog computing paradigm is

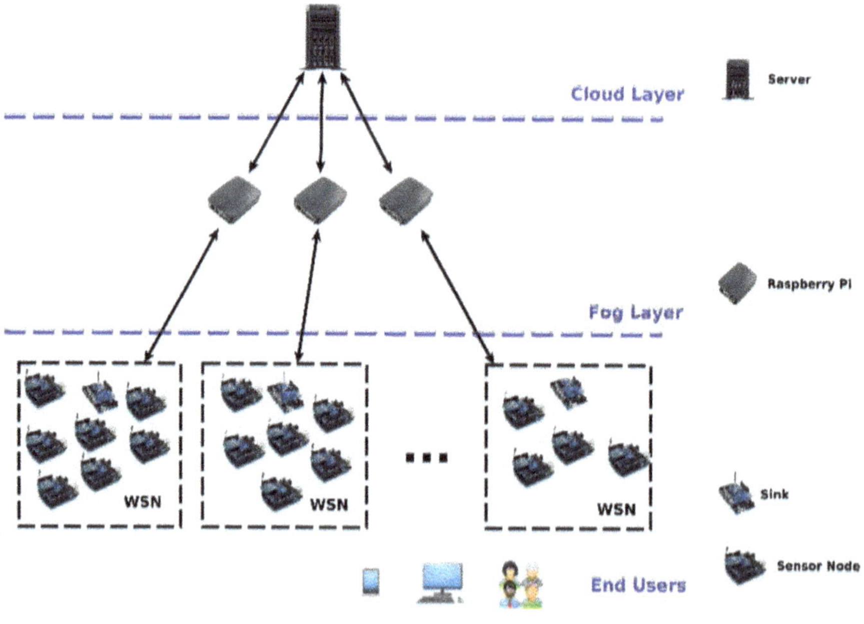

Fig. 7 The three-layered cloud/fog architecture: the server is on the first layer, the WNSs are located in the bottom layer, which consist of sink nodes and sensor nodes, and the fog devices are found in the middle one

adopted, creating an intermediate layer (Fog Layer) between the end users and the Cloud Layer.

Fog devices provide similar services with the cloud servers, but they usually offer less computational capabilities, which makes them cheaper and more flexible than their cloud counterparts. For the case at hand, the Fog Layer comprises small single-board computers, such as Raspberry Pis, and it is responsible for processing the data collected by the WSNs and/or relaying it to the Cloud Layer through wide area network connectivity.

The bottom layer (End Users), includes the deployed WSNs, along with the various people that manage them and exploit them, such as researchers, farmers, environmental agencies, etc., through various devices, like smart phones or computers. The WSNs, in turn, are synthesized by a plethora of sensor nodes. Section 2.3 presents some components commonly used for actuating the nodes, rendering them energy-efficient and low-cost. Furthermore, their installation does not harm or disturb the flora and the fauna of the ecosystem, making them an ideal hardware solution for hosting agricultural applications.

Since most of these systems are created for real-world applications, it is vital to evaluate their effectiveness, in regard to smart agricultural applications and reliable monitoring, by putting them to the test, under real use-case scenarios. One of the important aspects of an agricultural application architecture is the time needed by the system to respond in case of an emergency. In other words, the time needed for a data packet, encapsulating critical information regarding a serious ongoing event (e.g., wildfire or animal intrusion), to reach the fog device or the cloud, plus the time it takes for the data packet to be successfully processed and the generated system decision to travel back to the bottom layer. Clearly, this time must be kept as low as possible, in order for the end users or the corresponding WSN to proceed into swiftly dealing with the issue and launching appropriate countermeasures, whether these being a speaker making a loud noise to scare the animals away or a siren to alert the authorities about the fire outbreak. These are just two potential use case scenarios. Obviously, there exist many more. Therefore, it is vital to assess the system's performance, in dealing with similar affairs, based on a pertinent evaluation metric.

4.3 Evaluation

The evaluation process usually takes place by conducting experimental runs prior to large-scale deployment, to early detect faulty modules and optimize the functionality. In particular, the tests can be carried out either in a controlled environment using real equipment, such as a lab, where different environmental conditions can be simulated, or outdoors, where the WSNs can be implemented in real-world conditions to determine whether the selected architecture is indeed practical.

Since it is important to assess the time needed for the system to respond to an important issue, a performance evaluation metric based on the response time is exceptionally useful. One such metric is the Round Trip Time (RTT), which indicates the time needed for a data packet, generated at a WSN node, to reach the overseeing fog device or cloud data center, be processed accordingly, and travel back to the same node. The three-layered architecture was evaluated using the aforementioned metric, in order to determine the quality of services it can potentially provide and how it fairs when considering time-sensitive scenarios in olive grove applications [19].

The conducted experiments comprised a few WSNs, with each WSN's sink node being linked through a serial cable with a Raspberry Pi, acting as a fog device. The experiments took place in different buildings of the Ionian University's facilities, within closed, controlled laboratory conditions, emulating the considered three layered cloud/fog architecture. The response time varied based on different values of P, where P is the probability for a data packet to be processed either in a fog device or in the cloud server. In the former case, a system decision was generated in the Fog Layer, and was directly transmitted back towards the node, that initially generated the data, through the local area network. In contrast, during the latter case, where the data processing took place in the Cloud Layer, the generated system decision was first transmitted to the fog devices, which in turn relayed it to the appropriate WSN and the initial node.

A complete breakdown of the above behavior is presented in Fig. 8. Initially, a node of a WSN forwards a generated data packet to its sink node, which stores the data and forwards the data packet to the fog device, that manages the corresponding WSN. The fog device, upon reception, based on a random probability P, either processes the data packet locally (i.e., when P>p) or forwards it to the cloud (i.e., when P<p) respectively. The same route, inverted, is followed by the system to send the

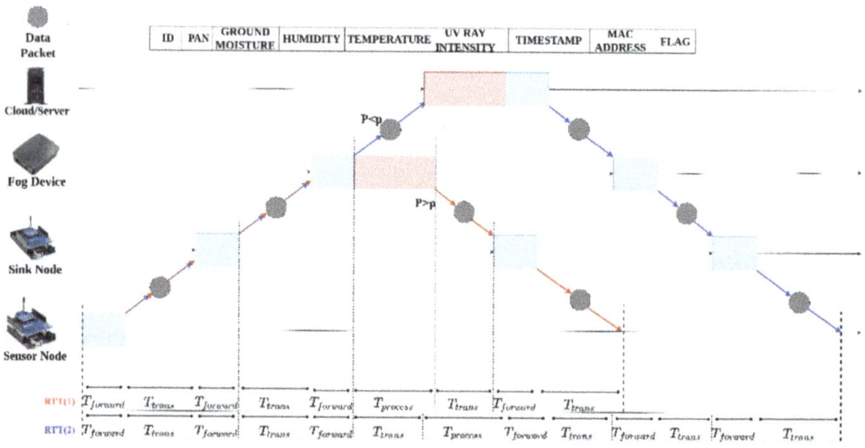

Fig. 8 The total time needed for an RTT of a packet transmission. T_{fo} represents the time required for a device to forward the data packet, T_{trans} represents the duration necessary for a data packet transmission from one device to another and lastly $T_{process}$ shows the necessary time for a data packet to be processed

processed data back to the WSN node, where the data originated, so it can be further managed by the end users' devices.

Understandably, RTT is a suitable performance evaluation metric for applications, that require fast responses and actions on special occasions. Agricultural applications are no exception, especially when it comes to prominent matters, such as the protection of crops against various threats. The following subsection focuses on the case study of olive groves to better contextualize the adjustable character, in terms of RTT, of the three-layered cloud/fog architecture, in order to deal with hazardous events.

To evaluate the system's effectiveness, a case study that includes multiple experimentation procedures, is revisited and examined in regard to the system's properties described earlier in the current section [19]. The system in this case has been adjusted to meet the requirements of an olive grove deployment. The main goal of the conducted experiments is to evaluate the system in terms of response time and effective load balancing. It is worth mentioning that, while highly accurate synchronization is not a prerequisite for a system designed for olive grove deployment, sufficient response time for hazardous events' detection (i.e., wildfires) is required, in order to prevent destruction as soon as possible.

The first experiment aims at evaluating the fog/cloud response time according to the proposed functionality. More specifically, the system includes six (6) different WSNs, which contain a total of 28 nodes. Three different fog devices handle the data packets (i.e., the measurements) created from all WSNs. When a fog device receives a data packet, it decides whether to proceed in the necessary calculations or relay the packet to the server. This decision is taken based on a probability P (probability for the packet to remain inside the fog device for processing), where $P \in [0, 1]$, incremented by 0.1. In the case where P is smaller than a randomly selected value from the fog device, the packet is calculated remotely in the server, otherwise it is processed inside the fog device.

With that said, the lowest RTT values (lower values are translated to faster response time) are observed for $P = 1$ (i.e., when all packets are processed by the fog device), while the highest RTT values are calculated for $P = 0$ (i.e., when all packets are relayed to the server). Figure 9 depicts the results for both 5 s and 25 s intervals between node measurement sensing. It is apparent, that the RTT's curve behaves similarly for both occasions, thus, the system is able to handle an increased number of packets without compromising the system's credibility, which means that the crop monitoring procedure can be carried out with low latency, even under stressful situations, requiring high frequency of environmental readings.

The next experiment concentrates on evaluating the system's load balancing mechanics. The latter aids in the proper adaptation of the system during peak times, when an increased number of packets, that require processing, are produced. The main aim is to determine if the fog devices are adequately equipped to handle such circumstances, by conforming the system's processing behavior to avoid congestion from building up. In order to achieve this, time is split into distinct time steps (t) and two packet boundaries are employed inside the fog devices. In the case where the fog devices acquire less than or equal to 10 packets per t (lower bound), they are

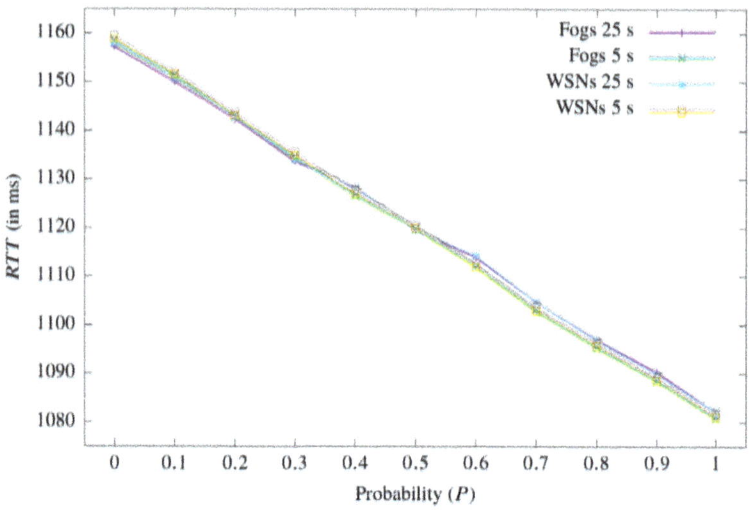

Fig. 9 The mean Round Trip Times for each WSN and each Fog Device are presented in a single diagram for both 5 s and 25 s experiments

considered underloaded, consequently, retaining the received packets for local processing, as opposed to having 15 or more packets (upper bound), in which case they considered overloaded, and they begin sending packets to the server for processing. In this way, the system is able to balance the packet load throughout between the Fog and the Cloud Layers (i.e., cloud server and fog devices).

Specifically, to achieve the two actions dynamically (i.e., server packet forwarding and local processing), each time the upper bound is violated, the probability P is decayed by 0.05, otherwise if the data packet generation rate infringes upon the lower bound, the server forwarding is cut down in order to free up server resources. Figure 10 depicts the packet load in all three fog devices and the server. It is observed that the configured window [10, 15] indeed manages to capture most of the data packets within the given thresholds, indicating that the employed fog mechanism is sufficiently driving the load balancing procedure. In fact, the results clearly visualize the system's ability to control the generated workload. Note that the spikes that appear on *Fog 3* are purely attributed to the fact that less nodes are connected to it (i.e., less packets are created for each time step), thus allowing it to distribute the load in a more controlled fashion compared to the rest.

4.4 Potential Future Applications: The Case of Wildfires

As already made clear, the issue of wildfires is a recurring threat to agriculture. To provide some context, Greece is heavily afflicted by wildfires during the summer seasons. In 2007, the country was hit by three consecutive tempestuous heat waves

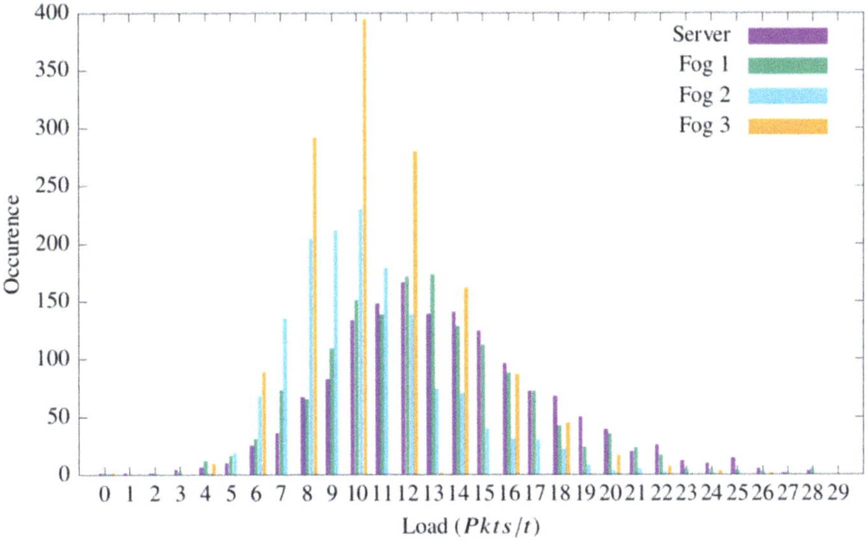

Fig. 10 The packet loads measured during the experiment on all fog devices and the Server while the former distributed the packet load to the latter according to the current requirements

with temperatures raging over 46 °C, which in conjunction with low relative humidity, resulted in wildfires spreading allover Greece. According to the European Space Agency[17], Greece's wildfire activity during the summer of 2007 was more intense than other European countries over the last decade. In fact, the number of wildfires rose up to 11,996, resulting in 675,000 acres of burned land. According to Greece's Fire Brigade (GFB)[18] recent official statistics, 8006 forest fires have been recorded in 2018 alone. To put this in more perspective, 174 wildfires were recorded just on Corfu Island, an area of barely 236 square miles.

As such, a large number of studies focuses on the issue of preventing, detecting, monitoring and managing wildfires using modern technological tools. For instance, methods have been developed for quantifying a region's vulnerability to fire, informing residents and farmers of fire risk [76, 77]. Most WSN-based applications specifically target sufficient monitoring and early detection, hoping to minimize damage [78–80]. WSNs have also been studied as a means of preventing and mitigating forest fires through irrigation scheduling, which requires an efficient WSN architecture to be effective [81].

Accordingly, an application combining most of these facets, while also fulfilling the duties of a typical agricultural application, is anything but unfathomable. Most applications, focusing on wildfires and forest fires, utilize, among others, relative humidity, soil moisture and temperature sensors for their purposes; sensors already ubiquitous in agricultural applications. By combining an efficient cloud/fog

[17] https://tinyurl.com/y86zoxy7

[18] https://tinyurl.com/yaberpnf

architecture with ideas and techniques presented throughout this entire chapter, including synchronization, an extremely capable and robust system can be developed, that is highly equipped to monitor the field, to yield improved product quality, as well as to minimize wildfire risk with appropriate detection or countermeasure methods. In the sequel, one such potential system is described, analyzing the different mechanisms that are integrated in each of the architecture's layers and their effects on the system's performance.

Starting from the top layer, maps outlining the sensing region's fire vulnerability are required. It is not uncommon for local authorities of particularly vulnerable regions to provide these maps. For instance, Greece's General Secretariat for Civil Protection (GSCP)[19] publishes a map depicting the risk of a fire outbreak for all regions of Greece on a daily basis during the firefighting season. A comprehensive fire risk metric can be estimated (e.g., through machine learning models) for each employed WSN, combining regional geographical and fire risk maps with other relevant metrics, such as time of day and previous measurements acquired by the WSNs themselves [77]. This metric, being frequently updated, can be included in the reply to each acquired data packet; i.e., the second half of the RTT described in the previous subsections.

Continuing with the bottom layer, it is extremely important for all acquired environmental metrics to be accurately timestamped, in order to improve the results of the fire risk estimation process. Synchronization can be used to provide timestamping, while also enabling sleep mode scheduling to minimize energy consumption, without sacrificing the system's sensing and routing capabilities, as argued in [81]. Furthermore, the fire risk metric received from the top layer can be used to adjust the measurement reading frequency of each WSN's nodes; an idea previously explored in [82]. When the fire risk is high, measurements should be more frequent to account for changes in the environment that contribute to fire spread, such as wind speed and direction. Actions aiming for fire prevention and mitigation, such as irrigation, can be performed as well, if the risk is extreme. On the other hand, when the fire risk is low, the measurement sensing period can be spread out to conserve energy levels.

Finally, the middle layer can be used to balance the load between the top and bottom layers. When the fire risk is great, a higher load is expected to originate from the WSNs, leading to lower responsiveness, at the time when it is most critical. To combat this, the probability P of each fog device forwarding acquired data packets to the cloud can be dynamically shifted in an inversely proportional manner, based on the estimated fire risk, as well. When fire risk is high, P should be lower to counteract the increased load, while still providing fire risk metrics through the fog devices, although perhaps less accurate, as only local WSN information is available. Conversely, if the risk is low and the expected load is low, P may be increased to improve future estimation accuracy. Adjusting this probability P has already proven

[19] https://www.civilprotection.gr/en/daily-fire-prediction-map

to be an effective mechanism at reducing RTT when the load is high, offering agility and flexibility.

The described system incorporates a state-of-the-art architecture and multiple techniques to form an overall well-rounded and fully fleshed-out application. It is capable of monitoring environmental conditions that affect crops and those that contribute to wildfire outbreak. As such, it can also be used for fire detection and prevention, while dynamically adjusting its behavior to improve performance with respect to multiple metrics, including energy efficiency, timestamping accuracy and RTT. All this can be accomplished using low-cost components, as previously outlined, highlighting the efficacy and potential of WSN applications in the domain of smart agriculture.

5 Conclusions and Future Directions

Following the advancements of WSNs in the field of smart agriculture, this chapter has focused on covering vital issues regarding the accountability and accuracy of systems that monitor agricultural environmental parameters.

Primarily, various wireless sensor applications and fundamental WSN technologies used in the field of smart agriculture were discussed in the respective literature review section. The low-cost equipment, utilized in frontier research, was then explicitly covered and the selected low-cost equipment used in the works that the current study focuses on, was thoroughly discussed.

Several schemes for synchronized monitoring were reviewed, which achieved synchronization either with the help of hardware or software components. Further, based on the literature, a particular simple synchronization scheme was selected and explored. This scheme achieves the correlation of its system measurements with actual physical time, by using the sink node's clock as a reference point to align sensors measurements.

The study, then, referenced the experimental evaluation of the described synchronization scheme, in which the considered WSN was deployed in an olive grove on the Island of Corfu, proving the simple synchronization scheme to be effective and accurate for monitoring. In the course of time, more sophisticated synchronization techniques are possible to be considered for these kinds of applications, in order to further improve the synchronization inaccuracies. One can also expect that future technological solutions will resolve to more capable low-cost equipment. In that vein, the applications will be able to incorporate more advanced antennas, which will further reduce energy consumption.

Additionally, to further explore the capacities of modern smart agriculture, the study also focused on covering various environmental monitoring architectures, which can collect sensor information from different locations, perform more advance operations and deliver the results to the appropriate parties, involved in the farming process. One such architecture employs the cloud/fog computing paradigm, which forms flexible WSNs that communicate with a main cloud infrastructure through a fog computing network (consisting of microcontrollers), allowing for real-time monitoring of environmental factors in time sensitive agricultural applications. It typically keeps the load at better capacity and has higher throughput. The evaluation metric, that is mostly used for determining its response time is RTT, as it measures the overall time needed for a packet to travel across the system.

To demonstrate the performance of such a system, a robust prototype demo of a three-layered cloud/fog architecture was also discussed, which was deployed in the facilities of the Ionian University to evaluate the system's performance [19]. This architecture features a homogeneous ability to dynamically control the network's latency and the load balance, which allows for the system's robust operation. Indeed, the results proved the system to be highly promising for agricultural applications. The average throughput showed to be consistently high and the average response time faster compared to cloud-only systems.

Following the olive grove solution, this cloud/fog architecture was further studied in the light of natural hazard prevention, another time sensitive application, which could be considered due to the system's fast response time. Wildfires, is a natural disaster that frequently afflicts large areas of Greece, causing indescribable damage. By incorporating the cloud/fog architecture and additional techniques, the study discusses the potential of an application for detecting wildfires. Such an application would be possible, due to the operational adaptability of the system, by dynamically changing the number of generated packages to reduce the network's load, in order to reserve energy, when the risk of a fire is estimated to be low, and respectively adjust it under the circumstances of danger.

It is a question of future research to investigate the efficiency and throughput of alternative sensor modules for various environments such as, for example, cameras that can capture audiovisual footage or even soil pH trackers. The intermediate goal would be the autonomous adjustability of the cloud/fog architectures to counterbalance implications and hazards imposed by system failures throughout the system's entities, e.g., the fog devices. Finally, a particularly useful feature would be web interfaces, to provide farmers with the ability to remotely manage the system, in a user-friendly way, via various platforms.

In closing, the direction of cloud/fog architectures holds significant value, thus, making this kind of implementations, highly promising for the future of smart agriculture. It is clear, that due to the lack of research in the field, actions towards the establishment of unified guidelines are necessary, to allow for a better understanding of the needs of the involved stakeholders. Such guidelines may include behavior optimization for the dynamic modification of the system's characteristics to deal with various other agricultural aspects, but also to support emerging technologies, such as future 5G standards, and bridge the gap among different agricultural scientific disciplines and environmental domains. In fact, the inherently generic nature of the cloud/fog system makes its future adaptation to numerous technological changes easily attainable.

Acknowledgments This work was supported in part by project "A Pilot Wireless Sensor Networks System for Synchronized Monitoring of Climate and Soil Parameters in Olive Groves,"

(MIS 5007309) which is partially funded by European and National Greek Funds (ESPA) under the Regional Operational Programme "Ionian Islands 2014-2020."

References

1. Klytchnikova II, Sadler MP, Townsend R, Edmeades S, Choudhary V, Hussain S, Kray HA, Fernandes EC, Moses E, Cantrell JT, et al. (2015) Future of food: shaping a climate-smart global food system. World Bank Group, Washington, DC
2. Popović T, Latinović N, Pešić A, Žarko Zečević, Krstajić B, Djukanović S (2017) Architecting an iot-enabled platform for precision agriculture and ecological monitoring: A case study. Computers and Electronics in Agriculture 140:255 – 265, https://doi.org/10.1016/j.compag.2017.06.008, URL http://www.sciencedirect.com/science/article/pii/S0168169916307566
3. Westermann O, Förch W, Thornton P, Körner J, Cramer L, Campbell B (2018) Scaling up agricultural interventions: Case studies of climate-smart agriculture. Agricultural Systems 165:283– 293, https://doi.org/10.1016/j.agsy.2018.07.007, URL http://www.sciencedirect.com/science/article/pii/S0308521X1830043X
4. Sørensen C, Fountas S, Nash E, Pesonen L, Bochtis D, Pedersen S, Basso B, Blackmore S (2010) Conceptual model of a future farm management information system. Computers and Electronics in Agriculture 72(1):37 – 47, https://doi.org/10.1016/j.compag.2010.02.003, URL http://www.sciencedirect.com/science/article/pii/S0168169910000396
5. Irvin R (2007) Information and communication technology (ict) literacy: Integration and assessment in higher education. Journal of Systemics, Cybernetics and informatics 5(4):50–55
6. Atzori L, Iera A, Morabito G (2010) The internet of things: A survey. Computer Networks 54(15):2787 – 2805, https://doi.org/10.1016/j.comnet.2010.05.010, URL http://www.sciencedirect.com/science/article/pii/S1389128610001568
7. Benlloch-González M, Sánchez-Lucas R, Benlloch M, Ricardo FE (2018) An approach to global warming effects on flowering and fruit set of olive trees growing under field conditions. Scientia Horticulturae 240:405 – 410, https://doi.org/10.1016/j.scienta.2018.06.054, URL http://www.sciencedirect.com/science/article/pii/S0304423818304461
8. Rodrigues N, Casal S, Peres AM, Baptista P, Bento A, Martín H, AsensioS-Manzanera MC, Pereira JA (2018) Effect of olive trees density on the quality and composition of olive oil from cv. arbequina. Scientia Horticulturae 238:222 – 233, https://doi.org/10.1016/j.scienta.2018.04.059, URL http://www.sciencedirect.com/science/article/pii/S030442381830308X
9. Pérez-Ruiz M, Agüera J, Gil J, Slaughter D (2011) Optimization of agrochemical application in olive groves based on positioning sensor. Precision Agriculture 12(4):564–575
10. Gebbers R, Adamchuk VI (2010) Precision agriculture and food security. Science 327(5967):828–831, https://doi.org/10.1126/science.1183899, URL https://science.sciencemag.org/content/327/5967/828, https://science.sciencemag.org/content/327/5967/828.full.pdf
11. Akyildiz I, Su W, Sankarasubramaniam Y, Cayirci E (2002) Wireless sensor networks: a survey. Computer Networks 38(4):393–422, https://doi.org/10.1016/S1389-1286(01)00302-4, URL http://www.sciencedirect.com/science/article/pii/S1389128601003024
12. Jiang R, Kasimov F, Zhang D, Kaliyeva K (2020) Development of intelligent agricultural machinery control device. In: Journal of Physics: Conference Series, IOP Publishing, vol 1617, p 012087
13. Gikas GD, Pérez-Villanueva M, Tsioras M, Alexoudis C, Pérez-Rojas G, MasísMora M, Lizano-Fallas V, Rodríguez-Rodríguez CE, Vryzas Z, Tsihrintzis VA (2018) Low-cost approaches for the removal of terbuthylazine from agricultural wastewater: constructed wetlands and biopurification system. Chemical Engineering Journal 335:647–656
14. Tavares Bruscato L, Heimfarth T, Pignaton de Freitas E (2017) Enhancing time synchronization support in wireless sensor networks. Sensors 17(12), https://doi.org/10.3390/s17122956, URL https://www.mdpi.com/1424-8220/17/12/2956

15. Zervopoulos A, Tsipis A, Alvanou AG, Bezas K, Papamichail A, Vergis S, Stylidou A, Tsoumanis G, Komianos V, Koufoudakis G, Oikonomou K (2020) Wireless sensor network synchronization for precision agriculture applications. Agriculture 10(3), https://doi.org/10.3390/agriculture10030089, URL https://www.mdpi.com/2077-0472/10/3/89
16. Bonomi F, Milito R, Zhu J, Addepalli S (2012) Fog computing and its role in the internet of things. In: Proceedings of the First Edition of the MCC Workshop on Mobile Cloud Computing, Association for Computing Machinery, New York, NY, USA, MCC '12, p 13–16, https://doi.org/10.1145/2342509.2342513, URL https://doi.org/10.1145/2342509.2342513
17. Guardo E, Di Stefano A, La Corte A, Sapienza M, Scatà M (2018) A fog computing-based iot framework for precision agriculture. Journal of Internet Technology 19(5):1401–1411
18. Dastjerdi A, Gupta H, Calheiros R, Ghosh S, Buyya R (2016) Chapter 4 – fog computing: principles, architectures, and applications. In: Buyya R, Dastjerdi, AV (eds) Internet of Things, Morgan Kaufmann, pp 61 – 75, https://doi.org/10.1016/B978-0-12-805395-9.00004-6, URL http://www.sciencedirect.com/science/article/pii/B9780128053959000046
19. Tsipis A, Papamichail A, Koufoudakis G, Tsoumanis G, Polykalas SE, Oikonomou K (2020) Latency-adjustable cloud/fog computing architecture for time-sensitive environmental monitoring in olive groves. AgriEngineering 2(1):175–205, https://doi.org/10.3390/agriengineering2010011, URL https://www.mdpi.com/2624-7402/2/1/11
20. Meyn A, White PS, Buhk C, Jentsch A (2007) Environmental drivers of large, infrequent wildfires: the emerging conceptual model. Progress in Physical Geography: Earth and Environment 31(3):287–312, https://doi.org/10.1177/0309133307079365, URL https://doi.org/10.1177/0309133307079365
21. Ray PP, Mukherjee M, Shu L (2017) Internet of things for disaster management: State-of-the-art and prospects. IEEE Access 5:18818–18835, https://doi.org/10.1109/ACCESS.2017.2752174
22. Visconti P, Primiceri P, Orlando C (2016) Solar powered wireless monitoring system of environmental conditions for early flood prediction or optimized irrigation in agriculture. Journal of Engineering and Applied Sciences (ARPN) 11(7):4623–4632
23. Alphonsa A, Ravi G (2016) Earthquake early warning system by iot using wireless sensor networks. In: 2016 International Conference on Wireless Communications, Signal Processing and Networking (WiSPNET), pp 1201–1205, https://doi.org/10.1109/WiSPNET.2016.7566327
24. Elijah O, Rahman TA, Orikumhi I, Leow CY, Hindia MN (2018) An overview of internet of things (iot) and data analytics in agriculture: Benefits and challenges. IEEE Internet of Things Journal 5(5):3758–3773, DOI https://doi.org/10.1109/JIOT.2018.2844296
25. Jawad HM, Nordin R, Gharghan SK, Jawad AM, Ismail M (2017) Energyefficient wireless sensor networks for precision agriculture: A review. Sensors 17(8), https://doi.org/10.3390/s17081781, URL https://www.mdpi.com/1424-8220/17/8/1781
26. Dlodlo N, Kalezhi J (2015) The internet of things in agriculture for sustainable rural development. In: 2015 International Conference on Emerging Trends in Networks and Computer Communications (ETNCC), pp 13–18, https://doi.org/10.1109/ETNCC.2015.7184801
27. de Lima GHEL, e Silva LC, Neto PFR (2010) Wsn as a tool for supporting agriculture in the precision irrigation. In: 2010 Sixth International Conference on Networking and Services, pp 137–142, https://doi.org/10.1109/ICNS.2010.26
28. Xuemei Li, Yuyan Deng, Lixing Ding (2008) Study on precision agriculture monitoring framework based on wsn. In: 2008 2nd International Conference on Anti-counterfeiting, Security and Identification, pp 182–185, https://doi.org/10.1109/IWASID.2008.4688381
29. Yoo S, Kim J, Kim T, Ahn S, Sung J, Kim D (2007) A2s: Automated agriculture system based on wsn. In: 2007 IEEE International Symposium on Consumer Electronics, pp 1–5
30. Srbinovska M, Gavrovski C, Dimcev V, Krkoleva A, Borozan V (2015) Environmental parameters monitoring in precision agriculture using wireless sensor networks. Journal of Cleaner Production 88:297 – 307, https://doi.org/10.1016/j.jclepro.2014.04.036, URL http://www.sciencedirect.com/science/article/pii/S0959652614003916, sustainable Development of Energy, Water and Environment Systems

31. Díaz SE, Pérez JC, Mateos AC, Marinescu MC, Guerra BB (2011) A novel methodology for the monitoring of the agricultural production process based on wireless sensor networks. Computers and Electronics in Agriculture 76(2):252 – 265, https://doi.org/10.1016/j.compag.2011.02.004, URL http://www.sciencedirect.com/science/article/pii/S0168169911000548
32. Shinghal D, Srivastava N, et al. (2017) Wireless sensor networks in agriculture: for potato farming. Neelam, Wireless Sensor Networks in Agriculture: For Potato Farming (September 22, 2017)
33. Bapat V, Kale P, Shinde V, Deshpande N, Shaligram A (2017) Wsn application for crop protection to divert animal intrusions in the agricultural land. Computers and Electronics in Agriculture 133:88 – 96, https://doi.org/10.1016/j.compag.2016.12.007, URL http://www.sciencedirect.com/science/article/pii/S0168169916303441
34. Friedrich T, Kienzle J (2007) Conservation agriculture: impact on farmers' livelihoods, labour, mechanization and equipment. In: Stewart, BI, Asfary, AF, Belloum, A. Steiner, K., and Friedrich, T., editors. Conservation Agriculture for Sustainable Land Management to Improve the Livelihood of People in Dry Areas. Proceedings of an international workshop, Citeseer, pp 25–36
35. Tzounis A, Katsoulas N, Bartzanas T, Kittas C (2017) Internet of things in agriculture, recent advances and future challenges. Biosystems Engineering 164:31–48
36. Tamura M, Nimura T, Naito K (2018) Development of field sensor network system with infrared radiation sensors. In: International Conference on Intelligent Interactive Multimedia Systems and Services, Springer, pp 74–83
37. Abbasi AZ, Islam N, Shaikh ZA, et al. (2014) A review of wireless sensors and networks' applications in agriculture. Computer Standards & Interfaces 36(2):263–270
38. ur Rehman A, Abbasi AZ, Islam N, Shaikh ZA (2014) A review of wireless sensors and networks' applications in agriculture. Computer Standards & Interfaces 36(2):263 – 270, https://doi.org/10.1016/j.csi.2011.03.004, URL http://www.sciencedirect.com/science/article/pii/S0920548911000353
39. Kiani F, Seyyedabbasi A (2018) Wireless sensor network and internet of things in precision agriculture. Int J Adv Comput Sci Appl 9(8):220–226
40. Yoon C, Huh M, Kang SG, Park J, Lee C (2018) Implement smart farm with iot technology. In: 2018 20th International Conference on Advanced Communication Technology (ICACT), IEEE, pp 749–752
41. Pei Z, Deng Z, Yang B, Cheng X (2008) Application-oriented wireless sensor network communication protocols and hardware platforms: A survey. In: 2008 IEEE International Conference on Industrial Technology, IEEE, pp 1–6
42. Aju OG (2015) A survey of zigbee wireless sensor network technology: Topology, applications and challenges. International Journal of Computer Applications 130(9):47–55
43. Vieira MAM, Coelho CN, Da Silva D, da Mata JM (2003) Survey on wireless sensor network devices. In: EFTA 2003. 2003 IEEE Conference on Emerging Technologies and Factory Automation. Proceedings (Cat. No. 03TH8696), IEEE, vol 1, pp 537–544
44. Nayyar A, Puri V (2016) Smart farming: Iot based smart sensors agriculture stick for live temperature and moisture monitoring using arduino, cloud computing & solar technology. In: Proc. of The International Conference on Communication and Computing Systems (ICCCS-2016), pp 9781315364094–121
45. Ahmad N, Hussain A, Ullah I, Zaidi BH (2019) Iot based wireless sensor network for precision agriculture. In: 2019 7th International Electrical Engineering Congress (iEECON), pp 1–4, https://doi.org/10.1109/iEECON45304.2019.8938854
46. Sahitya G, Balaji N, Naidu C (2016) Wireless sensor network for smart agriculture. In: 2016 2nd International Conference on Applied and Theoretical Computing and Communication Technology (iCATccT), IEEE, pp 488–493
47. Math RK, Dharwadkar NV (2017) A wireless sensor network based low cost and energy efficient frame work for precision agriculture. In: 2017 International Conference on Nascent Technologies in Engineering (ICNTE), IEEE, pp 1–6

48. Usha Rani M, Kamalesh S (2014) Web based service to monitor automatic irrigation system for the agriculture field using sensors. In: 2014 International Conference on Advances in Electrical Engineering (ICAEE), pp 1–5, https://doi.org/10.1109/ICAEE.2014.6838569
49. Sushanth G, Sujatha S (2018) Iot based smart agriculture system. In: 2018 International Conference on Wireless Communications, Signal Processing and Networking (WiSPNET), IEEE, pp 1–4
50. Vergis S, Komianos V, Tsoumanis G, Tsipis A, Oikonomou K (2020) A lowcost vehicular traffic monitoring system using fog computing. Smart Cities 3(1):138–156
51. Alvanou AG, Zervopoulos A, Papamichail A, Bezas K, Vergis S, Stylidou A, Tsipis A, Komianos V, Tsoumanis G, Koufoudakis G, et al. (2020) Cabius: Description of the enhanced wireless campus testbed of the ionian university. Electronics 9(3):454
52. Nikhade SG (2015) Wireless sensor network system using raspberry pi and zigbee for environmental monitoring applications. In: 2015 International Conference on Smart Technologies and Management for Computing, Communication, Controls, Energy and Materials (ICSTM), pp 376–381, https://doi.org/10.1109/ICSTM.2015.7225445
53. Zhang N, Wang M, Wang N (2002) Precision agriculture—a worldwide overview. Computers and Electronics in Agriculture 36(2):113 – 132, https://doi.org/10.1016/S0168-1699(02)00096-0, URL http://www.sciencedirect.com/science/article/pii/S0168169902000960
54. Sadler BM, Swami A (2006) Synchronization in sensor networks: an overview. In: MILCOM 2006 – 2006 IEEE Military Communications conference, pp 1–6, https://doi.org/10.1109/MILCOM.2006.302459
55. Sivrikaya F, Yener B (2004) Time synchronization in sensor networks: a survey. IEEE Network 18(4):45–50, DOI https://doi.org/10.1109/MNET.2004.1316761
56. Qun Li, Rus D (2006) Global clock synchronization in sensor networks. IEEE Transactions on Computers 55(2):214–226, DOI https://doi.org/10.1109/TC.2006.25
57. Elsts A, Fafoutis X, Duquennoy S, Oikonomou G, Piechocki R, Craddock I (2018) Temperature-resilient time synchronization for the internet of things. IEEE Transactions on Industrial Informatics 14(5):2241–2250, DOI https://doi.org/10.1109/TII.2017.2778746
58. Elson J, Römer K (2003) Wireless sensor networks: A new regime for time synchronization. SIGCOMM Comput Commun Rev 33(1):149–154, https://doi.org/10.1145/774763.774787, URL https://doi.org/10.1145/774763.774787
59. Mani SK, Durairajan R, Barford P, Sommers J (2018) An architecture for iot clock synchronization. In: Proceedings of the 8th International Conference on the Internet of Things, Association for Computing Machinery, New York, NY, USA, IOT '18, https://doi.org/10.1145/3277593.3277606, URL https://doi.org/10.1145/3277593.3277606
60. Maróti M, Kusy B, Simon G, Lédeczi A (2004) The flooding time synchronization protocol. In: Proceedings of the 2nd International Conference on Embedded Networked Sensor Systems, Association for Computing Machinery, New York, NY, USA, SenSys '04, p 39–49, https://doi.org/10.1145/1031495.1031501, URL https://doi.org/10.1145/1031495.1031501
61. Skiadopoulos K, Tsipis A, Giannakis K, Koufoudakis G, Christopoulou E, Oikonomou K, Kormentzas G, Stavrakakis I (2019) Synchronization of data measurements in wireless sensor networks for iot applications. Ad Hoc Networks 89:47 – 57, https://doi.org/10.1016/j.adhoc.2019.03.002, URL http://www.sciencedirect.com/science/article/pii/S1570870518308825
62. Koo J, Panta RK, Bagchi S, Montestruque L (2009) A tale of two synchronizing clocks. In: Proceedings of the 7th ACM Conference on Embedded Networked Sensor Systems, Association for Computing Machinery, New York, NY, USA, SenSys '09, p 239–252, https://doi.org/10.1145/1644038.1644062, URL https://doi.org/10.1145/1644038.1644062
63. Bezas K, Komianos V, Oikonomou K, Koufoudakis G, Tsoumanis G (2019) Structural health monitoring in historical buildings using a low cost wireless sensor network. In: 2019 4th South-East Europe Design Automation, Computer Engineering, Computer Networks and Social Media Conference (SEEDA-CECNSM), pp 1–4, https://doi.org/10.1109/SEEDA-CECNSM.2019.8908531

64. Capra A, Scicolone B (2018) Irrigation scheduling optimisation in olive groves. Journal of Experimental Agriculture International pp 1–19
65. Cruz MG (2010) Monte carlo-based ensemble method for prediction of grassland fire spread. International Journal of Wildland Fire 19(4):521–530
66. Theopoulos A, Boursianis A, Koukounaras A, Samaras T (2018) Prototype wireless sensor network for real-time measurements in hydroponics cultivation. In: 2018 7th International Conference on Modern Circuits and Systems Technologies (MOCAST), IEEE, pp 1–4
67. Chiang M, Zhang T (2016) Fog and iot: An overview of research opportunities. IEEE Internet of Things Journal 3(6):854–864
68. Noor TH, Zeadally S, Alfazi A, Sheng QZ (2018) Mobile cloud computing: Challenges and future research directions. Journal of Network and Computer Applications 115:70 – 85, https://doi.org/10.1016/j.jnca.2018.04.018, URL http://www.sciencedirect.com/science/article/pii/S1084804518301504
69. Osanaiye O, Chen S, Yan Z, Lu R, Choo KR, Dlodlo M (2017) From cloud to fog computing: A review and a conceptual live vm migration framework. IEEE Access 5:8284–8300, DOI https://doi.org/10.1109/ACCESS.2017.2692960
70. Castillo-Cara M, Huaranga-Junco E, Quispe-Montesinos M, Orozco-Barbosa L, Antúnez EA (2018) Frog: a robust and green wireless sensor node for fog computing platforms. Journal of Sensors 2018
71. Channe H, Kothari S, Kadam D (2015) Multidisciplinary model for smart agriculture using internet-of-things (iot), sensors, cloud-computing, mobilecomputing & big-data analysis. Int J Computer Technology & Applications 6(3):374–382
72. Hsu TC, Yang H, Chung YC, Hsu CH (2018) A creative iot agriculture platform for cloud fog computing. Sustainable Computing: Informatics and Systems p 100285, https://doi.org/10.1016/j.suscom.2018.10.006, URL http://www.sciencedirect.com/science/article/pii/S2210537918303275
73. Liyang Yu, Neng Wang, Xiaoqiao Meng (2005) Real-time forest fire detection with wireless sensor networks. In: Proceedings. 2005 International Conference on Wireless Communications, Networking and Mobile Computing, 2005., vol 2, pp 1214–1217, https://doi.org/10.1109/WCNM.2005.1544272
74. Anastasi G, Conti M, Francesco, MD, Passarella A (2009) Energy conservation in wireless sensor networks: A survey. Ad Hoc Networks 7(3):537 – 568, https://doi.org/10.1016/j.adhoc.2008.06.003, URL http://www.sciencedirect.com/science/article/pii/S1570870508000954
75. Hefeeda M, Bagheri M (2009) Forest fire modeling and early detection using wireless sensor networks. Ad Hoc & Sensor Wireless Networks 7(3-4):169–224
76. González JR, Kolehmainen O, Pukkala T (2007) Using expert knowledge to model forest stand vulnerability to fire. Computers and Electronics in Agriculture 55(2):107 – 114, https://doi.org/10.1016/j.compag.2006.12.005, URL http://www.sciencedirect.com/science/article/pii/S0168169906001207
77. de Bem PP, de Carvalho Júnior OA, Matricardi EAT, Guimarães RF, Gomes RAT (2019) Predicting wildfire vulnerability using logistic regression and artificial neural networks: a case study in brazil's federal district. International journal of wildland fire 28(1):35–45
78. Molina-Pico A, Cuesta-Frau D, Araujo A, Alejandre J, Rozas A (2016) Forest monitoring and wildland early fire detection by a hierarchical wireless sensor network. Journal of Sensors 2016
79. Abdulsahib GM, Khalaf OI (2018) An improved algorithm to fire detection in forest by using wireless sensor networks. International Journal of Civil Engineering & Technology (IJCIET)-Scopus Indexed 9(11):369–377
80. Abdullah S, Bertalan S, Masar S, Coskun A, Kale I (2017) A wireless sensor network for early forest fire detection and monitoring as a decision factor in the context of a complex integrated emergency response system. In: 2017 IEEE Workshop on Environmental, Energy, and Structural Monitoring Systems (EESMS), IEEE, pp 1–5

81. Vicente-Charlesworth L, Galmés S (2011) On the development of a sensor network-based system for wildfire prevention. In: Luo Y (ed) Cooperative Design, Visualization, and Engineering, Springer Berlin Heidelberg, Berlin, Heidelberg, pp 53–60
82. Manolakos ES, Logaras E, Paschos F (2009) Wireless sensor network application for fire hazard detection and monitoring. In: International Conference on Sensor Applications, Experimentation and Logistics, Springer, pp 1–15

Part III
Remote Sensing Applications

Potential of Sentinel-2 Satellite and Novel Proximal Sensor Data Fusion for Agricultural Applications

Miloš Pandžić, Aristotelis C. Tagarakis, Vasa Radonić, Oskar Marko, Goran Kitić, Marko Panić, Nataša Ljubičić, and Vladimir Crnojević

1 Introduction

Precision agriculture (PA) is often defined as a farming approach in which decisions are made at a high resolution according to the actual needs of the plants at each location. Precision agriculture is commercially practiced since the early 1990's [1] when the yield monitors with georeferencing capability became commercially available [2]. There are several definitions of PA. Fountas et al. [3] defined PA as *"the management of spatial and temporal variability in the fields using Information and Communications Technologies (ICT)"*. Khosla [4] referred to the five "R's" of PA which correspond to the application of inputs in agricultural systems at the "Right time", at the "Right amount" and to the "Right place" [5], using "the Right Source", and the "Right manner" [4].

In practice, in order to address all the aspects of PA, fields and plants are monitored using a variety of sensing technologies and decisions concerning fertilizer and pesticide application as well as irrigation, are adjusted accordingly. PA management systems show significant advantages compared to traditional farming such as increased application efficiency, and minimal environmental footprint of agricultural applications [6–9]. However, recent advances in technology provide an

M. Pandžić (✉) · V. Radonić · O. Marko · G. Kitić · M. Panić · N. Ljubičić · V. Crnojević
BioSense Institute, University of Novi Sad, Novi Sad, Serbia
e-mail: milos.pandzic@biosense.rs; vasarad@biosense.rs; oskar.marko@biosense.rs; gkitic@biosense.rs; panic@biosense.rs; natasa.ljubicic@biosense.rs; crnojevic@biosense.rs

A. C. Tagarakis
BioSense Institute, University of Novi Sad, Novi Sad, Serbia

Institute for Bio-Economy and Agri-Technology (iBO), Centre for Research and Technology Hellas (CERTH), Thessaloniki, Greece
e-mail: a.tagarakis@certh.gr

D. D. Bochtis et al. (eds.), *Information and Communication Technologies for Agriculture—Theme I: Sensors*, Springer Optimization and Its Applications 182,
https://doi.org/10.1007/978-3-030-84144-7_7

unprecedented opportunity for further development. Remote and proximal sensing using optical sensors are gaining popularity in crop production systems in the framework of PA applications. Nowadays, there is vast number of optical sensors recording the data about crops at different spatial, radiometric, and temporal resolutions. Both remote and in-field sensors are used for monitoring plant deficiency for nutrients and water, plant health status and soil condition [10]. Regarding the platform carrying the sensor and the sensors' proximity from the target (Fig. 1), they can be divided into three general categories, namely:

1. satellite-based sensors;
2. airborne and unmanned aerial vehicles (UAV)-mounted sensors;
3. ground-based proximal sensors.

1.1 Satellite-Based Sensors

The first and the uppermost level is remote sensing, where information is derived from satellite images. In recent studies, satellite data are increasingly used to monitor agricultural fields since they present a powerful tool for monitoring phenological trends and assessing the effect of climate variability. Compared to hand-held optical sensors, satellite images offer a valuable perspective at the field-scale, revealing

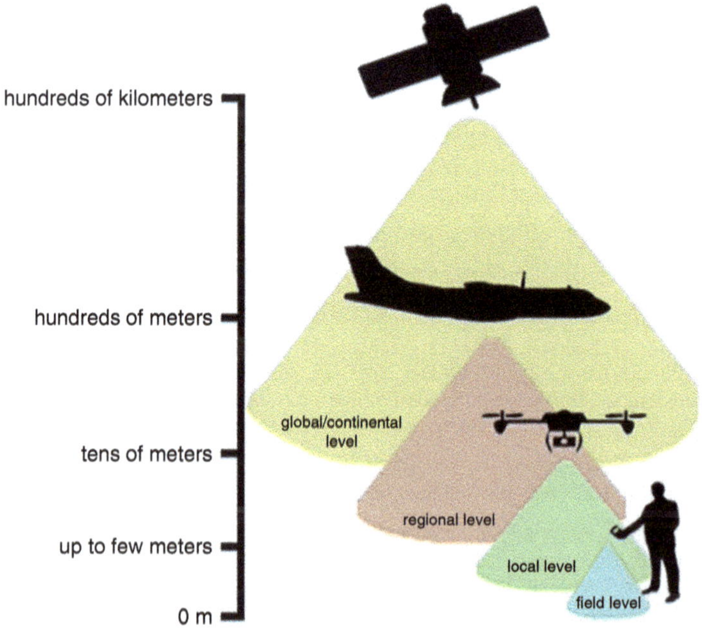

Fig. 1 Pictorial display of sensing platforms with regard to distance from target and coverage

regional crop conditions that are difficult to detect from ground manual measurements due to limited sampling capabilities [11]. This is by far the most scalable sensing technology, as satellites such as Sentinels by the European Space Agency (ESA) [12], Landsats by the National Aeronautics and Space Administration (NASA) [13] and many other commercial satellites, are covering the whole planet at different revisit frequencies providing images at spatial resolutions that may range from few decimeters to hundreds of meters (depending on the satellite and the sensor). Besides the visible range of the electromagnetic spectrum, many satellite sensors also employ red edge, near-infrared, infrared, and thermal domains, making them particularly valuable for crop monitoring [14]. Atzberger [15] pointed out that satellite remote sensing shows great potential for monitoring vegetation dynamics due to large spatial coverage and frequent revisit time. Satellite remote sensing has been used in agriculture since the 1970's when the first Landsat satellite was launched. Over the period of nearly half a century, the resolution of satellite images, as well as the revisit frequency, increased dramatically [1]. Today, a variety of different satellites are used for agricultural monitoring purposes, including Planet satellites with a resolution of up to 3 m, RapidEye with 5 m resolution and Pleiades with 50 cm resolution, each having a daily revisit time. However, a big drawback of the aforementioned commercial satellites is that the images are not free of charge and the cost of monitoring are thus prohibitively high for the majority of the farms.

In the last decade ESA and NASA changed their policies and made certain satellite imagery available to the public at no cost [16, 17]. Landsat's 40-year long archive is now freely available and the same holds for the state-of-the-art Earth observation program Copernicus operated by ESA on behalf of the European Commission. These led to an increased interest of the agricultural community towards satellite remote sensing in the previous years. Still, a few meters resolution of freely available satellite imagery may not be sufficient for precise monitoring of individual plants and could make the detection of fine differences within the field difficult. Another drawback of optical satellite sensors is the weather dependency. Clouds can heavily obscure images making them sometimes useless and that presents a serious challenge especially in rainy parts of Western Europe and North America. Satellites working in microwave domain like Sentinel-1 and TanDEM-X can, to a certain level, assist in this case since their signal penetrates clouds due to a different working principle. However, the information collected is different compared to the optical sensors and this is still an area of active research [18, 19].

1.2 Airborne- and Drone-Based Sensors

At the second level, when higher resolution is needed, sensors carried by aircrafts or, more recently, unmanned aerial vehicles (UAVs) can be used. They can provide precise measurements at high spatial resolution (a few cm) and are an ideal tool for crops that need detailed images for monitoring [20], such as orchards and vineyards, or any crops that do not fully cover the ground surface at the critical stage of image acquisition. The high cost of lifting an aircraft, the costs of equipment and operation, or the limited flight time of UAVs due to battery power limitations, are just some of the challenges in this category which make these solutions less scalable than satellite-based applications. However, there is an obvious and equally rising interest of the agricultural community in these techniques, with the UAV and the UAV-mounted sensors industry showing rapid growth [21].

1.3 Ground-Based Proximal Sensors

Ground-based proximal sensing is performed by sensors at a relatively short distance from the object of interest. Hand-held devices or sensors mounted on tractors and other vehicles are usually referred to as proximal sensors, here referred as ground-based proximal sensors. Their limitation is the small area coverage [22], but they also have significant advantages, such as high spatial resolution and independent choice of the time of acquisition. The typical spatial resolution of proximal sensing is in the range from millimeters to centimeters, as opposed to remote sensing that typically has resolution in the range from decimeters to hundreds of meters [23]. Another advantage is that their measurements are not compromised by cloudiness, in case of active sensors, and are ideal for practical applications such as on-the-go variable rate fertilization [24].

Active proximal sensors are independent of illumination conditions, since they emit their own light, and can operate under cloudy conditions or even at night. Furthermore, they do not require calibration to reflectance because the light source is known and constant [25]. However, the same is not applicable for passive sensors. Differences in environmental conditions between the start and end of measurement [26] occur due to time consuming character of ground-based sensing and this can reflect on losing precision in these types of measurements. Over the years, various different optical proximal sensors found practical applications, such as the Soil Plant Analysis Development (SPAD) chlorophyll meter (Konica Minolta Inc., Osaka, Japan), Hydro N-sensor (Yara International ASA, Oslo, Norway), GreenSeeker (Tribmle Inc., CA, USA), Crop Circle (Holland Scientific, NE, USA), CropScan (Next Instruments, Sydney, Australia), Analytical Spectral Devices (ASD) FieldSpec (ASD Inc., CO, USA), etc. Although they are least scalable, proximal sensors provide accurate assessment of the plants' growth, which is linked to photosynthetic activity and chlorophyll content [27], level of evapotranspiration [28], crops' nitrogen status [7, 29, 30], and yield [8, 31].

1.4 Vegetation Indices

The development of low-cost sensors, as well as the aforementioned liberalization of data access by data providers such as the ESA and NASA, have paved the way for the acquisition of vast amounts of sensor data. Compatibility studies between datasets acquired by different sensors are necessary prior to any kind of data fusion in practice. These studies are usually relying on the calculation of vegetation indices (VIs) that are the base for analyzing growth and vigor of vegetation. Vegetation index is a combination of different spectral bands in which information about reflected electromagnetic radiance from vegetation canopy is stored.

Satellite and hand-held sensors are usually multispectral sensing devices operating in the visible (red, green, and blue) and infrared wavelengths, which are especially interesting for agricultural monitoring. Due to variable lighting conditions or scanning angles among two or more measurement dates, absolute values of single band measurements are seldom used. On the other hand, relative difference between different bands is more stable as the additive effect of illumination on the response may be approximated as equal throughout the spectrum. For this reason, vegetation indices (VIs) have been developed in order to relate the reflected electromagnetic radiance from the canopy with the canopies' characteristics [32].

VIs are more widely used than absolute reflectance values in remote sensing algorithms for monitoring crop characteristics because of their simplicity and the ease of data processing [11]. Light of different parts of the electromagnetic (EM) spectrum acts differently when in contact with vegetation. In this regard, e.g., infrared light is reflected by mesophyll tissue whereas red light is absorbed by chlorophyll. Hence, their ratio will give high values for actively growing vegetation in contrast to stressed vegetation or other types of surfaces [33]. Due to the increase in biomass, changes of near-infrared radiation reflected from the canopy are the greatest during the growing season, whereas visible light shows lower variation mostly related to absorption of light by photosynthetic and photoprotective pigments [32]. Usage of VIs is very dependable on instruments and platforms that are recording this data [34].

Various VIs have been used for assessing different phenomena, such as plants' vegetation characteristics (NDVI, [35]), available soil moisture (NDWI, [36]), drought intensity (NMDI, [37] etc. Today, NDVI is most widely used, but it has its limitations with most important the fact that it reaches saturation when the canopy is dense, usually at the end of the growing season [38]. Generally, there is no perfect VI that could be used for all phenomena, all crops and all regions, and because of that, a variety of indices are used in scientific applications. The VIs can be affected by the canopy structure, leaf angles and their spatial distribution, plant row orientation, atmospheric conditions, which all together strongly influence canopy reflectance. They are generally mathematical expressions of the reflectance using just a few specific spectral bands and at a single angle of observation (usually nadir sensor orientation), which leads to an under-exploitation of the full spectral and directional ranges available when using new generation sensors [39].

This might, however, change in the future by utilizing other techniques that are rising in trend nowadays. As it is usually the case with data-driven decision-support systems and monitoring applications, the more data collected, the better. Nowadays, new approaches for data analysis are gaining popularity, providing the ability to analyze massive amounts of data and information. The essence of Big Data Analytics and Machine Learning is to acquire a huge quantity of data and leave it to the algorithm to find the hidden dependencies between them. Literature that convers machine learning for agriculture was recently reviewed by [40] who reported that over 60% of the studies were related to crop management. Hence, a likely pathway in Agriculture 4.0 is, among other, the utilization of VIs and Artificial Intelligence.

1.5 Inter-Comparison

Although proximal and remote sensing were extensively studied for assessing crop dynamics [41], direct inter-comparison between satellite remote sensing and proximal sensors with respect to the crop monitoring has rarely been discussed. Bausch and Khosla [42] compared QuickBird satellite-derived indices with ground-based Exotech radiometer-derived indices and found good correlation, with the highest agreement in green normalized difference vegetation index normalized for reference area (NGNDVI). Caturegli et al. [43] tested ground-based multispectral measurements (using Licor spectroradiometer and GreenSeeker) and GeoEye-1 satellite images for estimating nitrogen status of turfgrasses. Comparing NDVI values acquired from these instruments, the highest Pearson correlation coefficient was found between GreenSeeker and satellite derived NDVI ($r \approx 1$). Yang et al. [44] found moderate linear correlation ($r > 0.7$) between NDVI measured from Formosat-2 satellite images and ground portable spectroradiometer GER-2600. Wagner and Hank [45] revealed Pearson correlation coefficient of 0.85 between RapidEye and YARA-N sensor-derived Red Edge Inflection Point (REIP). Necessary modification was made in RapidEye measurements using YARA-N sensor-based model, so that the REIP could be calculated. Bu et al. [46] confirmed that yields of sugar beetroot, spring wheat, corn and sunflower can be predicted with GreenSeeker, Crop Circle and RapidEye red and red-edge imagery.

The use of VIs is of great importance in monitoring crop dynamics and predicting end-of-season yield from mid-season canopy reflectance measurements. Hence, it is essential to quantify the level of similarity between different sensor measurements prior to data fusion. Jackson and Huete [47] pointed that caution is needed when comparing VIs obtained by different sensors because of the differences between sensors' band-response functions, the differences in the fields of view, and the type of data, raw or transformed, acquired by each sensor.

In this chapter, 17 different indices were analyzed and their importance for crop monitoring, their stability and applicability on a large scale were discussed. VIs were derived from measurements made with a recently developed, active, multispectral proximal sensor named Plant-O-Meter (POM) and compared to VIs derived from Sentinel-2. The first practical value of this research lies in the fact that POM measurements could serve as the ground-truth for calibration of satellite images in large-scale applications. In this way, data with coarser spatial resolution and lower accuracy could be fine-tuned to fit the field measurements and provide higher accuracy results for broader areas swept by the satellite. Additionally, the motivation behind this research was to correlate measurements from a novel device with measurements widely used in modern research purposes. Nevertheless, both sensors represent modern optical instruments that are likely to find broader use in the near-term future.

2 Materials and Methods

The present study was carried out during the 2018 growing season at a commercial field located in Begeč (45° 14′ 32.712" N and 19° 36′ 21.486″ E), near Novi Sad, in Vojvodina (Serbian province) (Figs. 2 and 5). The study area covers a geographical area of 6 ha. The experimental field's soil properties were favorable for maize production showing high content of humus and nutrients, and neutral to slightly alkaline reaction, typical characteristics of the chernozem soils which are dominant in the area. The climate of Vojvodina is moderate continental, with cold winters and hot and humid summers with huge range of extreme temperatures and nonequal distribution of rainfall per month [48]. The mean annual air temperature is 11.1 $^{\circ}C$

Fig. 2 Location of the test site shown by the red dot

and the cumulative annual precipitation is about 606 mm according to climate data from the period 1949–2006 [49]. During maize growing season (from April 1 to September 30), the total cumulative precipitation was about 398 mm, which was 10% more than the 1970–2018 period average [50].

During maize sowing season, extremely dry and warm weather was recorded. Increased daily air temperatures in late April and early May accelerated maize development stages. However, the period from June to August was more humid and warmer compared to the average climatic conditions of the region. The soil conditions regarding soil moisture content were monitored using a wireless network of soil moisture sensors; three sensor nodes installed in specific locations according to soil apparent electrical conductivity zones. The crop was irrigated after seeding to enhance germination and support growth during the initial growth stages. Two additional irrigations were supplied in critical stages when the water demand is high; in June after V6 growth stage when rapid growth occurs and in August during grain filling stages. On the other hand, September was significantly dry, which was particularly favorable for maize crop allowing for the grain to mature and dry. The meteorological data, (air temperature and precipitation) were acquired from the official web portal of the Hydro-meteorological Service of the Republic of Serbia.[1] The field was sown with "Exxupery" (by RAGT Semences, France, FAO 560) maize hybrid (*Zea mays* L.) widely adopted by farmers in Europe. Maize field was sown on April 15, 2018 for the major cropping season. Seeding was done in 300 m long rows, at the plant distance of 0.2 m within rows and 0.7 m between rows. Maize was grown using common agronomic practices to avoid any nutrient deficiencies and other unfavorable conditions. A total of 300 kg ha^{-1} of composite granular synthetic fertilizer (15:15:15; N-P-K) was applied at planting. The field was harvested when the majority of plants reached full maturity, on September 30, 2018.

2.1 *Plant-O-Meter*

In-field proximal measurements were performed using POM sensor, an active crop sensor, recently developed by the BioSense Institute (Republic of Serbia). It is a multispectral sensing device emitting at four wavelengths of the electromagnetic spectrum: blue – 465 nm, green – 535 nm, red – 630 nm and near-infrared – 850 nm and measuring the amount of energy reflected by the target. Based on these measurements, a number of vegetation indices can be calculated. As previously explained, these indices represent combinations of surface reflectance at two or more wavelengths and they have been adopted with the aim to highlight particular properties of the scanned plants [51]. The multispectral source is developed using four super bright LED dices that emit radiation at the defined wavelengths and have been attached on a surface mounted technology (SMT) chip, Fig. 3.

[1] www.hidmet.gov.rs

Fig. 3 SMT multispectral source with four wavelengths: 850, 630, 535 and 465 nm

The reflected light is detected by a silicon PIN photodiode BPX 61 [52] with supporting electronics for filtering and amplifying the useful signal as well as for providing immunity to ambient light up to 10,000 lux. This photodiode-based detector module sequentially receives the reflected spectrum and sends the raw data to the microcontroller which averages the measurements and sends the data via Bluetooth to Android-operated device such as tablet, smartphone or similar. All further processing is performed using the processor of the Android device through a specially designed Android application. Each measurement is georeferenced using the internal antenna of the Android device (which typically support GPS, GLONASS, or Beidou systems). Based on these inputs, VIs are calculated and displayed on the screen through the POM Android interface. The application has two operating modes:

1. stationary measurements used for recording unique georeferenced measurements of specific locations in the field, and
2. continuous measurement intended for mapping the plants' canopy or soil reflectance in the field.

More detailed information about POM sensor (Fig. 4) and its operating principles can be found in the dedicated paper [53].

During the field ground-based proximal measurements, the POM was used to scan the whole length of every tenth row by walking along the rows, holding the sensor directly on top of the crop row with the scanning footprint perpendicular to the row direction. The measuring frequency was 1 Hz, which roughly corresponded to 1 m distance between the POM record points along the row. POM measurements were performed at four different dates and were carried out in the following stages

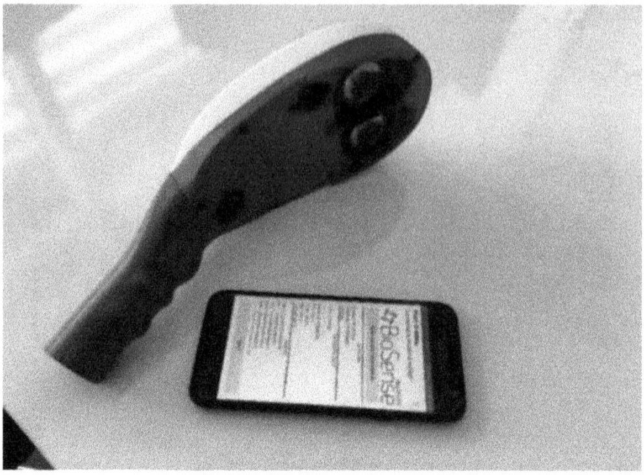

Fig. 4 Plant-O-Meter and specially developed Android application

Table 1 Corresponding acquisition dates for POM and Sentinel-2 and development stage of maize

POM date	Sentinel-2 date	Crop development stage
01.06.2018	30.05.2018	6-leaf (V6)
21.06.2018	24.06.2018 (cloudy)	Tassel (VT)
04.07.2018	14.07.2018	Silking (R1)
26.07.2018	29.07.2018	Blister (R2)

of maize development: 6-leaf growth stage (V6), beginning of tasseling (VT), silking (R1) and at the end of blister stage (R2), (Table 1).

2.2 *Sentinel-2*

Sentinels are series of space missions of the European Union's Copernicus programme which was created for monitoring the Earth's environment [54]. Each mission is specifically designed and is using different technology to collect various type of information about Earth's land, water, and atmosphere. Sentinels −1, −2, −3 and -5P have currently been launched. Sentinel-2 is a constellation of two identical satellites A and B, launched respectively on June 23, 2015, and March 7, 2017. They fly at the average altitude of 786 km, in the same orbit phased at 180° to each other [55], thus having joint revisit time of 5 days at the equator. Sentinel-2's swath width is 290 km, and it images the Earth's surface between 56° S and 84° N latitude. Mission lifespan is designed to last for 7 years.

Each Sentinel-2 satellite carries an optical multispectral instrument that provides images in 13 spectral bands with spatial resolutions of either 10, 20, or 60 m [56]. ESA delivers Sentinel-2 images either as Level-1C or Level-2A products, which both represent radiometrically and geometrically corrected images, where Level-2A in addition to

Table 2 Operating wavelengths for POM and Sentinel-2

	Plant-O-Meter	Sentinel-2 [nm]		
Band name	Wavelengths range [nm]	Central wavelength [nm]		Bandwidth [nm]
Blue	450–465	490		65
Green	520–535	560		35
Red	620–630	665		30
NIR	830–870	**Wide**	842	115
		Narrow	865	20

Level-1C includes atmospheric correction. Products used in our study, are Bottom-of Atmosphere Level-2A products. The bands used in the study are the blue (490 nm), green (560 nm), red (665 nm) and near infra-red (NIR) (842 nm), with a 10 m resolution, and the narrow NIR (865 nm) with a 20 m resolution. With respect to POM measurement dates, corresponding cloud-free satellite images were downloaded and processed. Atmospherically corrected images were downloaded from the official Copernicus Open Access Hub[2] and processed using the official Sentinel-2 Toolbox [57] and QGIS software [58]. Acquisition dates for Sentinel-2 images are given in Table 1 and operating wavelengths are given in Table 2.

2.3 *Data Analysis*

Since the narrow NIR band of Sentinel-2 images was only available at 20 m resolution, all images were resampled using the nearest neighbor method. Thus, the blue, green, red and NIR bands from Sentinel-2 images were down-sampled from 10 m to 20 m resolution.

Due to the higher resolution of POM measurements compared to Sentinel-2 images, i.e. several POM measurements points fell within a single Sentinel-2 image pixel (Fig. 5), all POM measurements inside a Sentinel-2 pixel were averaged. By employing this, there was only one corresponding value per POM spectral band for each single image pixel. Hence, 1-to-1 comparison between measurements of the two sensors was achieved. Using different spectral band combinations, various indices were calculated (Table 3).

Linear regression analysis was elaborated, using the Statistica 13 [71] statistical software, to define the relationship between the VIs calculated from the POM measurements and from the Sentinel-2 satellite images. Since Sentinel-2 datasets include two separate bands for the NIR spectrum, the effect of using either one of them in the calculation of the indices was studied. Therefore, the analysis was performed twice: (1) using the wide range NIR band from Sentinel-2 datasets, and (2) using the narrow NIR band.

At the second phase of data analysis, pixels that were confirmed as outliers were manually excluded from further analysis. Those were either border pixels,

[2] https://scihub.copernicus.eu/

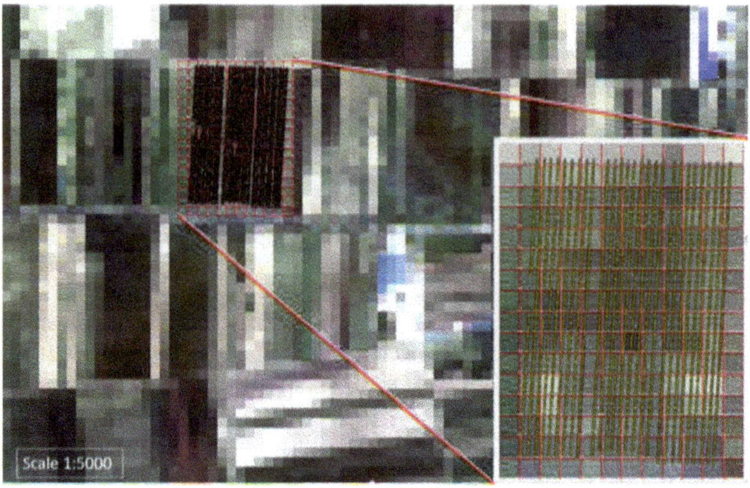

Fig. 5 Sentinel-2 image of the experimental field in Begeč at 20 m resolution where yellow dots represent POM measurement points

contaminated by the features outside the field, or pixels contaminated by other objects located inside the parcel (Fig. 6). Linear regression analysis was also performed in the outlier-free dataset (after removing the contaminated pixels) to export the final results.

3 Results and Discussion

The analysis of the Sentinel-2 image acquired on 24 June 2018, provided poor results due to the significant effect of a layer of clouds over the experimental field. Therefore, this date was excluded from the analysis. This is a good example of the constraints of the use of optical satellite images as they highly depend on the weather [1]. Sentinel-2 provides two measurements in the NIR channel: wide (785–900 nm) and narrow (855–875 nm) range. Therefore, the analysis was performed twice, using the wide range NIR band and narrow NIR band, for the calculation of indices acquired from Sentinel-2 images. In the initial analysis the full dataset regarding the study area, for each of the three useable measurement dates, was used. Regression analysis revealed significant positive correlations between the indices derived from Sentinel-2 satellite images and Plant-O-Meter measurements acquired at V6 growth stage (01-06-2018) before the plants entering the reproductive stages. However, the coefficient of determination (r^2) was considerably low ranging from 0.22 (weakest relationship for EVI) to 0.46 (strongest relationship for $NDVI_{narrow}$, NDVI calculated using narrow NIR band from Sentinel-2, and $PNDVI_{narrow}$, PNDVI calculated using narrow NIR band from Sentinel-2). This was attributed to the existence of contaminated pixels within the dataset that behaved as outliers in the analysis. Those pixels

Table 3 List of the vegetation indices used in the analysis with given formulas, short description, and further reference

VI	Formula	Description	References
Normalized Difference Vegetation Index	$NDVI = \frac{R_{NIR} - R_{Red}}{R_{NIR} + R_{Red}}$	Most popular vegetation index; indicates plant health status and photosynthetic activity.	[59]
Simple ratio	$SR = \frac{R_{NIR}}{R_{red}}$	Indicates greenness of vegetation.	[60]
Infrared Percentage Vegetation Index	$IPVI = \frac{R_{NIR}}{R_{NIR} + R_{red}}$	Functionally equivalent to NDVI, but computationally faster.	[61]
Green NDVI	$NDVI_g = \frac{R_{NIR} - R_{Green}}{R_{NIR} + R_{Green}}$	More sensitive to chlorophyll concentration variations than NDVI.	[62]
Blue NDVI	$NDVI_b = \frac{R_{NIR} - R_{Blue}}{R_{NIR} + R_{Blue}}$	Less sensitive to crop stress than NDVI, due to absorption of blue light by anthocyanins and its scattering in the atmosphere.	[63, 64]
Structure Insensitive Pigment Index	$SIPI = \frac{R_{NIR} - R_{Blue}}{R_{NIR} - R_{Red}}$	Relates Carotenoids to Chlorophyll a	[65]
Enhanced Vegetation Index	$EVI = \frac{2.5\left(R_{NIR} - R_{red}\right)}{R_{NIR} + 6R_{red} - 7.5R_{blue} + 1}$	Sensitive to canopy variations in regions with high biomass, where NDVI saturates.	[38]
Green Soil Adjusted Vegetation Index	$GSAVI = 1.5\frac{\left(R_{NIR} - R_{green}\right)}{R_{NIR} + R_{green} + 0.5}$	Both indices show sensitivity to nitrogen concentration change.	[66]
Green Optimized Soil Adjusted Vegetation Index	$GOSAVI = \frac{R_{NIR} - R_{green}}{R_{NIR} + R_{green} + 0.16}$		

(continued)

Table 3 (continued)

VI	Formula	Description	References
Green Chlorophyll Index	$GCI = \frac{R_{NIR}}{R_{green}} - 1$	Chlorophyll content estimation	[67]
Non-Linear Index	$NLI = \frac{NIR^2 - R_{red}}{NIR^2 + R_{red}}$	Minimizes impact of leaf angle distribution, view azimuth and soil brightness	[68]
Transformed Difference Vegetation Index	$TDVI = 1.5 \frac{R_{NIR} - R_{red}}{\sqrt{R_{NIR}^2 + R_{red} + 0.5}}$	Does not saturate like NDVI; minimizes the effect of bare soil beneath vegetation cover	[69]
Wide Dynamic Range Vegetation Index	$WDRVI = \frac{0.2\,R_{NIR} - R_{red}}{0.2\,R_{NIR} + R_{red}}$	Higher sensitivity than NDVI to high values of LAI	[70]
Visible Normalized Difference Vegetation Indices	$GRNDVI = \frac{R_{NIR} - (R_{green} + R_{red})}{R_{NIR} + R_{green} + R_{red}}$	VIs constructed with red or blue band are moderately corelated with Leaf Area Index (LAI), whereas those constructed with green band are highly corelated with LAI.	[63]
	$GBNDVI = \frac{R_{NIR} - (R_{green} + R_{blue})}{R_{NIR} + R_{green} + R_{blue}}$		
	$RBNDVI = \frac{R_{NIR} - (R_{red} + R_{blue})}{R_{NIR} + R_{red} + R_{blue}}$		
	$PNDVI = \frac{R_{NIR} - (R_{green} + R_{red} + R_{blue})}{R_{NIR} + R_{green} + R_{red} + R_{blue}}$		

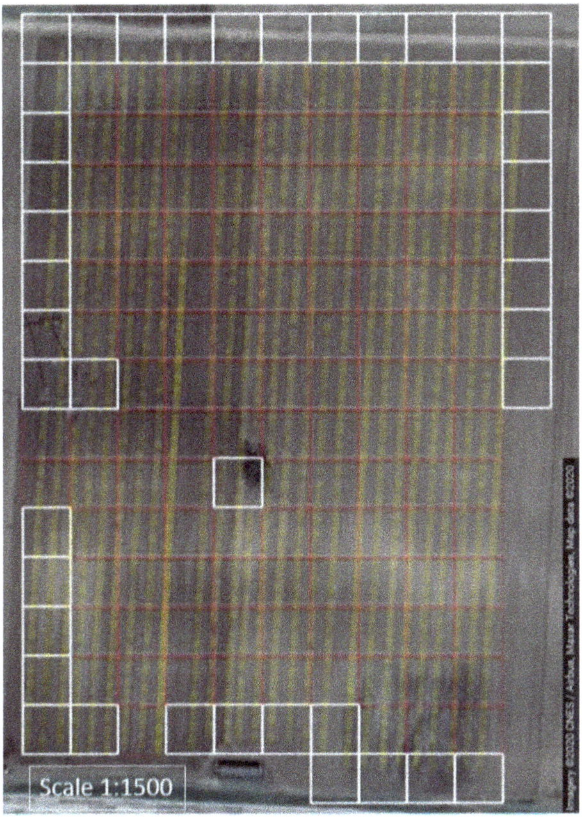

Fig. 6 Study field with contaminated pixels (white), pixels used in the analysis (pale red) and POM measurement points (pale yellow); (Google Maps, Imagery ©2020 CNES/Airbus, Maxar Technologies, Map data ©2020)

were therefore removed, and the regression analysis was repeated with the outlier-free datasets (Fig. 6).

The final linear regression analysis provided an insight of which indices calculated using POM are in better agreement with the same ones calculated using Sentinel-2 satellite images. The differences were mainly due to the deviations in the operating wavelengths for the two sensors and in the different sensitivity of each sensor at different bands. According to the statistical analysis, using the narrow range NIR in calculations of Sentinel-2 indices provided better correlation to the POM indices (Table 5; Fig. 7b) as compared to the results using the wide range NIR (Table 4; Fig. 7a). In the regression between indices calculated from Sentinel-2, using the wide range NIR band (Table 4) for the first date of measurement (V6 growth stage; 01-06-2018), the coefficient of determination (r^2) ranged between 0.325 (for EVI) and 0.700 (for RBNDVI) while the root mean square error (RMSE) ranged from 0.905 (for SR) to 0.014 (for NLI). In the analysis using the Sentinel-2's narrow band NIR, the coefficient of determination (r^2) was better, ranging between

Table 4 Coefficient of determination (r^2) and Root Mean Square Error (RMSE) from the regression between indices calculated from Sentinel-2, using the wide range NIR band, and POM

Date	01-06-2018		04-07-2018		26-07-2018	
	r²	RMSE	r²	RMSE	r²	RMSE
NDVI	0.680	0.075	0.162	0.093	0.036	0.093
SR	0.612	0.905	0.147	4.283	0.045	4.616
IPVI	0.668	0.045	0.162	0.047	0.036	0.046
NDVIg	0.616	0.058	0.102	0.209	0.008	0.210
NDVIb	0.652	0.103	0.050	0.096	0.000	0.134
SIPI	0.546	0.527	0.059	0.266	0.000	0.267
EVI	0.325	0.182	0.002	0.719	0.000	1.327
GSAVI	0.614	0.091	0.105	0.319	0.008	0.318
GOSAVI	0.615	0.058	0.103	0.210	0.008	0.210
GCI	0.574	0.500	0.087	4.625	0.028	4.611
NLI	0.478	0.014	0.124	0.006	0.060	0.004
TDVI	0.672	0.115	0.167	0.091	0.034	0.089
WDRVI	0.648	0.120	0.156	0.226	0.041	0.230
GRNDVI	0.659	0.062	0.177	0.236	0.022	0.239
GBNDVI	0.676	0.061	0.097	0.244	0.008	0.274
RBNDVI	0.700	0.131	0.143	0.152	0.010	0.184
PNDVI	0.686	0.082	0.155	0.259	0.017	0.288
Average	**0.580**	**0.179**	**0.111**	**0.677**	**0.020**	**0.736**

Table 5 Coefficient of determination (r^2) and Root Mean Square Error (RMSE) from the regression between indices calculated from Sentinel-2, using the narrow range NIR band, and POM

Date	01-06-2018		04-07-2018		26-07-2018	
	r²	RMSE	r²	RMSE	r²	RMSE
NDVI	0.710	0.069	0.162	0.100	0.045	0.103
SR	0.644	0.863	0.135	4.791	0.058	5.615
IPVI	0.696	0.042	0.162	0.050	0.045	0.052
NDVIg	0.630	0.059	0.106	0.217	0.012	0.224
NDVIb	0.676	0.099	0.044	0.101	0.002	0.141
SIPI	0.579	0.522	0.057	0.265	0.001	0.264
EVI	0.344	0.176	0.003	0.715	0.000	1.312
GSAVI	0.627	0.093	0.108	0.331	0.012	0.339
GOSAVI	0.629	0.060	0.107	0.218	0.012	0.225
GCI	0.595	0.515	0.094	5.015	0.032	5.349
NLI	0.493	0.014	0.116	0.006	0.097	0.004
TDVI	0.702	0.106	0.167	0.096	0.043	0.098
WDRVI	0.677	0.114	0.153	0.243	0.052	0.261
GRNDVI	0.683	0.057	0.178	0.248	0.029	0.260
GBNDVI	0.696	0.056	0.095	0.255	0.017	0.293
RBNDVI	0.728	0.124	0.136	0.162	0.022	0.200
PNDVI	0.710	0.076	0.151	0.274	0.028	0.313
Average	**0.603**	**0.175**	**0.110**	**0.733**	**0.028**	**0.842**

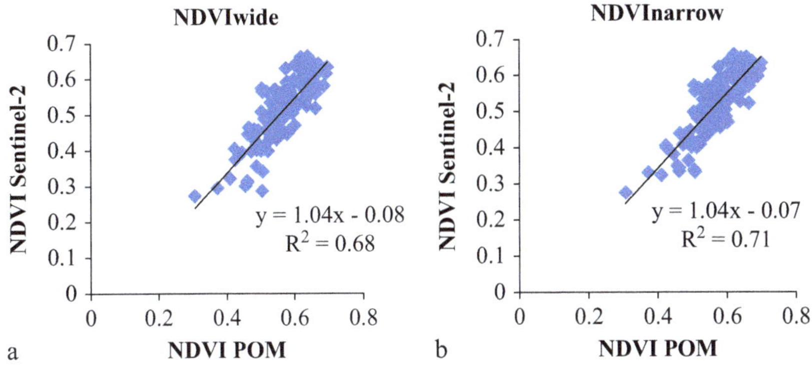

Fig. 7 Linear regression between the NDVI calculated from POM measurements and Sentinel-2 satellite images using wide (a) and narrow (b) NIR bands, at V6 maize growth stage

0.344 (for EVI) and 0.728 (for RBNDVI) while the root mean square error (*RMSE*) ranged from 0.863 (for SR) to 0.014 (for NLI). This was expected since the measuring range of the narrow NIR band of Sentinel-2 and NIR band of POM, are much closer.

The results for the two measurements during the reproductive stages, silking stage (R1) and blister stage (R2), provided poor results. This is attributed to the fact that after tasselling, the measurements showed considerably lower correlation between the two sensors explained by the mixture of colours after the tassels appear, and the different shades of the canopy from green to yellow as the plants approach maturity. Due to the large difference in the spatial resolution of the measurements of the two sensors in the study, this random mixture of colours affected the results of each sensor differently. The regression results showed low correlation between indices derived by the two sensing systems, showing the coefficient of determination (r^2) lower than 0.178 (Tables 4 and 5). Therefore, these results are not discussed further.

The POM measurements acquired at the V6 growth stage showed good correlation with Sentinel-2 results, mainly due to the uniformity of the colour of the area scanned in the field, which at this stage consists of crop canopy leaves (green colour). With 6 leaves fully developed, the leaf area covered a significant proportion of the ground without significant overlapping of the vegetation. According to literature, this is the most appropriate stage for remotely sensing the maize crop as this is when the highest spatial variability in the reflectance measurements is observed [72].

Concerning the NDVI, which is the most widely used vegetation index [31, 32, 73], the linear regression showed significant correlation between the POM and Sentinel-2 derived datasets at V6 growth stage for both using the narrow NIR band (r^2 =0.710, RMSE=0.069; Table 5, Fig. 7b) and the wide NIR band (r^2 =0.680, RMSE=0.075; Table 4, Fig. 7a) showing an almost 1:1 relationship; the slope of the linear model approached 1 and the constant approached 0 (Fig. 7). The correlation would probably increase if the resolution of the satellite imagery was greater and if

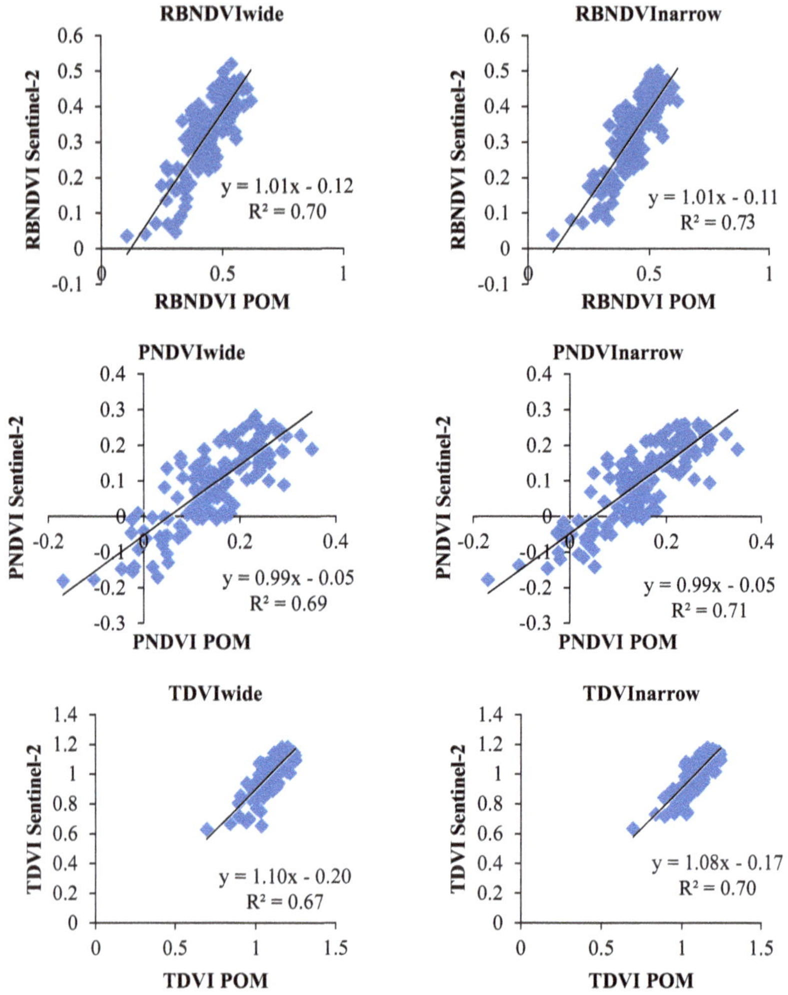

Fig. 8 Linear regression between the RBNDVI, TDVI and PNDVI calculated from POM measurements and Sentinel-2 satellite images using wide (left) and narrow (right) NIR bands, at V6 maize growth stage

the different crop type was scanned which allows minimal reflectance of soil, such as turfgrasses, as reported in [43]. Apart from NDVI, most of the indices showed significant correlations with RBNDVI showing the highest coefficient of determination (r^2 =0.728) followed by NDVI and PNDVI (r^2 =0.710), and TDVI (r^2 =0.702), (Table 5, Fig. 7b and Fig. 8 – right). This result suggests that RBNDVI, NDVI, PNDVI, and TDVI calculated using POM are more similar to the Sentinel-2 derived image results and are therefore more preferable to be used in studies that involve both POM (ground-based proximal) and Sentinel-2 (satellite remote) sensors.

The overall results of this study suggest that the POM active proximal multispectral sensor can serve as a good alternative to the Sentinel-2 satellite sensor, having the benefits that the active proximal sensors offer, high spatial resolution, flexibility in the measurement timing and independence from cloudiness [46]. In addition, satellite data often need auxiliary ground-truth data for correcting the image interpretation [23], and therefore combined use of Sentinel-2 imagery and POM measurements could provide added value in contrast to single source derived information. Attention should be paid to the time of sensing as the measurements acquired by the two different sensing systems (ground-based proximal and satellite-based remote) are dissimilar when the canopy is irregular and the cops' surface shows variability in structure and color, due to the large difference of the spatial resolution.

4 Conclusions

The present work shows a comparative study among a ground-based proximal multispectral sensor (named Plant-O-Meter) and satellite images from Sentinel-2 mission. Both instruments operate in similar range of specific wavelengths, enabling the calculation of various spectral indices which could be valuable for monitoring crop properties. According to the results, ground-based proximal sensing provides comparable results to the indices calculated from Sentinel-2 satellite images at certain growth stages of maize. Therefore, Plant-O-Meter active proximal sensor can be an alternative to satellite images, providing measurements at high spatial resolution. This system operates independently of weather and illumination conditions overcoming the limitations that passive optical sensors are facing, such as cloudiness, an important limiting factor of satellite remote sensing. Also, since satellite data usually need auxiliary ground-truth data for image interpretation, data fusion of Sentinel-2 imagery and POM measurements may provide added value in contrast to single source derived information.

On the other hand, POM provides information on plant level, while Sentinel-2 is more scalable and can provide information on field or regional level. The plant development stage plays an important role in the agreement between the indices derived by POM and Sentinel-2 due to the large difference in spatial resolution of the measurements. Satisfactory agreement between the indices calculated from the two sensing systems was achieved at progressed growth stages (V6, with 6 fully developed leaves emerged) of maize and before entering the reproductive stages. After the emergence of tassels, the regression between the two datasets provided poor results due to the uneven spatial distribution of the canopy growth and color mixture of leaves and tassels. The results can also apply to other crops; however, additional studies are needed to validate the relationship at the different development stages according to the physiological specifications of each individual crop.

Further use cases may involve the POM proximal crop sensor mounted on tractors or pivot irrigation systems which would, independently or in combination with Sentinel-2 data, enable decision making for variable rate irrigation, fertilization, or fertigation. In addition, the data provided by the two sensing systems will be used for developing and testing algorithms for early- and late-season yield prediction. With an estimated market price below 500 € and a planned commercialization for POM, and in combination with free access to Sentinel-2 data and web applications that utilize satellite data [74], farmers will have an ultimate possibility to apply precision agriculture regardless the size and market share of their business.

Acknowledgments This work was supported by the Ministry of Education, Science and Technological Development of the Republic of Serbia (Grant No. 451-03-68/2020-14/200358), European Union's Horizon 2020 research and innovation programme (Grant No. 739570 – ANTARES) and Provincial Secretariat for Higher Education and Scientific Research of Vojvodina through the project "Sensor technologies for integrated monitoring of agricultural production".

References

1. Mulla, D. J. (2013) 'Twenty five years of remote sensing in precision agriculture: Key advances and remaining knowledge gaps', Biosystems Engineering, Special Issue: Sensing in Agriculture, pp. 358–371. https://doi.org/10.1016/j.biosystemseng.2012.08.009.
2. Yang, C., Everitt, J. H., Du, Q., Luo, B. and Chanussot, J. (2013) 'Using High-Resolution Airborne and Satellite Imagery to Assess Crop Growth and Yield Variability'. Proceedings of the IEEE, 101 (3), pp. 582-592. https://doi.org/10.1109/JPROC.2012.2196249.
3. Fountas, S., Aggelopoulou, K., and Gemtos, T. A. (2016) 'Precision Agriculture', In: 'Supply Chain Management for Sustainable Food Networks', (pp. 41–65), Chichester, UK: John Wiley & Sons, Ltd. https://doi.org/10.1002/9781118937495.ch2.
4. Khosla, R. (2008) 'The 9th International Conference on Precision Agriculture opening ceremony presentation'. July 20-23rd, 2008.ASA/CSSA/SSSA.
5. Robert, P., Rust, R. and Larson, W. (1994) 'Site-specific Management for Agricultural Systems', Proceedings of the 2nd International Conference on Precision Agriculture, 1994, Madison, WI.
6. Auernhammer H. (2001) 'Precision farming — the environmental challenge', Computers and Electronics in Agriculture, 30, pp. 31–43. https://doi.org/10.1016/S0168-1699(00)00153-8.
7. Raun, W.R., Solie, J.B., Johnson, G.V., Stone, M.L., Mullen, R.W., Freeman, K.W., Thomason, W.E, and Lukina, E.V. (2002) 'Improving nitrogen use efficiency in cereal grain production with optical sensing and variable rate application', Agron. J. 94:815–820. https://doi.org/10.2134/agronj2002.8150.
8. Tagarakis A. C., Ketterings Q. M., Lyons S. and Godwin G. (2017) 'Proximal sensing to estimate yield of brown midrib forage sorghum', Agronomy Journal, 109(1), 107-114. https://doi.org/10.2134/agronj2016.07.0414.
9. Tagarakis A., Liakos V., Fountas S., Koundouras S. and Gemtos T. (2013) 'Management zones delineation using fuzzy clustering techniques in grapevines', Precision Agriculture 14(1), 18-39. https://doi.org/10.1007/s11119-012-9275-4.
10. Lee, W.S., Alchanatis, V., Yang, C., Hirafuji, M., Moshou, D. and Li, C. (2010) 'Sensing technologies for precision specialty crop production', Computers and Electronics in Agriculture, 74, pp. 2–33. https://doi.org/10.1016/j.compag.2010.08.005.
11. Wang, R., Cherkauer, K. and Bowling, L. (2016) 'Corn Response to Climate Stress Detected with Satellite-Based NDVI Time Series', Remote Sens., 8, 269. https://doi.org/10.3390/rs8040269.
12. PwC Advisory France for the European Commission (2019) 'Copernicus Market Report', Issue 2, ISBN 978-92-79-98973-5. https://doi.org/10.2873/011961.

13. Wulder, M.A.; Loveland, T.R.; Roy, D.P.; Crawford, C.J.; Masek, J.G.; Woodcock, C.E.; Dwyer, J.; Hermosilla, T.; Hipple, J.D.; Hostert, P.; et al. Current status of Landsat program, science, and applications. Remote Sens. Environ. 2019, 225, 127–147. https://doi.org/10.1016/j.rse.2019.02.015.
14. Tucker, C. J. (1979) 'Red and photographic infrared linear combinations for monitoring vegetation', Remote sensing of Environment, 8(2), 127-150. https://doi.org/10.1016/0034-4257(79)90013-0.
15. Atzberger, C. (2013) 'Advances in Remote Sensing of Agriculture: Context Description, Existing Operational Monitoring Systems and Major Information Needs', Remote Sens., 5(2), 949-981; https://doi.org/10.3390/rs5020949.
16. Aschbacher, J. and Milagro-Pérez, M. P. (2012) 'The European Earth monitoring (GMES) programme: Status and perspectives', Remote Sensing of Environment, 120, pp. 3-8. https://doi.org/10.1016/j.rse.2011.08.028.
17. Woodcock, C. E., Allen, R., Anderson, M., Belward, A., Bindschadler, R., Cohen, W., Gao, F., Goward, S. N., Helder, D., Helmer, E., Nemani, R., Oreopoulos, L., Schott, J., Thenkabail, P. S., Vermote, E. F., Vogelmann, J., Wulder, M. A. and Wynne, R. (2008) 'Free access to Landsat imagery', Science, 320, pp. 1011. https://doi.org/10.1126/science.320.5879.1011a.
18. Joshi, N., Baumann, M., Ehammer, A., Fensholt, R., Grogan, K., Hostert, P., Jepsen, M.R., Kuemmerle, T., Meyfroidt, P. Mitchard, E.T.A., Reiche, J., Ryan, C. M. and Waske, B. (2016) 'A review of the application of optical and radar remote sensing data fusion to land use mapping and monitoring', Remote Sensing, 8, 70. https://doi.org/10.3390/rs8010070.
19. McNairn, H., Champagne, C., Shang, J., Holmstrom, D. and Reichert, G., (2009) 'Integration of optical and Synthetic Aperture Radar (SAR) imagery for delivering operational annual crop inventories', ISPRS J. Photogramm. Remote Sens., 64 (5), pp. 434-449. https://doi.org/10.1016/j.isprsjprs.2008.07.006.
20. Wójtowicz, M., Wójtowicz, A., Piekarczyk, J. (2016) Application of remote sensing methods in agriculture. Commun. Biometry Crop Sci. 2016, 11, 31–50.
21. Puri, V., Nayyar, A., Raja, L. (2017) Agriculture drones: A modern breakthrough in precision agriculture. J. Stat. Manag. Syst., 20, 507–518. https://doi.org/10.1080/09720510.2017.1395171.
22. Jackson, R. D. (1986) 'Remote Sensing of Biotic and Abiotic Plant Stress', Annual review of Phytopathology, 24, pp. 265-287. https://doi.org/10.1146/annurev.py.24.090186.001405.
23. Oerke, E.C., Mahlein, A.K., and Steiner, U. (2014) 'Proximal sensing of plant diseases' In: Gullino, M.L., Bonants, P.J.M. (eds) 'Detection and diagnostics of plant pathogens', Springer, Dordrecht, pp. 55–68. https://doi.org/10.1007/978-94-017-9020-8_4.
24. Shanahan, J.F., Kitchen, N.R., Raun, W.R. and Schepers, J.S. (2008) 'Responsive in-season nitrogen management for cereals', Computers and Electronics in Agriculture, 61, pp. 51-62. https://doi.org/10.1016/j.compag.2007.06.006.
25. Fitzgerald, G. J. (2010) 'Characterizing vegetation indices derived from active and passive sensors', International Journal of Remote Sensing, 31:16, 4335-4348. https://doi.org/10.1080/01431160903258217.
26. Tattaris, M., Reynolds, M. P. and Chapman, S. C. (2016) 'A direct comparison of remote sensing approaches for high-throughput phenotyping in plant breeding', Front. Plant Sci. 7:1131. https://doi.org/10.3389/fpls.2016.01131.
27. Adamsen, F.J., Pinter Jr., P.J., Barnes, E.M., LaMorte, R.L., Wall, G.W., Leavitt, S.W. and Kimball, B.A. (1999) 'Measuring wheat senescence with a digital camera', Crop Science, 39, pp. 719-724. https://doi.org/10.2135/cropsci1999.0011183X003900030019x.
28. Helman, D., Bonfil, D. J., and Lensky, I. M. (2019) 'Crop RS-Met: A biophysical evapotranspiration and root-zone soil water content model for crops based on proximal sensing and meteorological data', Agricultural Water Management, 211, 210–219. https://doi.org/10.1016/j.agwat.2018.09.043.
29. Stone, M.L., Solie, J.B., Raun, W.R., Whitney, R.W., Taylor, S.L., and Ringer, J.D., (1996) 'Use of spectral radiance for correcting in-season fertilizer nitrogen deficiencies in winter wheat', Trans. ASAE 39, 1623–1631. https://doi.org/10.13031/2013.27678.

30. Tagarakis A. C. and Ketterings Q. M. (2018) 'Proximal sensor-based algorithm for variable rate nitrogen application in maize in northeast U.S.A.', Computers and Electronics in Agriculture, 145, p. 373-378. https://doi.org/10.1016/j.compag.2017.12.031.
31. Tagarakis, A. C. and Ketterings, Q. M. (2017) 'In-Season Estimation of Corn Yield Potential Using Proximal Sensing', Agronomy Journal, 109, pp. 1323-1330. https://doi.org/10.2134/agronj2016.12.0732.
32. Hatfield, J.L., Gitelson, A.A, Schepers, J.S. and Walthall C.L. (2008) 'Application of Spectral Remote Sensing for Agronomic Decisions', Agronomy Journal, 100, pp. 117–131. https://doi.org/10.2134/agronj2006.0370c.
33. Campbell, J.B. and Wynne, R.H. (2011) 'Introduction to Remote Sensing', Guilford Press, New York, Fifth edition, p. 483. ISBN 978-1-60918-176-5.
34. Xue, J. and Su, B. (2017) 'Significant remote sensing vegetation indices: A review of developments and applications', Journal of Sensors, Volume 2017, Article ID 1353691. https://doi.org/10.1155/2017/1353691.
35. Rouse, J.W., Haas, R.H., Schell, J.A. and Deering, D.W. (1973) 'Monitoring vegetation systems in the great plains with ERTS', Third ERTS Symposium, NASA SP-351, I, pp. 309-317.
36. Gao, B. C. (1996) 'NDWI—A normalized difference water index for remote sensing of vegetation liquid water from space', Remote sensing of environment, 58(3), 257-266. https://doi.org/10.1016/S0034-4257(96)00067-3.
37. Wang, L. and Qu, J. J. (2007) 'NMDI: A normalized multi-band drought index for monitoring soil and vegetation moisture with satellite remote sensing', Geophysical Research Letters, 34(20). org/https://doi.org/10.1029/2007GL031021.
38. Huete, A., Didan, K., Miura, T., Rodriguez, E.P., Gao, X. and Ferreira L.G. (2002) 'Overview of the radiometric and biophysical performance of the MODIS vegetation indices', Remote Sensing of Environment, 83, pp. 195-213. https://doi.org/10.1016/S0034-4257(02)00096-2.
39. Pasqualotto, N., D'Urso, G., Bolognesi, S.F., Belfiore, O.R., Van Wittenberghe, S., Delegido, J., Pezzola, A., Winschel, C. and Moreno, J. (2019) 'Retrieval of Evapotranspiration from Sentinel-2: Comparison of Vegetation Indices, Semi-Empirical Models and SNAP Biophysical Processor Approach', Agronomy, 9, 663. https://doi.org/10.3390/agronomy9100663.
40. Liakos, K., Busato, P., Moshou, D., Pearson, S. and Bochtis, D. (2018) 'Machine Learning in Agriculture: A Review', Sensors, 18, 2674. https://doi.org/10.3390/s18082674.
41. Corti, M., Cavalli, D., Cabassi, G., Gallina, P. M. and Bechini, L. (2018) 'Does remote and proximal optical sensing successfully estimate maize variables? A review', European Journal of Agronomy, 99, pp. 37-50. https://doi.org/10.1016/j.eja.2018.06.008.
42. Bausch, W.C. and Khosla, R. (2010) 'QuickBird satellite versus ground-based multi-spectral data for estimating nitrogen status of irrigated maize', Precision Agriculture, 11, pp. 274-290. https://doi.org/10.1007/s11119-009-9133-1.
43. Caturegli, L., Casucci, M., Lulli, F., Grossi, N., Gaetani, M., Magni, S., Bonari, E. and Volterrani, M. (2015) 'GeoEye-1 satellite versus ground-based multispectral data for estimating nitrogen status of turfgrasses', International Journal of Remote Sensing, 36, pp. 2238–2251. https://doi.org/10.1080/01431161.2015.1035409.
44. Yang, C.-M., Liu, C.-C. and Wang, Y.-W. (2008) 'Using Formosat-2 Satellite Data to Estimate Leaf Area Index of Rice Crop', Journal of Photogrammetry and Remote Sensing, 13, pp. 253–260.
45. Wagner, P. and Hank, K. (2013) 'Suitability of aerial and satellite data for calculation of site-specific nitrogen fertilisation compared to ground based sensor data', Precision Agriculture, 14, pp. 135–150. https://doi.org/10.1007/s11119-012-9278-1.
46. Bu, H., Sharma, L. K., Denton, A. and Franzen, D. W. (2017) 'Comparison of Satellite Imagery and Ground-Based Active Optical Sensors as Yield Predictors in Sugar Beet, Spring Wheat, Corn, and Sunflower', Agronomy Journal, 109, pp. 299–308. https://doi.org/10.2134/agronj2016.03.0150.
47. Jackson, R.D. and Huete, A.R. (1991) 'Interpreting vegetation indices', J. Preventative Vet. Med., 11, pp. 185-200. https://doi.org/10.1016/S0167-5877(05)80004-2.

48. Hrnjak, I., Lukić, T., Gavrilov, M.B., Marković, S.B., Unkašević, M., and Tošić, I. (2014) 'Aridity in Vojvodina, Serbia', Theor. Appl. Climatolgy, 115: 323. https://doi.org/10.1007/s00704-013-0893-1.
49. Gavrilov, M.B., Lazić, L., Pešić, A., Milutinović, M., Marković, D., Stanković, A. and Gavrilov, M.M. (2010) 'Influence of hail suppression on the hail trend in Serbia', Phys. Geogr., 31, 5, 441–454. https://doi.org/10.2747/0272-3646.31.5.441.
50. Radičević, Z., Džingalašević, Lj., Bojović, J., Milakara, S. and Radević, S. (2018) 'Agrometeorološki uslovi u proizvodnoj 2017/2018. godini na teritoriji Republike Srbije', Republic Hydrometeorological Service of Serbia, Division for Applied Climatology and Agrometeorology, QF-E-012, http://www.hidmet.gov.rs/podaci/agro/ciril/AGROveg2018.pdf (accessed January 14, 2020).
51. Bannari, A., Morin, D., Bonn, F. and Huete, A.R. (1995) 'A review of vegetation indices', Remote Sensing Reviews 13, 95–120. https://doi.org/10.1080/02757259509532298.
52. OSRAM Opto Semiconductors (2015) 'Silicon PIN Photodiode Version 1.3.' https://www.osram.com/media/resource/hires/osram-dam-2495839/BPX%2061.pdf (accessed December 17, 2018).
53. Kitić, G., Tagarakis, A., Cselyuszka, N., Panić, M., Birgermajer, S., Sakulski, D. and Matović, J. (2019) 'A new low-cost portable multispectral optical device for precise plant status assessment', Computers and Electronics in Agriculture 162, 300–308. https://doi.org/10.1016/j.compag.2019.04.021.
54. Aschbacher, J. (2017) 'ESA's Earth Observation Strategy and Copernicus' In: Onoda, M. and Young, O. R. (eds) 'Satellite Earth Observations and Their Impact on Society and Policy', Springer Singapore, pp. 81-86. https://doi.org/10.1007/978-981-10-3713-9.
55. Pandžić, M., Mihajlović, D., Pandžić, J., and Pfeifer, N. (2016) 'Assessment of the geometric quality of Sentinel-2 data', International Archives of the Photogrammetry, Remote Sensing & Spatial Information Sciences, 41, 489–494. https://doi.org/10.5194/isprsarchives-XLI-B1-489-2016.
56. European Space Agency (2015) 'Sentinel-2 User Handbook', Issue 1, pp. 1-64. https://sentinel.esa.int/documents/247904/685211/sentinel-2_user_handbook (accessed July 25, 2018)
57. European Space Agency (2018) 'Sentinel Application Platform – SNAP', v6.0.5, http://step.esa.int/main/toolboxes/snap/ (accessed July 25, 2018)
58. QGIS Development Team (2018) 'QGIS Geographic Information System', Open Source Geospatial Foundation Project. http://qgis.osgeo.org (accessed December 28, 2018).
59. Teal, R.K., Tubana, B., Girma, K., Freeman, K.W., Arnall, D.B., Walsh, O. and Ruan, W.R. (2006) 'In-season prediction of corn grain yield potential using normalized difference vegetation index', Agron. J., 98, 1488–1494. https://doi.org/10.2134/agronj2006.0103.
60. Rouse, J.W., Haas, R.H., Schell, J.A., Deering, D.W. and Harlan, J.C. (1974) 'Monitoring the vernal advancement and retrogradation (greenwave effect) of natural vegetation', NASA/GSFC Type III Final Report, p. 371 Greenbelt, Md.
61. Crippen R.E. (1990) 'Calculating the vegetation index faster', Remote Sens. Environ., 34, pp. 71-73. https://doi.org/10.1016/0034-4257(90)90085-Z.
62. Gitelson, A.A., Kaufman, Y.J. and Merzlyak, M.N. (1996) 'Use of a green channel in the remote sensing of global vegetation from EOS-MODIS', Remote Sens. Environ., 58, pp. 289-298. https://doi.org/10.1016/S0034-4257(96)00072-7.
63. Wang, F., Huang, J., Tang, Y. and Wang, X. (2007) 'New vegetation index and its application in estimating leaf area index of rice', Science, 14 (3), pp. 195-203. https://doi.org/10.1016/S1672-6308(07)60027-4.
64. Sentek Systems (2015) 'NDVI DEFINITIONS (RED, BLUE, ENHANCED)' http://www.senteksystems.com/2015/11/23/ndvi-definitions-red-blue-enhanced/ (accessed October 23, 2019)
65. Peñuelas, J. and Filella, I. (1998) 'Visible and near-infrared reflectance techniques for diagnosing plant physiological status', Trends Plant Sci, 3, pp. 151-156. https://doi.org/10.1016/S1360-1385(98)01213-8.

66. Sripada, R. (2005) 'Determining In-Season Nitrogen Requirements for Corn Using Aerial Color-Infrared Photography', Ph.D. dissertation, North Carolina State University. http://www.lib.ncsu.edu/resolver/1840.16/4200 (accessed January 10, 2020)
67. Gitelson, A., Gritz, Y. and Merzlyak, M. (2003) 'Relationships Between Leaf Chlorophyll Content and Spectral Reflectance and Algorithms for Non-Destructive Chlorophyll Assessment in Higher Plant Leaves', Journal of Plant Physiology 160, 271-282. https://doi.org/10.1078/0176-1617-00887.
68. Goel, N. and Qin, W. (1994) 'Influences of Canopy Architecture on Relationships Between Various Vegetation Indices and LAI and Fpar: A Computer Simulation', Remote Sensing Reviews 10, pp. 309-347. https://doi.org/10.1080/02757259409532252.
69. Bannari, A., Asalhi, H. and Teillet, P. (2002) 'Transformed Difference Vegetation Index (TDVI) for Vegetation Cover Mapping', In Proceedings of the Geoscience and Remote Sensing Symposium, IGARSS '02, IEEE International, Volume 5. /https://doi.org/10.1109/igarss.2002.1026867.
70. Gitelson, A.A., (2004) 'Wide Dynamic Range Vegetation Index for Remote Quantification of Biophysical Characteristics of Vegetation', Journal of Plant Physiology 161, 165–173. https://doi.org/10.1078/0176-1617-01176.
71. TIBCO Statistica (2017) 'Statistica – version 13.3, TIBCO Data Science', TIBCO Software Inc., Palo Alto, CA, USA, www.tibco.com.
72. Raun, W. R., Solie, J.B., Martin, K.L., Freeman, K.W., Stone, M. L., Johnson, G.V. and Mullen, R.W. (2005) 'Growth stage, development, and spatial variability in corn evaluated using optical sensor readings', J. Plant Nutr., 28, pp. 173-182. https://doi.org/10.1081/PLN-200042277.
73. Ljubičić, N., Kostić, M., Marko, O., Panic, M., Brdar, S., Lugonja, P., Knežević, M., Minić, V., Ivošević, B., Jevtic, R. and Crnojević, V. (2018) 'Estimation of aboveground biomass and grain yield of winter wheat using NDVI measurements', Proceedings of the IX International Agricultural Symposium "Agrosym 2018", 390-397. ISBN 978-99976-718-8-2.
74. AgroSense (2020) 'Digital Agriculture of Serbia', v3.5.0 https://agrosens.rs/ (accessed January 10, 2020)

Trends in Satellite Sensors and Image Time Series Processing Methods for Crop Phenology Monitoring

Luca Pipia, Santiago Belda, Belen Franch, and Jochem Verrelst

1 Introduction

Agricultural production undergoes increasing pressure from anthropogenically-induced and natural changes, including rising population, conversion of food (cereals) into biofuels, increased protein demands and climatic extremes [1]. Through a fleet of Earth Observation (EO) satellites, National and International space agencies are determined to keep their fingers on the pulse of agricultural land and crop growth [2]. Among the objectives of the multiple EO satellite missions launched in the last five decades, primary importance has been given to observe agricultural and natural vegetation land covers [3–7]. The strong correlation between the response of vegetation in the visible and near-infrared spectrum and its biophysical activities led the preference towards optical sensors for crop growth monitoring [8].

Optical data from EO image time series at high temporal resolution can effectively assist in vegetation monitoring over time as they provide key information about vegetation status over large areas. However, imagery acquired at a high temporal resolution goes traditionally at the expense of a low spatial resolution, and EO missions dedicated to time series studies have long been restricted to the domain of wide swath that achieve global coverage on a near daily basis. For instance, the

L. Pipia
Institut Cartogràfic i Geològic de Catalunya (ICGC), Parc de Montjüic,
Barcelona, Spain

Image Processing Laboratory (IPL), Universitat de València. C/Catedrático José Beltrán,
Paterna, Valencia, Spain

S. Belda · B. Franch · J. Verrelst (✉)
Image Processing Laboratory (IPL), Universitat de València. C/Catedrático José Beltrán,
Paterna, Valencia, Spain
e-mail: jochem.verrelst@uv.es

D. D. Bochtis et al. (eds.), *Information and Communication Technologies for Agriculture—Theme I: Sensors*, Springer Optimization and Its Applications 182,
https://doi.org/10.1007/978-3-030-84144-7_8

Advanced Very High-Resolution Radiometer (AVHRR) was pioneering in time series studies for vegetation monitoring studies at regional to global scales for more than 25 years. AVHRR has been collecting a near-daily global coverage of coarse-to-moderate spatial resolution (1 km and 8 km) providing a consistent time-series of temporally-composited observations [9–11]. As a marked improvement, the Moderate Resolution Imaging Spectrometer (MODIS) has provided, since the early 2000s, an improved times-series of multispectral observations, acquiring a global coverage of multispectral imagery with a high temporal (daily) resolution, a higher spatial resolution (250–500 m) and seven land-related spectral bands for vegetation detection. MODIS data have become increasingly used for vegetation growth monitoring over large geographic regions [12, 13]. Yet, probably the most noteworthy pioneering mission for land applications is the Landsat series of satellite-based sensors. Landsat has long been appropriate for many landscape characterization applications such as land cover classification, change detection and vegetation monitoring. It has a nominal 16-day temporal resolution and up to 30 m spatial resolution, with a data archive extending from the early 1970s to present. However, the usage of Landsat time series for crop growth monitoring has limitations because vegetation changes may occur more rapidly than the 16-day revisit time of Landsat. In addition, cloud cover contamination of the optical satellite observations further reduces the number of Landsat images available to adequately detect many seasonal events [8].

These pioneering monitoring missions paved the path for a diversity of dedicated EO land missions initiated by National and International space agencies with emphasis in exploiting the spatial, spectral, or temporal domain. With current and upcoming EO satellite missions, an ever-increasing amount of optical EO satellites are orbiting around the Earth, such as the Sentinel constellations on behalf of the joint ESA/European Commission initiative Copernicus and the NASA A-Train satellite constellations. With the operational super-spectral Copernicus' Sentinel-2 (S2) [14] and Sentinel-3 missions [15], as well as the recently launched and upcoming imaging spectrometer missions [16–19], an unprecedented data stream for vegetation mapping and monitoring becomes available. For instance, the unprecedented frequency of S2 multispectral observations (every five days) with a spatial resolution of 20 m (up to 10 m for specific bands) captures rapid changes of agricultural land-cover from national to field scale, serving as a major support for environmental monitoring and agricultural subsidy control [14]. Hence, S2 time series allows for high-resolution coverage of large areas with systematic data acquisition with high-frequency sampling during critical phases of the crop growth cycle [20]. The Sentinel-3 satellites even enable a short revisit time of less than two days for the optical sensor OLCI (Ocean and Land Colour Instrument), but it is a medium-resolution imaging spectrometer as it provides a spatial resolution of 300 m [15], and thus is less suited for crop monitoring at field scale.

Having an unprecedented influx of optical time series data at disposal, an essential condition for using image data for further processing is that it requires to be spatially and temporally continuous, i.e., gap-free data. Unfortunately, in reality this need is often unfulfilled, due to multiple causes: (1) inadequate climatic conditions (clouds, snow, dust and aerosols), (2) instrumentation errors, (3) losses of data during data transmission or (4) low temporal resolution (i.e., long time needed to revisit and acquire data for the exact same location), among others. The causes above degrade the availability of spatial and temporal information required to retrieve land surface properties. Therefore, the impact of missing data on quantitative research can be serious, leading to biased estimates of parameters, loss of information, decreased statistical power, increased standard errors, and weakened findings [21]. For this reason, spatiotemporal reconstruction of gapped areas from satellite imagery is becoming crucial for monitoring purposes [22], including the knowledge of the life cycle of vegetation, i.e., vegetation phenology [23].

Another important remark is that, from an EO perspective, specific plant seasonal events such as budbreak, leaf out, land leaf senescence, flowering and maturity of cereal crops cannot be directly detected at the spatial resolution of satellite imagery. Instead, more general descriptors of vegetation dynamics termed 'land surface phenology (LSP)' are calculated [8]. LSP refers to the seasonal pattern of variation in vegetated land surfaces observed from remote sensing [24]. This is distinct from observations of individual plants or species, as space-based observations aggregate information on the timing of heterogeneous vegetation development over pixel-sized areas. This aggregation often disassociates the response signal of the landscape from that of the individual species; yet is important for representing landscape scale processes in biosphere atmosphere interaction and crop monitoring models [24]. LSP metrics are typically associated with general inter-annual vegetation changes interpretable from spectral remote sensing imagery such as start of greening/season (SOS), the peak of growing season, onset of senescence or end of the season (EOS), and growing season length [24, 25], as well as other transition stages (e.g., maturity and senescence) [5]. Therefore, this chapter provides an overview of the possibilities for calculation of these LSP metrics from time series images for crop monitoring purposes.

Altogether, when aiming to process time series data for calculation of LSP metrics and agricultural monitoring purposes, a critical aspect to deal with is that EO data is spatially and temporally discontinuous. This implies that the ability to process irregular time series becomes indispensable for studying seasonal vegetation patterns. In this respect, this chapter aims to provide a general overview on agricultural land monitoring by means of EO image time series analysis and subsequent LSP calculation. To do so, first an historical overview of EO satellites with optical sensors that are designed to monitor the phenology of agricultural lands is given. Second, solutions are offered on how to gap-fill time series image data and then to calculate LSP metrics. Third, the calculation of LSP metrics from MODIS and 1 km aggregated S2 data is presented for two demonstration areas characterized by different dominant crop: corn and winter wheat. Finally, trends in EO missions and image time series processing are being discussed in the broader context of monitoring croplands' phenology.

2 Satellite Sensors for Crop Phenology Monitoring

Although in the current era of EO missions time series processing has become standard practice in agriculture monitoring, it only recently reached maturity. Almost half a century was dedicated to overcoming challenges related to EO technology and optimizing for ideal temporal and spatial resolution. In this respect, this section intends to give a brief historical overview about EO satellite missions for agricultural monitoring purposes. Afterwards, time series data from NASA and ESA flagship missions for land applications are used for presenting crop monitoring demonstration cases.

When EO satellites were first available in the 60s, it was recognized that the technology held considerable promise for agricultural monitoring [26]. NASA was pioneering with EO programs for agricultural monitoring purposes. Initial efforts involved the NASA LACIE and AgriSTARS programs in the 70s. They made significant advances in crop monitoring but were seriously constrained by satellite data availability. At the beginning of EO missions, satellite optical data have been primarily provided globally at coarse-resolution (c. 250 m–8 km) by systems specifically developed for land applications. This is especially true for the AVHRR sensors, launched back in the 80s. AVHRRs provided daily global observations, which represent one of the most critical features needed for agriculture monitoring, but they were limited by their low spatial resolution (1 km). It has long been recognized that when working on agriculture applications, a good temporal resolution is required, given that the crop phenology and conditions (e.g., water supply, pests, environmental) can change very quickly. To this end, the NASA Long Term Data Record (LTDR) contains gridded daily surface reflectance and brightness temperatures derived from processing of the data acquired by the AVHRR sensors onboard four NOAA polar-orbiting satellites: NOAA-7, -9, -11 and -14. The Version 4 contains improvements to geolocation, cloud masking and calibration, making the data record suitable for crop monitoring [27]. This product is still operational, and its usefulness has been demonstrated for a wide variety of applications such as snow cover estimation [28], agricultural modeling [27], Leaf Area Index (LAI) and Fraction of Absorbed Photosynthetically Active Radiation (FAPAR) retrieval [29, 30], global vegetation monitoring [31, 32], burned area mapping [33] and albedo estimation products [34].

A next milestone involved the MODIS sensor on the Terra satellite. Since its launch in 2000, observations from the EOS/MODIS sensors have several of the key qualities needed for global agriculture monitoring such as global, daily coverage at coarse spatial resolution (250 m) and a suite of validated products. With MODIS onboard Terra (morning satellite) and Aqua (afternoon satellite) getting to the end of its operational life, it was high time to transition into new satellites. The Visible Infrared Imaging Radiometer Suite (VIIRS) onboard the Suomi National Polar-orbiting Partnership (S-NPP) satellite provided continuity with MODIS from 2012 [35, 36]. It overpasses once a day and during the afternoon, which decreases the chance of getting cloud-free observations, especially in the tropical regions.

However, the combination with the ESA Sentinel-3 satellite, that overpasses during the morning and has similar characteristics to MODIS, provides continuity to the successful and still valuable (due to its high temporal resolution) coarse resolution missions. Despite the advantage of the high revisit time, the main disadvantage of coarse to moderate resolution sensors is the spatial resolution that often mixes, in a given pixel, signals from different land cover types and crops. Stratifying a region into different crop types (commonly termed as crop masking) is an important step in developing EO-based agriculture models [37]. Such masks enable the isolation of the remotely sensed, crop-specific signal throughout the growing season, reducing the noise on the signal from other land cover or crop types [38]. In the United States (US), the US Department of Agriculture (USDA) generates a yearly national Cropland Data Layer (CDL) since 2007 [39] and Canada provides yearly national Annual Crop Inventory Maps (ACIM) since 2009 [40]. These masks are provided at the end of the growing season and no crop type masks are available for other countries. Therefore, generally EO-based agriculture models use static cropland or crop type masks.

The flagship moderate resolution NASA mission Landsat, with data going back to the 70s, was long used for agriculture monitoring, but with limitations mainly due to its low temporal resolution of 16 days. Coupled with the frequency of cloud cover, the revisit time for some regions is often worse. A number of studies have fused Landsat with MODIS data [41–43], and combined Landsat data with biophysical models [44, 45], leading to varying results in terms of errors as they are still constrained by the low temporal frequency of Landsat imagery. The launch of the ESA optical moderate resolution missions Sentinel-2A in 2015 and Sentinel-2B in 2017 have been revolutionary for the moderate agriculture monitoring. The increased temporal coverage and the new technologies offered by the Sentinel systems and their combination with NASA sensors, provides new opportunities for high temporal frequency moderate resolution remote sensing, enabling a new generation of agriculture products to be generated. Specifically, with the Sentinel-2A and -2B fusion with Landsat, it is now possible to achieve a temporal resolution of three to five days globally. In fact, recent studies leverage the combination of these satellites to address crop yield assessment at field scale [46–48]. Yet, simply having synergistic sensors on orbit is not sufficient for end users; the data products themselves must also be processed in such a way as to ease preprocessing and analysis burden. The Harmonized Landsat/Sentinel-2 (HLS) project [49] developed by NASA provides a surface reflectance product that combines observations from USGS/NASA's Landsat-8 (LS8) and ESA's Sentinel-2 (S2) satellites at moderate spatial resolution (30 m). The main goal is to provide a unique dataset based on both satellites' data to improve the revisit time to three to five days depending on the latitude. Along with a common atmospheric correction algorithm [50], geometric resampling to 30 m spatial resolution and geographic registration [49], the product is also corrected for Bidirectional Reflectance Distribution Function (BRDF) effects and band pass adjustment. Besides, the Sen2like tool [51] developed by ESA will provide analysis ready Harmonized LS8 and S2 data/products to the user. Using the S2 tiling system, the sen2like tool processes S2 Level-1 products and LS8 Level-1 products and create a harmonized surface reflectance data stack at 10 m spatial resolution. Working on the same baseline principles as NASA HLS initiative, geometric, radiometric and image processing algorithms are applied. Recent studies took advantage of Landsat and S2 data to address crop yield assessment at a moderate spatial resolution [52, 53].

Recent advances in data acquisition and processing (e.g., cloud computing) are making possible the development of global high-to-moderate resolution data sets (10–30 m). Such global time series data will permit improved mapping of crop type, crop area and vegetation properties essential for regional implementation of monitoring strategies. Higher temporal frequency from multiple high-to-moderate resolution satellites will also provide a better characterization of agronomic growth stages, with the consequent improvement of crop production modeling accuracy.

3 Time Series Processing for Crop Seasonality Monitoring

3.1 Gap-Filling

An essential step for being able to use EO data for further processing such as LSP calculation, is converting raw data time series into spatio-temporal continuous datasets. To ensure this, gaps mostly provoked by clouds must be filled. Time series gap filling essentially refers to the prediction of missing values in time. Mostly, these missing values are located within the dataset time series, so in principle interpolation methods to fill them up would suffice. It is therefore no surprise that interpolations and fitting methods are commonly used as a first step in the time series processing. According to the recent review by [8], gap filling methods can be categorized into: (1) smoothing and empirical methods, (2) data transformations, and (3) curve fitting methods. From these three categories, the curve fitting methods are the most commonly used, with double logistic curves being a popular method for seasonality estimation [54–56]. This family of methods has expanded rapidly in the last few years with the emergence of adaptive machine learning regression algorithms [57]. See also [57, 58] for a quantitative evaluation of these methods. Some machine learning methods proved to be particularly attractive; not only because of achieving higher accuracies when validated against a reference image, but also because of additional properties such as delivering uncertainty estimates (e.g. Gaussian processes regression: GPR). Most of these methods have been recently implemented into an in-house developed graphical user interface (GUI) toolbox, named DATimeS (Decomposition and Analysis of Time Series software) [57]. DATimeS has been developed to generate cloud-free composite maps from regular or irregular satellite time series. The novelty of the toolbox lies in expanding established time series gap-filling methods with a diversity of advanced machine learning fitting algorithms. An overview of the gap-filling methods is provided in Table 1.

Here, a brief description of the toolbox is provided, as it will be used in subsequent calculation of phenology indicators. In short, DATimeS is developed as a

Table 1 Interpolation methods used for gap-filling

	Gap-filling methods
Smoothing and Interpolation Methods	Linear, Polynomial, Nearest, Next, Previous, Pchip, Spline
Data Transformation	Offset + Harmonic analysis Offset + Harmonic analysis + Linear Term Offset + Harmonic analysis + Linear Term Offset + Harmonic Analysis using Sliding Window
Fitting methods (e.g., machine learning)	Bagging trees (BAGTREE) Adaptive Regression Splines (ARES) Boosting trees (BOOST) k-nearest neighbors regression (KNNR) Gaussian Process Regression (GPR) Kernel Ridge Regression (KRR) Locally-Weighted Polynomials (LWP) Support Vector Regression (SVR) Neural networks (NNIPL) Random forests (RF2) Boosting random trees (RF1) Structured Kernel Ridge Regression with linear Kernel (SKRRlin) Relevance Vector Machine (RVM) Sparse Spectrum Gaussian Process Regression (SSGPR) Structured Kernel Ridge Regression with RBF kernel (SKRRrbf) Decision trees (TREE) Variational Heteroscedastic Gaussian Process Regression (VHGPR) Double Logistic curve Whittaker

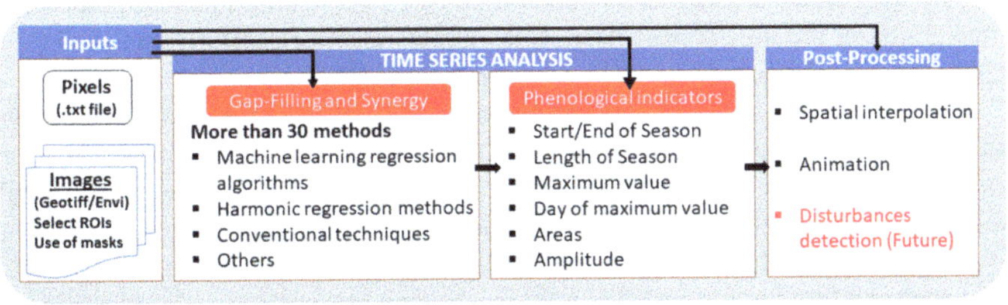

Fig. 1 Hierarchical design of DATimeS

modular toolbox that can be applied to both, set of images and discrete time series data stored in a text file. An overview of the DATimeS' modules contained in this first version (v.1.06) is shown in Fig. 1. The core machinery of DATimeS is the "Time Series Analysis" module, where the gap-filling methods can be selected, and subsequent phenology indicators can be computed. The user may choose whether to incorporate the smoothing function prior to the parameter estimation. Although a prior smoothing step may help in finding general patterns, it must be remarked that most fitting methods perform a time series anyhow, smoothing along with the fitting prediction. Before starting the gap-filling procedure, a compulsory step is to define the output time settings, i.e., the days to which cloud-free interpolated images will be generated (e.g., every 10 days). These composite products from operational land missions (e.g., AVHRR, MODIS, SPOT Vegetation (VGT)) are commonly used for subsequent LSP calculation [25, 59–61].

3.2 LSP Calculation

A next step involves the calculation of the phenology indicators from the prepared cloud-free time series data, i.e., the LSP metrics. Numerous studies have dealt with the retrieval of phenological phases from remotely sensed data [55, 62–65]. LSP metrics quantification over croplands is widely used for yield estimation, or to improve management and timing of field works (planting, fertilizing, irrigating, crop protection or harvesting) [66, 67]. Distinct LSP metrics may be of interest to the scientific community, private companies and farmers, such as dates of start and end of the growing season (SOS and EOS, respectively), maximum peak, seasonal amplitude defined between the base level and the maximum value for each individual season, length of the season, etc. [38, 68]. These LSP metrics are extremely sensitive to changes in vegetation cycles related to multiple factors such as climate anomalies or extreme weather events, which can have a profound impact in the agricultural production [69–71]. Hence, estimating LSP metrics is a convenient way to summarize seasonal information in a few comprehensive quantitative descriptors. However, it must be taken into account that these metrics are sensitive to the processing data characteristics or methods used (e.g., gap-filling method, pixel size, time period of the time series). Therefore, outputs must be carefully analyzed (see also review in [8]), as will be further demonstrated in the case study.

In practice, LSP metrics are recommended to be derived after the interpolation step so that cloud-free composite images are created, and trends become evident for easy phenological metrics derivation. For this reason, DATimeS recommends LSP estimation as the next logical processing module after the gap-filling module, even if going directly to this step is also possible. In this module, the whole time series is first analyzed looking for possible multiple growing seasons.

Then, each individual season is processed separately to estimate the phenological indicators (e.g., SOS and EOS) based on conventional threshold methods, analogous to [68, 72–75]. The computational routine for LSP calculation follows multiple steps. It runs pixelwise, and for each pixel it: (1) extracts the time series, (2) identifies automatically individual growing seasons within each year, (3) locates specific points within the growing season (e.g., SOS, EOS, peak), (4) computes seasonal integrals (area under the curve between SOS and EOS) and (5) stores the estimates in output ENVI or Tiff files. Three alternative methods have been implemented to calculate the SOS/EOS: (1) seasonal, (2) relative and (3) absolute amplitude. In the former case, the SOS/EOS are identified where the left/right part of the curve reaches a fraction of the seasonal maximum amplitude along the rising/decaying part of the curve. The second approach is similar to the previous one, but now a mean amplitude is estimated considering the minimum/maximum values of all seasons. Consequently, the SOS/EOS correspond to dates where the curve reaches a specific percentage of the reference amplitude. In the latter method, the SOS/EOS are determined when each growing season reaches the same fixed value.

4 Demonstration Cases Time Series Processing

Having outlined the main principles of (1) EO missions dedicated to crop monitoring, (2) gap-filling methods, and (3) LSP calculation, this section provides some time series demonstration cases with temporal data coming from the currently most successful optical missions at low and high spatial resolution, i.e. MODIS and S2 acquisitions. The study focuses on the trade-off between revisit time and spatial resolution of each sensor and is carried out over two agricultural landscapes of the US, each one characterized by the presence of a different dominant crop type with specific phenological dynamics: winter wheat and corn.

4.1 Study Area and Data Acquisition

The US is one of the main producers and exporters of corn and wheat globally. In 2016 the US was the leading wheat exporting country, shipping 14.8% of global wheat exports[1]. Wheat is produced in almost every state in the United States and winter wheat varieties dominate US production, representing between 70% and 80% of the total wheat production. The winter wheat is planted in the fall and harvested during June-July. Generally, wheat is rain-fed and just 7% of the national production is irrigated. The main wheat class is Hard Red Winter Wheat, which is grown primarily in the Great Plains, with Kansas being the largest producing state. Besides, the US is a major player in the world corn trade market, with between 10% and 20% of its corn crop exported to other countries. Corn is grown in most U.S. States, but production is concentrated in the Heartland region (including Illinois, Iowa, Indiana, eastern portions of South Dakota and Nebraska, western Kentucky and Ohio, and the northern two-thirds of Missouri). Iowa and Illinois, the top corn-producing States, typically account for about one-third of the U.S. crop. The corn is planted during April-June and is harvested during September-November.

With the aim of performing a fair comparison of multispectral spatiotemporal information carried by high- and low spatial resolution multispectral imageries, i.e. S2 and MODIS, over corn and winter wheat, the Crop Data Layer (CDL) yearly

[1] http://www.worldstopexports.com/wheat-exports-country/

produced by the National Agricultural Statistics Service (NASS) of the US Department of Agriculture (USDA) was analyzed, and selected two S2 tiles representative of each crop type were selected.

4.1.1 Crop Data Layer

The Crop Data Layer (CDL) is distributed by NASS since 2008 at 30 m as part of the official archive of county-level statistics on yield, area harvested and production that are available from the USDA National Agricultural Statistics Service (NASS) Quick Stats database[2]. It is a rasterized land cover map using field level training data from extensive ground surveys, farmer reports provided to the US Farm Service Agency (FSA), and remotely sensed data from Landsat Thematic Mapper (TM), Landsat Enhanced Thematic Mapper (ETM+) and Advanced Wide Field Sensor(AWiFS). These data are used in a decision tree classifier in order to produce a land cover classification that distinguishes between different crop types, including winter wheat [39, 76].

4.1.2 MODIS and Sentinel-2 Surface Reflectance Time-Series

The time span chosen for the study was the year 2019 due to the availability in Google Earth Engine (GEE) [77] of S2 surface reflectance images over the US from December 2018 on. The first tile chosen is 11TLM, which is located in North West of US, in the South of Washington and contains mainly winter wheat cultivated areas. The second one is 15TVH, centered in North Iowa, with essentially corn and soybean crops. S2 data were downloaded from GEE in UTM projective coordinates. Limited by the spatial resolution of CDL, S2 information distributed at 20 m were gathered. Aside from the crop-of-interest spatial density, a second criterion for the selection of the two areas was their medium frequency level of cloudiness estimated by analyzing MODIS daily cloud mask. This way, the main advantages and drawbacks of the shorter revisit time of coarse resolution MODIS imagery against the longer revisit time of high resolution S2 acquisitions can be assessed. Details about the spatial properties of the test sites are summarized in Fig. 2. The nine classes in the legend correspond to the most frequent classes within the two test-sites, among the 134 provided by USDA [78]. Tile 15TVH is essentially made up of two main classes, corn and soybeans; tile 11TLM presents a more heterogeneous scenario, with winter wheat being the dominant crop class after pasture. The landscape is further characterized by shrubland, but crops as spring wheat, alfalfa and potatoes are also cultivated. Grey and blue colors indicate urban and water areas, respectively.

The analysis of MODIS time series was based on MODIS daily surface reflectance Collection 6 data (MOYD09GQ) distributed by the Land Processes Distributed

[2] http://www.nass.usda.gov/Quick_Stats/

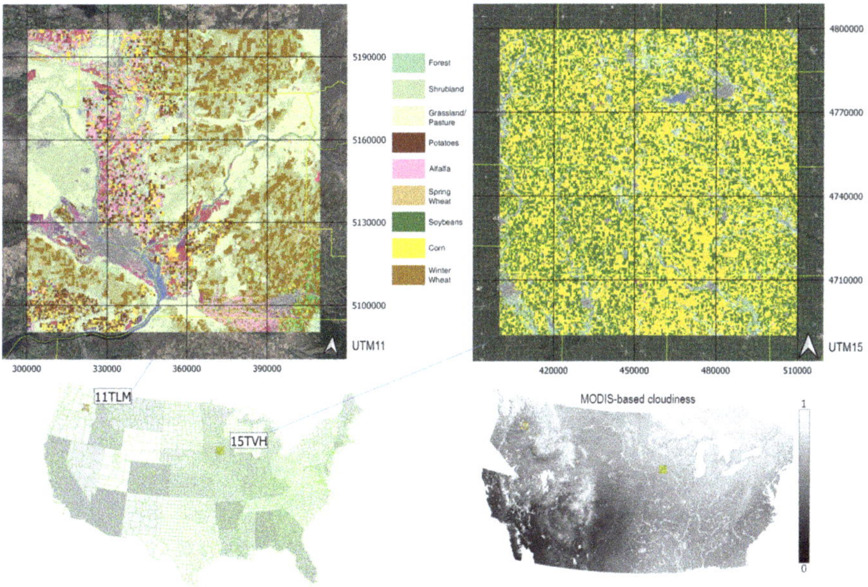

Fig. 2 USDA land cover map of 2019 over S2 footprint @30m for tiles 11TLM (left) of Washington and 15TVH (right) of Iowa. The legend details the main classes within the two tiles, among the 134 defined by USDA. Green lines and greyish areas define US Counties and States limits, respectively. The cloudiness map along 2019 was estimated as percentage of per pixel MODIS cloudy acquisitions

Active Archive Center[3] (LPDAAC), which are gridded in the sinusoidal projection at 250m resolution. Additionally, the product MOYD0 was used 9GA to extract the geometry of observation illumination of each image. Since the nominal 250 m MODIS resolution decreases for the off-nadir observations and due to inaccurate registration [79], the 250 m surface reflectance was re-scaled to 1 km spatial resolution to mitigate that effect by aggregating 4 × 4 pixels. The wide swath MODIS sensor allows for near global coverage of the Earth every day. However, it has a 16-day repeat cycle, which means that every day the geometry of observation is different and can include View Zenith Angles (v) of up to 65 degrees. As a consequence, the surface reflectance that is defined for a given geometry of observation-illumination has different values every day. In order to normalize the BRDF effects on the surface reflectance, we used the VJB method [80, 81]. This method uses longer compositing periods (five years in [80]), than the MCD43 product (16 days) [82], which reduces the noise in the normalized reflectance time series [83]. In this study, the nadir BRDF parameters at 1 km spatial resolution using the most recent five years (2012–2016) were derived.

[3] https://lpdaac.usgs.gov/dataset_discovery/modis/modis_products_table

Table 2 MODIS Terra/Aqua and S2A/B imagery information over the two test-sites

Sensor	Start date	End date	Number of images	Full coverage orbit	Partially coverage orbit	Spatial resolution
MODIS T/A	1/1/2018	31/12/2019	730	–	–	1 km
S2A/B 11TLM	13/12/2018	10/01/2020	244	113	70/13	20 m
S2A/B 15TVH	13/12/2018	10/01/2020	166	69	112	20 m

By using the daily surface reflectance (from both Aqua and Terra) and its angular conditions during the five-year period considered, the variables that define the BRDF shape (V and R in Equation (1)) are derived using the approach proposed by [81]. The MODIS dataset consists of daily acquisitions covering the whole US territory during 2018 and 2019. For S2, a total amount of 237 and 158 partially cloudy or cloud-free images was collected for 11TLM and 15TVH, respectively. Despite the nominal revisit time of S2 being five days, 11TLM is fully covered by orbit 113 and partially covered by orbits 70 and 13, whereas 15TVH is fully covered by orbit 69 and partially covered by only orbit 112. This explains the different number S2 images. Details about the dataset are reported in Table 2.

4.2 *Time Series Processing Over Croplands*

As pointed out in Sect. 2, the main disadvantage of coarse to moderate resolution sensors is the spectral mixing from different land cover types and crops. In order to characterize the degree of homogeneity of coarse resolution pixels, the higher spatial resolution information provided by CDL map and S2 imagery can be exploited. First, a common coordinates' reference must be defined to allow establishing a pixel-to-pixel correspondence among the different information sources. MODIS data were cropped over S2 tiles 11TLM and 15TVH, projected to their corresponding UTM reference at 20 m using the nearest-neighbor interpolation, and finally aggregated at 1 km. Similarly, CDL maps were projected onto S2 UTM reference at 20 m.

The interpolated CDL was then used to calculate the percentage of each land cover class within each MODIS 1 km pixel. A qualitative description of the homogeneity of MODIS pixels is given in the 1 km land cover maps shown in images (a) and (b) of Fig. 3. For their generation, the 3 most likely classes at pixel level were taken into account. Denoting them ordered by probability as Cl_1, Cl_2 and Cl_3, the RGB composite was obtained by weighing the color coding of the three classes with the corresponding percentages. The visual comparison of the land cover maps at 20 m (Fig. 2) and 1 km indicates that a dominant class can be still identified at MODIS scale. Yet, the less saturated colors point out the presence of a non-negligible class mixing. The higher the mixing, the more relevant the difference between

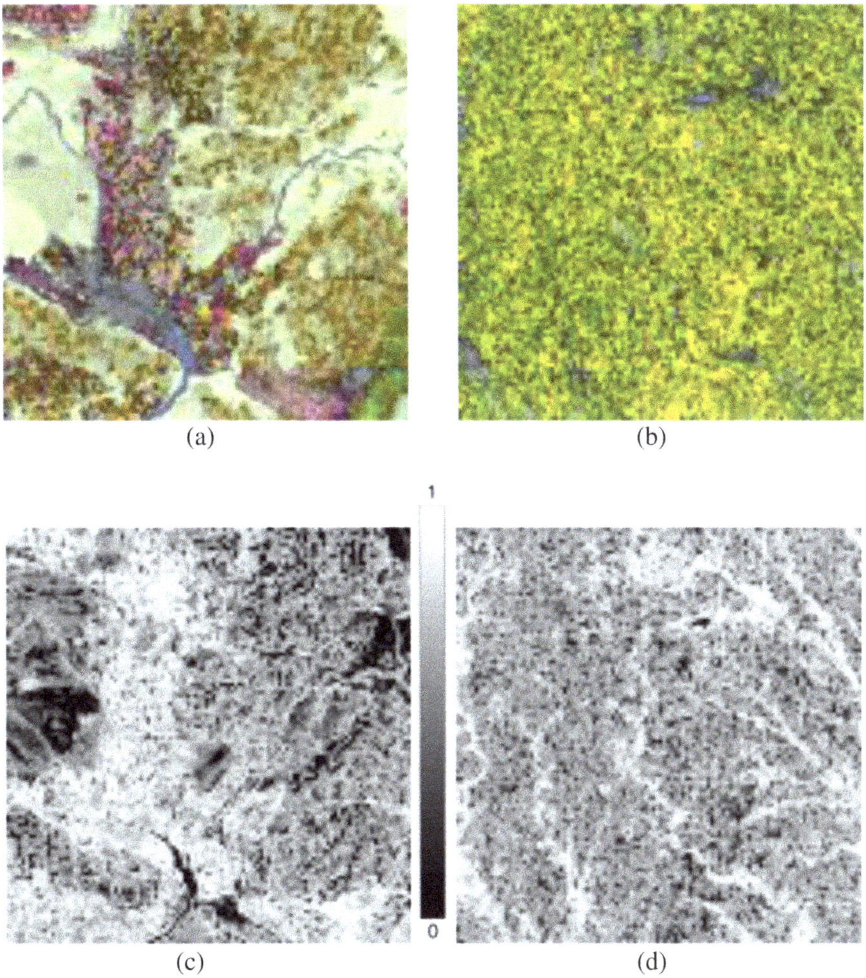

Fig. 3 Synthetic Land cover map @1km of 11TLM (**a**) and 15TVH (**b**) tiles based on USDA land cover product @30m weighted by the probability of classes Cl_1, Cl_2 and Cl_3 within MODIS pixels. In (**c**) and (**d**) the corresponding GSI maps are showcased

MODIS and S2 spectra is. A quantitative estimation of this mixing effect is provided by the Gini-Simpson Index (GSI) [84, 85]. The GSI essentially quantifies how many different types of classes the pixel of interest contains, and is computed as the complement of the sum of squared N-member fractions of classes:

$$GSI(x,y) = \sum_{i=1}^{N_c} 1 - p_i(x,y)^2 \tag{1}$$

where (x, y) denotes the coordinates of MODIS pixel's center, N_c is the total number of CDL classes within the pixel and p_i is the fraction of the area covered by the ith

class. The result obtained for the area corresponding to the two S2 tiles are shown in Fig. 3c, d. The closer GSI to zero, the purer the pixel is, i.e. a dominant class characterizes the pixel. Conversely, a higher GSI denotes a heterogeneous 1 km pixel where multiple classes are present with comparable percentages. Over the latter ones, the interpretation of coarse resolution imagery deserves special attention, as the information they contain cannot be transferred directly to individual classes. To clarify this concept, we use the Difference Vegetation Index (DVI).

DVI is a non-normalized parameter simply defined as the difference between the near-infrared and the red bands, with the main advantage to describe the evolution in time of crop phenology avoiding saturation effects often detected with other normalized indexes such as NDVI [86]. Accordingly, five DVI time series at 1 km from the two imagery sources were generated. The first two ones are the MODIS DVI from the BDRF-corrected MODIS, and the S2 DVI obtained by simply upscaling the S2 product to 1 km. Besides, for each 1 km pixel the three classes with the highest probability were selected and the S2 DVI value of 20 m pixels belonging to each of them separately was averaged. The corresponding DVI at 1 km for pixel (x,y) at time t was hence obtained as follows:

$$DVI_{CLk}\left(x,y,t\right)=\frac{1}{N_{pClk}}\sum_{i=1}^{N_{pClk}} DVI_{S2_{20m}}\left(x_i,y_i,t\right) \tag{2}$$

where N_{pClk} indicates the number of 20 m pixels (x_i, y_i) within the 1 km pixel centered in (x, y) and belonging to the class C_{lk}, with $k = 1, 2, 3$. Examples of the five-time series obtained for almost pure and heterogeneous pixels of corn and winter wheat are shown in Figs. 4 and 5, respectively.

In general, an analogous temporal evolution of DVI from MODIS (blue triangle) and upscaled S2 (magenta circle) images for the two crop types can be observed on both homogeneous and heterogeneous pixels, confirming both the effectiveness of the BDRF correction and the accuracy of the datasets spatial alignment. In terms of time sampling, the lower sampling rate of S2 does not seem to affect the reconstruction of the overall shape of vegetation dynamics significantly. Yet, quantitative assessments of phenology descriptors are required to estimate the real effect on vegetation characterization. As expected, over pixels characterized by GSI close to zero the coarse resolution imagery mimics faithfully the evolution of the dominant classes (blue asterisks). There, the 1 km information can be used directly to infer crop properties, being spurious contributions from the rest of classes negligible. On the contrary, pixels characterized by higher degrees of heterogeneity are not able to provide a direct description of the crop-type of interest, being the information drifted apart from the pure time series as far as it becomes less dominant within the pixels. For them, unmixing approaches are mandatory if reliable vegetation evolution is to be retrieved, and if only coarse resolution imagery is available the solution comes with accepting an additional loss of spatial details. A successful solution has been put forward in [87], where spectral unmixing is carried out with an Ordinary Least Square method at US County level and provides a unique crop-type time

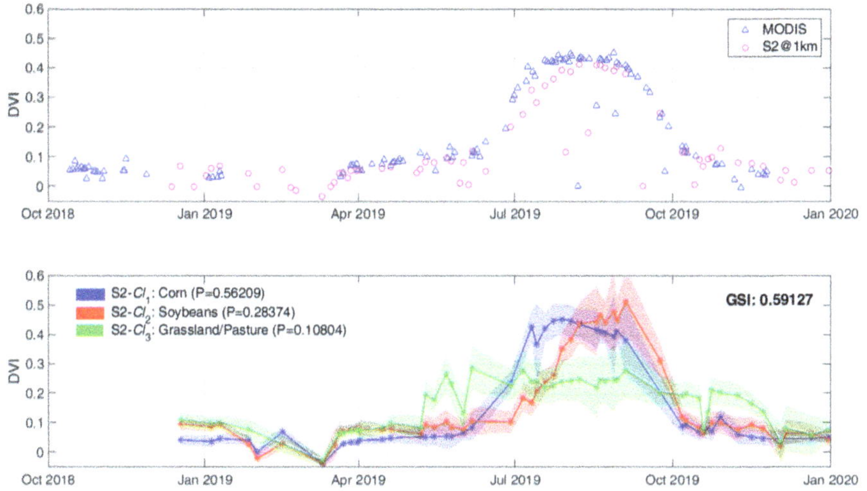

(a)

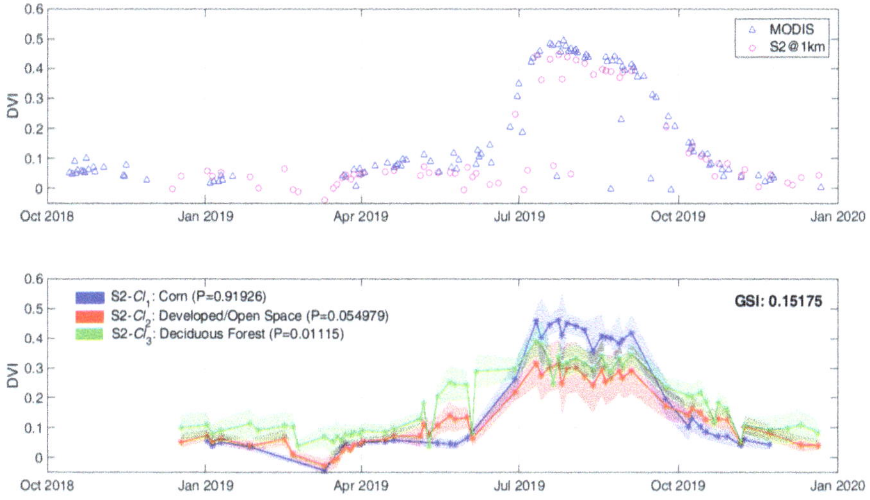

(b)

Fig. 4 DVI Time series over almost pure corn pixel (**a**), more abundant but not dominant corn pixel (**b**, **c**) @1km from MODIS, S2 and S2-based pure classes (mean value ± 1 standard deviation with the 1 km pixel)

series at US county level. Overall, this hypothesis is fully satisfied for irrigation crops such as corn, and the county-level characterization is also representative of crop behaviors at 1 km. This can be observed in the normalized 2D histograms of DVI time series at 1 km for the tile 15TVH, shown in Fig. 6.

The corn region time series at 1 km was obtained by averaging at 1 km scale only S2 pixels labeled as corn in the USDA land cover map. A minimum crop-type

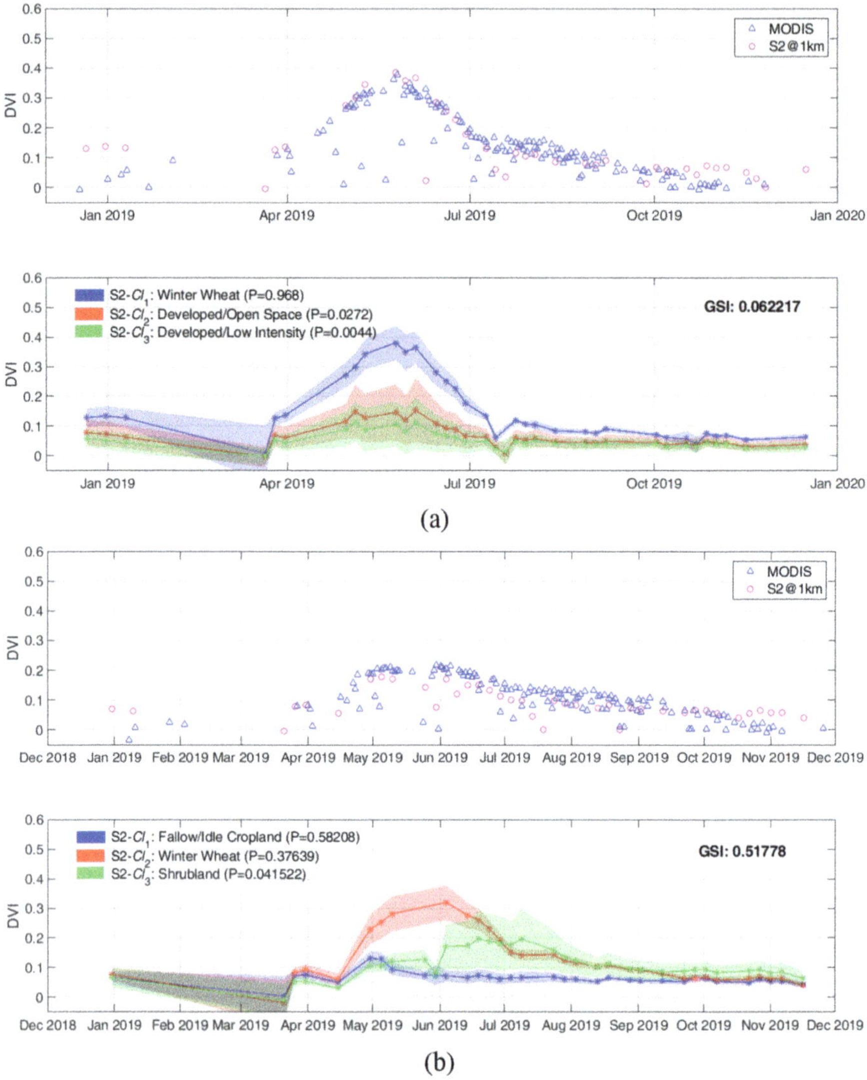

Fig. 5 DVI Time series over almost pure winter wheat pixel (**a**), more abundant but not dominant Winter Wheat pixel (**b**) @1km from MODIS, S2 and S2-based pure classes (mean value ± 1 standard deviation with the 1km pixel)

percentage threshold of 20% was also applied to filter out noisy information. The results over the four counties entirely covered by the tile 15TVH of Iowa (Hancock, Cerro Gordo, Franklin, and Wright) show that minimum differences are detectable in the temporal evolution of the DVI, being the time sample dispersion slightly higher just during the start and end of season. Because of the corn dominance, smooth temporal profiles with a clear phenology can be detected.

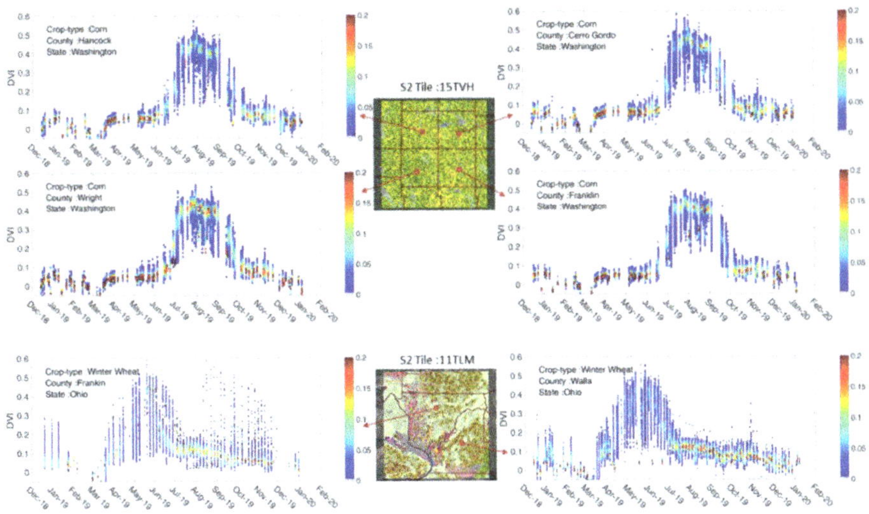

Fig. 6 Normalized 2D histograms of corn (tile: 15TVH) and winter wheat (tile: 11TLM) DVI time series at 1 km grouped by County

Conversely, when applying the same analysis to winter wheat fields of tile 11TLM, a significant spreading of time series during the whole evolution of the crop in the two counties of Washington (Frankin and Walla Walla) can be observed. Whereas the bare soil period before seeding and after harvest are stable overall the tile, the magnitude of the phenological evolution of this crop type turns out to be dependent on the specific 1 km pixel selected for the analysis. The larger spread suggests a more heterogeneous land cover with variations in phenology due to different crop types and natural vegetation. These two contrasting land covers show the case for an in depth systematic and quantitative analysis, i.e., as done by the LSP calculation.

4.3 LSP Calculation Over Croplands

The two test cases presented in the section above (e.g. see Fig. 6) have been processed by DATimeS in order to estimate the LSP metrics. To do so, first gaps due to cloud cover were filled by means of a machine learning (ML) fitting method over the temporal data. The ML algorithm Gaussian processes regression (GPR) was chosen because of excellent fitting performances (see [57, 58] for a quantitative analysis of over 20 gap-filling algorithms). As such, cloud-free DVI maps were reconstructed on a five-days basis for the year 2019. Subsequently, the LSP metrics can be reliably calculated.

The LSP metrics were calculated for the following three time series products:

Table 3 Mean and standard deviation (SD) of phenological indicators estimated from S2, S2Cl_1 and MODIS over 1km-pixel whose Cl_1 corresponded to corn (tile 15TVH). SOS, EOS and Day MV are in DOY 2019

	SOS	EOS	LOS	MV	Amp.	Day MV	Area
S2Cl_1	168.7±8.0	278.0±8.5	110.5±11.6	0.3±0.1	0.3±0.1	222.0±8.6	25.6± 4.3
S2	167.8±10.0	278.7±8.2	109.8±13.1	0.4±0.1	0.4±0.1	224.2±8.9	28.5±4.1
MODIS	164.9±11.4	279.7±5.2	113.8±12.7	0.4±0.1	0.4±0.1	225.1±6.1	31.1± 3.3

Table 4 Mean absolute deviation (MAD) and standard deviation (MSD) of MAD estimated from S2, S2Cl_1 and MODIS over 1km-pixel whose Cl_1 corresponded to corn (tile 15TVH). SOS, EOS and Day MV are in DOY 2019

	SOS	EOS	LOS	MV	Amp.	Day MV	Area
S2Cl_1 vs. S2	5.3±6.8	5.6±7.1	7.5±9.0	0.1±0.1	0.0±0.1	5.6±7.0	3.8±3.7
S2Cl_1 vs. MODIS	6.7±8.1	6.9±8.3	8.5±10.0	0.1±0.1	0.1±0.1	7.2±8.2	6.1±5.2
S2 vs. MODIS	6.4±7.9	5.6±7.0	8.0±9.5	0.0±0.1	0.0± 0.1	5.5±7.0	4.2±4.8

- **S2Cl_1**: Sentinel-2 data at 1 km obtained by averaging only pixels labeled as the dominant crop at MODIS scale, according to the CDL map. Thus, this represents the time series of pure dominant crop within each MODIS pixel;
- **S2**: Sentinel-2 data at 1 km resolution;
- **MODIS**: MODIS data at the nominal 1 km resolution.

Starting with the homogeneous corn fields dataset, general LSP results are reported in Table 3. The mean values for all the pixels are provided, as well as the associated standard deviation (SD). Considering the pure corn crop S2CL_1 as reference, it can be noticed that the S2 and MODIS data at 1km provide similar statistics, with especially the S2 product providing analogous values as the S2CL_1. The consistency can be explained by the dominance of corn fields in the S2 tile. The consistency of the LSP metrics among the three time series products can probably be better expressed by calculating the mean absolute deviation (MAD) and its dispersion, as displayed in Table 4. Differences are low, especially when comparing the S2CL1 against the S2 product, meaning that for this more homogeneous region both S2 and MODIS datasets provide consistent temporal information.

When repeating the same exercise for the more heterogeneous landscape with winter wheat as dominant crop (Table 5), it becomes apparent that the consistency among the S2 and MODIS information somewhat degrades. This especially holds for EOS and consequent LOS with more than a month difference. On the other hand, the MV, Amp and day MV seem more robust, suggesting that the mismatch took only place in identifying the EOS. In general, the S2 dataset resembles closer the S2Cl_1 dataset, as is also quantified by the Area between SOS and EOS. The differences between S2 and MODIS are also revealed by calculating the mean absolute deviation against S2Cl_1 (Table 6); the differences with MODIS are up to two twice as large as compared to S2. Altogether, it suggests that the MODIS dataset is harder

Table 5 Mean and standard deviation (SD) of phenological indicators estimated from S2, S2Cl1 and MODIS over 1km-pixel whose Cl1 corresponded to winter wheat (tile 11TLM). SOS, EOS and Day MV are in DOY 2019

In order to display and interpret specific spatial patterns, the maps of the LSP metrics over the two study sites turn out to be very useful	SOS	EOS	LOS	MV	Amp.	Day MV	Area
S2Cl$_1$	90.6±10.0	235.7±27.8	144.2±31.0	0.3±0.1	0.3±0.1	145.0±11.8	22.2±6.3
S2	88.2±8.6	220.4±30.3	131.7±31.5	0.3±0.1	0.2±0.1	145.7±14.2	20.8±7.5
MODIS	84.7±17.4	261.9±18.4	172.3±29.2	0.3±0.1	0.3±0.1	148.4±13.8	31.1±8.8

Table 6 Mean absolute deviation (MAD) and standard deviation (MSD) of MAD estimated from S2, S2Cl$_1$ and MODIS over 1km-pixel whose Cl$_1$ corresponded to winter wheat (tile 11TLM). SOS, EOS and Day MV are in DOY 2019

	SOS	EOS	LOS	MV	Amp.	Day MV	Area
S2Cl$_1$ vs. S2	6.0±7.8	14.3±19.7	15.7±20.3	0.0±0.1	0.1±0.1	6.6±10.2	4.7±6.2
S2Cl$_1$ vs. MODIS	12.6±16.4	27.5±20.1	27.1±19.4	0.1±0.1	0.1±0.1	9.3±12.7	10.7±10.8
S2 vs. MODIS	11.8±16.5	24.5±22.7	24.1±23.2	0.0±0.1	0.1±0.1	7.2±10.2	11.7±11.0

to interpret in view of the phenology of the dominant crop, winter wheat, due to the larger heterogeneity in croplands and patches of natural vegetation.

Figure 7 shows the maps for the more homogeneous region dominated by corn fields (tile 15TVH), and Fig. 8 shows the maps for the more heterogeneous landscape dominated by winter wheat (tile 11TLM). Masked areas correspond to water or urban pixels.

Starting with the corn field maps, LSP metrics maps reveal that the region is highly spatially and temporally homogeneous. This is probably best visible in the SEOS and EOS maps. All three maps show the same pattern with a pronounced SOS around DOY 165–169 (half of June) and EOS around DOY 278 (beginning of October). These numbers are in agreement with the typical corn growing patterns in the Corn belt region [88]. The maximum DVI value (MV) and amplitude show some more variation. Here slight discrepancies between S2 and MODIS can be noticed, with S2 closer to the reference maps of S2Cl$_1$. The thin blue line in some of the S2 maps is due to border artifacts of those S2 captures covering the tile only partially, which generate local discontinuities in time that ripple along the pixel time series and affect LSP estimation. In order to eliminate these effects, these partial acquisitions should be either filtered out from the collection or processed with morphological erosion operators to modify the boundary contours.

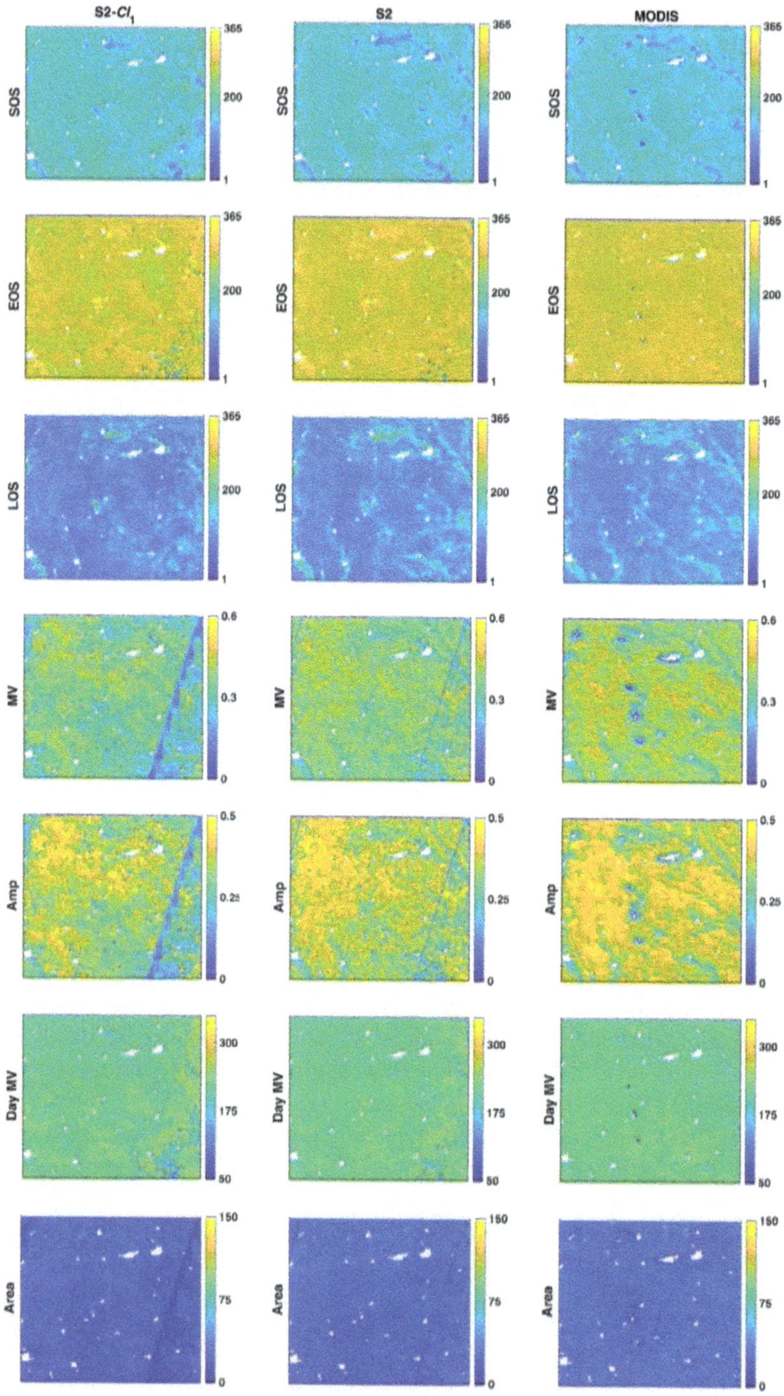

Fig. 7 LSP indicators for the year 2019 estimated from S2Cl_1, S2 and MODIS at 1km-pixel over a more homogeneous agricultural region (tile 15TVH). SOS, EOS and Day MV are in DOY 2019. Masked areas correspond to water or urban pixels

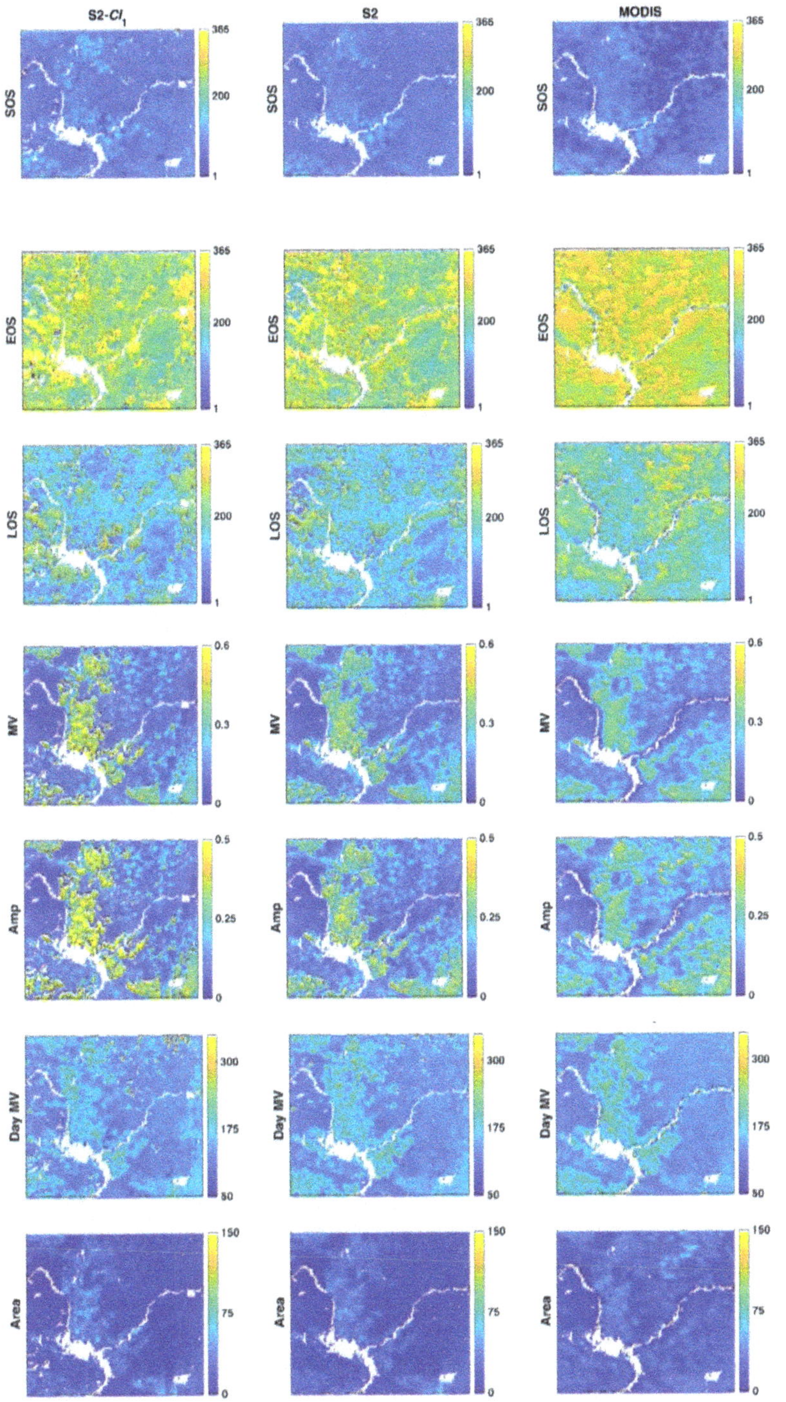

Fig. 8 Phenological indicators for the year 2019 estimated from S2Cl_1, S2 and MODIS at 1km-pixel over a more heterogeneous agricultural region (tile 11TLM). SOS, EOS and Day MV are in DOY 2019. Masked areas correspond to water or urban pixels

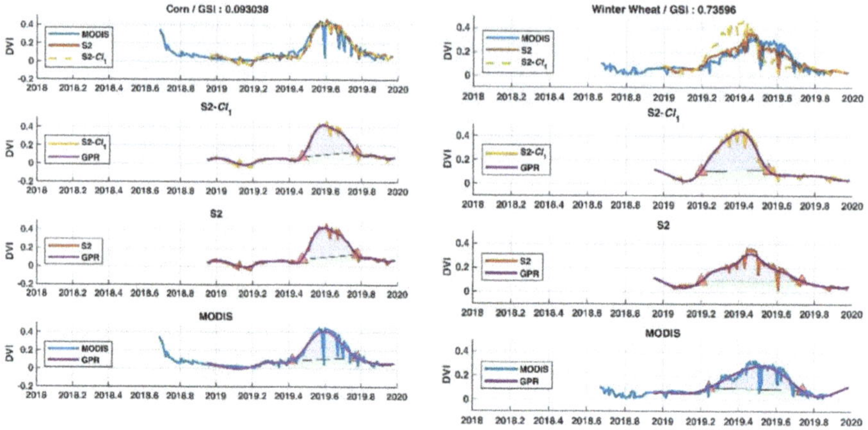

Fig. 9 Phenological indicators estimated from S2Cl$_1$, S2 and MODIS for corn (low-GSI) [left] and for winter wheat (high GSI) [right]

Conversely, the more heterogeneous landscape with croplands of winter wheat but also grasslands and shrubland, display more inconsistencies in the LSP metrics maps among the three data sources. While SOS still provides consistent patterns, with a SOS around DOY 85–91 (end of March), the EOS map is remarkably less consistent. Here, S2 still provides the same patterns as S2Cl$_1$ (EOS half of September), while the MODIS data shows a systematic later EOS (end of September). Noteworthy is that S2Cl$_1$ maximum values (MV) and amplitude (Amp) maps provide regions with more pronounced higher values than S2 and MODIS. Both S2 and MODIS deliver smoother, more blurred maps, which again must be attributed to the greater heterogeneity in vegetation cover.

Finally, in order to improve the understanding of the LSP maps, it is worth inspecting the temporal profiles of the three data sources more closely, and relating them to the land cover heterogeneity, i.e., as expressed by the Gini-Simpson index (GSI). Figure 9 shows the temporal profiles of the three data sources for two pixels with contrasting GSA values: low for a corn field pixel and high for a winter wheat pixel. These temporal profiles help also to understand how the LSP indicators are calculated.

When having a closer look to the homogeneous corn fields (Fig. 9, left), the temporal profiles for S2Cl$_1$, S2 and MODIS are shown in the top. A first observation is that the MODIS dataset is generally spikier, which is likely due to the higher temporal resolution, with more chances of observing inconsistencies, e.g., due to undetected cloud issues such as partial cloud cover. Regardless of the noise, the general temporal patterns of the three data products resemble closely. Accordingly, when the phenology indicators are calculated, they are alike. That is also shown in the individual calculation of SOS and EOS for each data source (see Fig. 9 underneath). For three data sources the SOS and EOS were identified at about the same dates. Conversely, for the more heterogeneous winter wheat landscape with a high GSI

(Fig. 9, right) the temporal profile of the wheat crop ($S2Cl_1$) follows a distinct pattern when compared to MODIS and S2 at 1 km patterns. Winter wheat has an earlier peak as opposed to the other phenology patterns. A closer inspection of the individual SOS and EOS calculations reveals that both MODIS and S2 express a smoother and longer pattern due to mixture of vegetation types (summer crops, grasslands or shrublands) with subsequent similar identification of SOS and EOS. The result suggests that independently of source, dataset at 1 km should be less related to crop phenology quantification and thus more care is required when interpreting this information towards crop monitoring.

5 Discussion

Having outlined a general overview of EO missions and time series processing technique applied to crop monitoring, this section provides a brief overview of a few ongoing trends with respect to satellite-based crop monitoring. They are summarized into the following topics: (1) trends in EO missions; (2) trends in gap-filling methods; (3) trends in time series data fusion, and (4) trends in time series software.

When it comes to EO imagery for crop monitoring purposes, a trade-off has to be made between spatial and temporal resolution. It does not come as a surprise that spatial resolution is a key factor to consider in phenology detection, given that medium to coarse spatial resolution imagery from sensors such as MODIS or Sentinel-3 are comprised of pixels containing a heterogeneous mosaic of multiple land cover types with varying phenological signals [8]. The impact of heterogeneity has been demonstrated here for the winter wheat case within MODIS pixels. Hence, coarse resolution data limits the extraction of specific phenological stages for specific land cover types given this sub-pixel land cover heterogeneity [89, 90]. However, in the extensive review by [8], it was also argued that the spectral-temporal signal at the coarse spatial sale is more stable over longer periods of time because the land cover composition within pixels at a resolution of 1 km or lower remains relatively static from year to year compared to higher spatial resolution pixels (e.g., S2) that detect common short-term land cover changes such as crop rotations. The study presented here just analyzes one growing season for one year, and therefore that statement cannot be confirmed, yet it is true that nominal S2 resolution (20 m) is well able to capture crop rotations (see also [57, 58]). At the same time, there is an ongoing tendency to move towards maximizing spatial and temporal resolution by making use of multiple satellites. i.e., constellations. This was first initiated with the two similar NASA satellites (Terra and Aqua) that both are equipped with the MODIS sensor [91]. The same concept of launching multiple satellites was repeated with the two S2 and Sentinel-3 constellations [92]. Progressing further along this trend, worth noting is the recent CubeSat initiative from Planet Labs, a private Earth imaging company[4]. For the last few years Planet Labs designed and launched a constellation of CubeSats of more than 100 units. It forms the largest satellite constellation in the world that provides a complete image of Earth once per day at 3–5 m spatial resolution. Their goal is to image the entirety of the planet daily to monitor changes and pinpoint trends. With such an unprecedented richness of spatiotemporal information, first initiatives are underway to estimate phenology stages at fine spatial resolution over the US Corn Belt and so provide significant advancement to crop monitoring and precision agriculture [93].

When it comes to EO imagery time series processing, there is a strong ongoing trend towards embracing artificial intelligence methods. Particularly the machine learning (ML) fitting algorithms entered as attractive alternatives of conventional gap-filling functions. Not only may ML methods lead to more accurate reconstructions (see [57, 58] for a quantitative comparison), but they are also adaptive towards unevenly spaced data over multiple seasons. The GPR used in this chapter is of special interest, as its associated uncertainty estimate provides per-pixel information of the gap-filling confidence. Typically, the longer the gap between two consecutive input samples, the higher the uncertainty. Another interesting method is Whittaker smoother, being almost as accurate as GPR and much faster (results not shown). Its adaptive fitting performance was already earlier reported [56, 94]. It must also be remarked that the multiple provided gap-filling techniques offer, to a greater or lesser extent, different performances. Each method has its own advantages and drawbacks, which depend strongly on the characteristics of the input time series [8, 94], i.e., a method that fits well with some data can be unsuited for a different set of data points. Concerning the appropriate length of time series, even if there is no limit of amount of data, the accuracy of the time series reconstruction increases with the data size. The main limitation of the interpolation module is the high time consuming and computational cost of specific algorithms. Although not the slowest method within the family of ML fitting methods (see [57]), also GPR becomes computationally inefficient in its standard per-pixel usage when processing time series of full images, mainly due to GPR training rather than fitting step. To mitigate this computational burden, it was recently proposed to substitute the per-pixel optimization step with the creation of a cropland-based pre-calculations for the GPR hyperparameters θ [95], which basically rule the way training samples contribute to time series reconstruction depending on their distribution along the time axis. The results of this optimized approach showed that accuracies were on the same order (at most 12% RMSE degradation), whereas processing time accelerated about 90 times. The alternative option of using the same hyperparameters for all the pixels within the complete scene was further evaluated. It led to similar overall accuracies over crop areas and computational performance. Hence, it means that calculating in advance and fixing θ substantial gain in run-time can be achieved in time series reconstruction while maintaining the advantages of GPR, i.e., a high accuracy and provision of associated uncertainties.

[4] https://www.planet.com/

While in this chapter only single-source imagery time series datasets were addressed, among the most exciting progress in time series analysis involves multi-source data fusion. Data fusion is being increasingly used to generate time series with high temporal and spatial resolutions [41, 96]. Data fusion algorithms are expected to generate fine resolution synthetic images based on infrequent observations at fine resolution and relatively frequent coarse remote sensing data with relatively higher temporal resolution [8]. ML methods are particularly promising for data fusion, and one of the most attractive fusion methods involves the multi-output (MO) version of GPR (MOGPR). This MOGPR approach was firstly introduced in [58] to fuse optical (S2) and radar (Sentinel-1) data for improved spatiotemporal reconstruction of vegetation products such as leaf area index (LAI). This approach proved to be particularly advantageous for long gapped time series, such as prolonged cloud clover, where optical data alone notoriously fails. Hence, the data from cloud-penetrating radar technology kicks in as complementary information, although the relationship between radar and vegetation phenology is less obvious, and strongly depends on structural properties. The absolute novelty of the solution proposed in [58] is that the parameters of the trained model implicitly predict the meaningfulness of any fusion approach: they quantify the amount of information shared between the two-time series and rule the interaction of low- and high-frequency GPs for output reconstruction. Moreover, the LAI data gap filling described in [58] is only one example of MOGPR possible applications. In fact, with MOGPR multiple datasets can be fused, so to say, that is not restricted to two data sources. Any set of time series collection can be entered into the MOGPR, i.e., the use of variables from multiple optical and radar data sources, coming from multiple satellite missions, e.g., Landsat, SPOT, the Sentinels, MODIS, can be envisaged, as long as they all share a certain amount of information and are georeferenced on a common grid. This data is nowadays easily accessible on cloud-based platforms such as the Google Earth Engine. Accordingly, in the present era of freely available, continuous multi-source satellite data streams, there is no doubt that fused time series processing will become indispensable in producing accurate cloud-free data and subsequent vegetation phenology monitoring.

Finally, to the benefit of the broader community and users in the agricultural sector, another interesting trend is that increasingly dedicated software packages become available for image time series processing and phenology-related studies. As reviewed by [8], the best known, and first software package is TIMESAT [68]. Subsequent software packages are variations and extensions of it or written in other (open-source) languages, such as: Phenological Parameters Estimation Tool (PPET [97]), enhanced TIMESAT [98], TimeStats [99], Phenosat [100], HANTS [101], CropPhenology [102] and QPhenoMetrics [103]. These software tools provide free functionalities for the reconstruction of time series data and extraction of phenological information customized with a number of user-defined input parameters based on time series data (e.g., vegetation indices). They are applicable in data reconstruction providing multiple common data gap-filling methods like logistic models, Savitzky–Golay, asymmetric Gaussian functions, piecewise regression, Fourier transforms etc. and generally perform well in general LSP extraction (e.g.,

SOS, EOS) providing common extraction methods, e.g., threshold method and inflection method [8]. It must hereby be remarked that all these software packages include the same established gap-filling algorithms. Apart from being equipped with these algorithms, the newly presented DATimeS software package [57] complements with a suite of versatile ML fitting algorithms. In comparison to other time series software packages, DATimeS is state of the art, through the: (1) ability to process unevenly spaced satellite image time series; (2) possibility to select over multiple ML fitting methods for time series prediction (some methods include associated uncertainties, e.g., GPR); (3) option to fuse multiple data sources with MOGPR, and (4) provision and analysis of phenological indicators over multiple growing seasons.[5]

6 Conclusions

Satellite imagery has become an essential source of information to enable monitoring agricultural lands. Specifically, optical data from EO image time series at high temporal resolution can assist in seasonal crop monitoring, as it provides key information about vegetation growing stages over large areas. In this chapter, the ongoing trends in image time series processing for the extraction of information about land surface phenology (LSP) metrics to quantify the key moments of the crop growing season is discussed. Identified trends go in the directions of: (1) a tendency towards constellation of multiple satellites to reach both a high spatial and temporal resolution; (2) adopting machine learning algorithms for fitting multi-year and irregular time series data sources; (3) time series fusion of multiple data sources, and (4) development of dedicated software packages. With the unprecedented availability of EO data and advanced image processing methods, these trends eventually lead to improved quantification of LSP metrics, e.g., start and end of season, but also metrics more related to crop biomass or yield, such as amplitude and area. By making use of the newly developed DATimeS toolbox, the LSP calculation for time series of MODIS and S2 data at 1 km resolution over predominantly (1) homogeneous and (2) heterogeneous agricultural landscapes has been carried out. It is concluded that LSP metrics can be consistently calculated and related to the dominant crop type over a homogeneous landscape. Conversely, heterogeneous regions show some discrepancies in the LSP metrics, which may be a consequence of the more complex landscape with varying phenological behaviors of croplands and natural vegetation, combined with the different temporal resolution of the two sensors analyzed and the role of cloud cover herein. Altogether, given the extraordinary flexibility of current processing algorithms and toolboxes, it can be safely concluded that the same level of maturity is reached in exploiting optical EO data in the temporal domain as in the spatial and spectral domains.

[5] The toolbox can be freely downloaded at https://artmotoolbox.com/

Acknowledgments This chapter was funded by the European Research Council (ERC) under the ERC-2017-STG SENTIFLEX project (grant agreement 755617) and supported by the Action CA17134 SENSECO (Optical synergies for spatiotemporal sensing of scalable ecophysiological traits) funded by COST (European Cooperation in Science and Technology, www.cost.eu)." Belen Franch and Santiago Belda were also supported by the program Generacio Talent of Generalitat Valenciana (CIDEGENT/2018/009 and SEJIGENT/2021/001, respectively), and Jochem Verrelst by Ramón y Cajal Contract (Spanish Ministry of Science, Innovation and Universities).

References

1. Thomas F Stocker, Dahe Qin, Gian-Kasper Plattner, M Tignor, Simon K Allen, Judith Boschung, Alexander Nauels, Yu Xia, Vincent Bex, and Pauline M Midgley. Climate change 2013: The physical science basis, 2014.
2. ESA. ESA's living planet programme: Scientific achievements and future challenges. Scientific Context of the Earth Observation Science Strategy for ESA, 2015.
3. Michele Chevrel, MICHEL Courtois, and G Weill. The SPOT satellite remote sensing mission. *Photogrammetric Engineering and Remote Sensing*, 47:1163–1171, 1981.
4. Alfredo Huete, Kamel Didan, Tomoaki Miura, E Patricia Rodriguez, Xiang Gao, and Laerte G Ferreira. Overview of the radiometric and biophysical performance of the MODIS vegetation indices. *Remote sensing of environment*, 83(1–2):195–213, 2002.
5. Xiaoyang Zhang, Mark A Friedl, Crystal B Schaaf, Alan H Strahler, John CF Hodges, Feng Gao, Bradley C Reed, and Alfredo Huete. Monitoring vegetation phenology using modis. *Remote sensing of environment*, 84(3):471–475, 2003.
6. Stephen H Boles, Xiangming Xiao, Jiyuan Liu, Qingyuan Zhang, Sharav Munkhtuya, Siqing Chen, and Dennis Ojima. Land cover characterization of Temperate East Asia using multi-temporal VEGETATION sensor data. *Remote Sensing of Environment*, 90(4):477–489, 2004.
7. Wouter Dierckx, Sindy Sterckx, Iskander Benhadj, Stefan Livens, Geert Duhoux, Tanja Van Achteren, Michael Francois, Karim Mellab, and Gilbert Saint. PROBA-V mission for global vegetation monitoring: standard products and image quality. *International Journal of Remote Sensing*, 35(7):2589–2614, 2014.
8. Linglin Zeng, Brian D Wardlow, Daxiang Xiang, Shun Hu, and Deren Li. A review of vegetation phenological metrics extraction using time-series, multispectral satellite data. *Remote Sensing of Environment*, 237:111511, 2020.
9. Sophie Moulin, Laurent Kergoat, Nicolas Viovy, and Gerard Dedieu. Global-scale assessment of vegetation phenology using NOAA/AVHRR satellite measurements. *Journal of Climate*, 10(6):1154– 1170, 1997.
10. Aaron Moody and David M Johnson. Land-surface phenologies from AVHRR using the discrete fourier transform. *Remote Sensing of Environment*, 75(3):305–323, 2001.
11. Benjamin W Heumann, JW Seaquist, Lars Eklundh, and Per Jönsson. AVHRR derived phenological change in the Sahel and Soudan, Africa, 1982–2005. *Remote sensing of environment*, 108(4):385–392, 2007.
12. Douglas E Ahl, Stith T Gower, Sean N Burrows, Nikolay V Shabanov, Ranga B Myneni, and Yuri Knyazikhin. Monitoring spring canopy phenology of a deciduous broadleaf forest using MODIS. *Remote Sensing of Environment*, 104(1):88–95, 2006.
13. Brian D Wardlow and Stephen L Egbert. Large-area crop mapping using time-series MODIS 250 m NDVI data: An assessment for the us central great plains. *Remote sensing of environment*, 112(3):1096–1116, 2008.
14. M. Drusch, U. Del Bello, S. Carlier, O. Colin, V. Fernandez, F. Gascon, B. Hoersch, C. Isola, P. Laberinti, P. Martimort, A. Meygret, F. Spoto, O. Sy, F. Marchese, and P. Bargellini.

Sentinel2: ESA's Optical High-Resolution Mission for GMES Operational Services. *Remote Sensing of Environment*, 120:25–36, 2012.
15. C. Donlon, B. Berruti, A. Buongiorno, M.-H. Ferreira, P. Féménias, J. Frerick, P. Goryl, U. Klein, H. Laur, C. Mavrocordatos, J. Nieke, H. Rebhan, B. Seitz, J. Stroede, and R. Sciarra. The Global Monitoring for Environment and Security (GMES) Sentinel-3 mission. *Remote Sensing of Environment*, 120:37–57, 2012.
16. D. Labate, M. Ceccherini, A. Cisbani, V. De Cosmo, C. Galeazzi, L. Giunti, M. Melozzi, S. Pieraccini, and M. Stagi. The PRISMA payload optomechanical design, a high performance instrument for a new hyperspectral mission. *Acta Astronautica*, 65(9–10):1429–1436, 2009.
17. T. Stuffler, C. Kaufmann, S. Hofer, K.P. Farster, G. Schreier, A. Mueller, A. Eckardt, H. Bach, B. Penn´e, U. Benz, and R. Haydn. The EnMAP hyperspectral imager-An advanced optical payload for future applications in Earth observation programmes. *Acta Astronautica*, 61(1–6):115–120, 2007.
18. D.A. Roberts, D.A. Quattrochi, G.C. Hulley, S.J. Hook, and R.O. Green. Synergies between VSWIR and TIR data for the urban environment: An evaluation of the potential for the Hyperspectral Infrared Imager (HyspIRI) Decadal Survey mission. *Remote Sensing of Environment*, 117:83–101, 2012.
19. S. Kraft, U. Del Bello, M. Bouvet, M. Drusch, and J. Moreno. Flex: Esa's earth explorer 8 candidate mission. pages 7125–7128, 2012.
20. D.K. Bolton and M.F. Friedl. Forecasting crop yield using remotely sensed vegetation indices and crop phenology metrics. *Agricultural and Forest Meteorology*, 173:74–84, 2013.
21. Y. Dong and C.Y. Peng. Principled missing data methods for researchers. *Springerplus*, 2(1):222:1– 17, 2013.
22. M. Schwartz. *Phenology: An Integrative Environmental Science*, volume 2. Springer Netherlands, 2013.
23. Michael A White, Forrest Hoffman, William W Hargrove, and Ramakrishna R Nemani. A global framework for monitoring phenological responses to climate change. *Geophysical Research Letters*, 32(4), 2005.
24. B.C. Reed, M.D. Schwartz, and X. Xiao. Remote sensing phenology: Status and the way forward. *Phenology of Ecosystems Processes*, pages 231–246, 2009.
25. Kirsten M De Beurs and Geoffrey M Henebry. Land surface phenology, climatic variation, and institutional change: Analyzing agricultural land cover change in kazakhstan. *Remote Sensing of Environment*, 89(4):497–509, 2004.
26. Robert B Macdonald. A summary of the history of the development of automated remote sensing for agricultural applications. *IEEE Transactions on Geoscience and Remote Sensing*, (6):473–482, 1984.
27. Belen Franch, Eric F Vermote, Jean-Claude Roger, Emilie Murphy, Inbal Becker-Reshef, Chris Justice, Martin Claverie, Jyoteshwar Nagol, Ivan Csiszar, Dave Meyer, et al. A 30+ year AVHRR land surface reflectance climate data record and its application to wheat yield monitoring. *Remote Sensing*, 9(3):296, 2017.
28. Siyuan Wang, Hang Yin, Qichun Yang, Hui Yin, Xiaoyue Wang, Yaoyao Peng, and Ming Shen. Spatiotemporal patterns of snow cover retrieved from NOAA-AVHRR LTDR: a case study in the tibetan plateau, china. *International Journal of Digital Earth*, 10(5):504–521, 2017.
29. Martin Claverie, Jessica L Matthews, Eric F Vermote, and Christopher O Justice. A 30+ year AVHRR LAI and fAPAR climate data record: Algorithm description and validation. *Remote Sensing*, 8(3):263, 2016.
30. Aleixandre Verger, Frédéric Baret, Marie Weiss, Iolanda Filella, and Josep Peñuelas. Geoclim: A global climatology of LAI, fAPAR, and fcover from vegetation observations for 1999–2010. *Remote Sensing of Environment*, 166:126–137, 2015.
31. Yves Julien and José A Sobrino. The yearly land cover dynamics (YLCD) method: An analysis of global vegetation from NDVI and LST parameters. *Remote sensing of environment*, 113(2):329–334, 2009.

32. Xiao-Peng Song, Matthew C Hansen, Stephen V Stehman, Peter V Potapov, Alexandra Tyukavina, Eric F Vermote, and John R Townshend. Global land change from 1982 to 2016. *Nature*, 560(7720):639–643, 2018.
33. Jose A Moreno Ruiz, David Riaño, Manuel Arbelo, Nancy HF French, Susan L Ustin, and Michael L Whiting. Burned area mapping time series in Canada (1984–1999) from NOAA-AVHRR LTDR: A comparison with other remote sensing products and fire perimeters. *Remote Sensing of Environment*, 117:407–414, 2012.
34. Zhen Song, Shunlin Liang, Dongdong Wang, Yuan Zhou, and Aolin Jia. Long-term record of top-of-atmosphere albedo over land generated from AVHRR data. *Remote Sensing of Environment*, 211:71–88, 2018.
35. Christopher O Justice, Eric Vermote, Jeff Privette, and Alain Sei. The evolution of US moderate resolution optical land remote sensing from AVHRR to VIIRS. pages 781–806, 2010.
36. Sergii Skakun, Christopher O Justice, Eric Vermote, and Jean-Claude Roger. Transitioning from MODIS to VIIRS: an analysis of inter-consistency of NDVI data sets for agricultural monitoring. *International journal of remote sensing*, 39(4):971–992, 2018.
37. Jude H Kastens, Terry L Kastens, Dietrich LA Kastens, Kevin P Price, Edward A Martinko, and Re-Yang Lee. Image masking for crop yield forecasting using AVHRR NDVI time series imagery. *Remote Sensing of Environment*, 99(3):341–356, 2005.
38. Belen Franch, Eric F Vermote, Sergii Skakun, Jean-Claude Roger, Inbal Becker-Reshef, Emilie Murphy, and C Justice. Remote sensing based yield monitoring: Application to winter wheat in United States and Ukraine. *International Journal of Applied Earth Observation and Geoinformation*, 76:112–127, 2019.
39. David M Johnson, Richard Mueller, et al. The 2009 cropland data layer. *Photogramm. Eng. Remote Sens*, 76(11):1201–1205, 2010.
40. T Fisette, P Rollin, Z Aly, L Campbell, B Daneshfar, P Filyer, A Smith, A Davidson, J Shang, and I Jarvis. AAFC annual crop inventory. In *2013 Second International Conference on AgroGeoinformatics (Agro-Geoinformatics)*, pages 270–274. IEEE, 2013.
41. Feng Gao, Martha C Anderson, Xiaoyang Zhang, Zhengwei Yang, Joseph G Alfieri, William P Kustas, Rick Mueller, David M Johnson, and John H Prueger. Toward mapping crop progress at field scales through fusion of Landsat and MODIS imagery. *Remote Sensing of Environment*, 188:9–25, 2017.
42. Liang Sun, Feng Gao, Martha C Anderson, William P Kustas, Maria M Alsina, Luis Sanchez, Brent Sams, Lynn McKee, Wayne Dulaney, William A White, et al. Daily mapping of 30 m LAI and NDVI for grape yield prediction in California vineyards. *Remote Sensing*, 9(4):317, 2017.
43. Lin Yan and David P Roy. Spatially and temporally complete landsat reflectance time series modelling: The fill-and-fit approach. *Remote Sensing of Environment*, 241:111718, 2020.
44. David B. Lobell. The use of satellite data for crop yield gap analysis. *Field Crops Research*, 143:56–64, 2013. Crop Yield Gap Analysis – Rationale, Methods and Applications.
45. David B Lobell, Gregory P Asner, J Ivan Ortiz-Monasterio, and Tracy L Benning. Remote sensing of regional crop production in the Yaqui valley, Mexico: estimates and uncertainties. *Agriculture, Ecosystems & Environment*, 94(2):205–220, 2003.
46. Marie-Julie Lambert, Pierre C Sibiry Traoré, Xavier Blaes, Philippe Baret, and Pierre Defourny. Estimating smallholder crops production at village level from Sentinel-2 time series in Mali's cotton belt. *Remote Sensing of Environment*, 216:647–657, 2018.
47. YR Lai, MJ Pringle, Peter M Kopittke, Neal W Menzies, Tom G Orton, and Yash P Dang. An empirical model for prediction of wheat yield, using time-integrated Landsat NDVI. *International journal of applied earth observation and geoinformation*, 72:99–108, 2018.
48. Sergii Skakun, Eric Vermote, Belen Franch, Jean-Claude Roger, Nataliia Kussul, Junchang Ju, and Jeffrey Masek. Winter wheat yield assessment from Landsat 8 and Sentinel-2 data: Incorporating surface reflectance, through phenological fitting, into regression yield models. *Remote Sensing*, 11(15):1768, 2019.

49. Martin Claverie, Junchang Ju, Jeffrey G Masek, Jennifer L Dungan, Eric F Vermote, Jean-Claude Roger, Sergii V Skakun, and Christopher Justice. The harmonized Landsat and Sentinel-2 surface reflectance data set. *Remote sensing of environment*, 219:145–161, 2018.
50. Eric Vermote, Chris Justice, Martin Claverie, and Belen Franch. Preliminary analysis of the performance of the Landsat 8/OLI land surface reflectance product. *Remote Sensing of Environment*, 185:46–56, 2016.
51. Śebastien Saunier, Jérôme Louis, Vincent Debaecker, Thomas Beaton, Enrico Giuseppe Cadau, Valentina Boccia, and Ferran Gascon. Sen2like, a tool to generate Sentinel-2 harmonised surface reflectance products-first results with Landsat-8. In *IGARSS 2019-2019 IEEE International Geoscience and Remote Sensing Symposium*, pages 5650–5653. IEEE, 2019.
52. F. Gao, M. Anderson, C. Daughtry, A. Karnieli, D. Hively, and W. Kustas. A within-season approach for detecting early growth stages in corn and soybean using high temporal and spatial resolution imagery. *Remote Sensing of Environment*, 242, 2020.
53. R. Fieuzal, V. Bustillo, D. Collado, and G. Dedieu. Combined use of multi-temporal Landsat-8 and Sentinel-2 images for wheat yield estimates at the intra-plot spatial scale. *Agronomy*, 10(3), 2020.
54. Pieter S.A. Beck, Clement Atzberger, Kjell Arild Høgda, Bernt Johansen, and Andrew K. Skidmore. Improved monitoring of vegetation dynamics at very high latitudes: A new method using MODIS NDVI. *Remote Sensing of Environment*, 100(3):321–334, 2006.
55. Y. Julien and J. A. Sobrino. Global land surface phenology trends from GIMMS database. *International Journal of Remote Sensing*, 30(13):3495–3513, 2009.
56. Peter M. Atkinson, C. Jeganathan, Jadu Dash, and Clement Atzberger. Inter-comparison of four models for smoothing satellite sensor time-series data to estimate vegetation phenology. *Remote Sensing of Environment*, 123:400–417, 2012.
57. DATimeS: a machine learning time series GUI toolbox for gap-filling and vegetation phenology trends detection], author=Belda, Santi and Pipia, Luca and Rivera-Caicedo, Juan Pablo and Amin, Eatidal, de Grave, Charlotte and Verrels, Jochem, Environmental Modelling and Software Journal, volume=submitted, publisher=Elsevier.
58. Luca Pipia, Jordi Muñoz-Marí, Eatidal Amin, Santiago Belda, Gustau Camps-Valls, and Jochem Verrelst. Fusing optical and SAR time series for LAI gap filling with multioutput gaussian processes. *Remote Sensing of Environment*, 235:111452, 2019.
59. Jin Chen, Per Jönsson, Masayuki Tamura, Zhihui Gu, Bunkei Matsushita, and Lars Eklundh. A simple method for reconstructing a high-quality NDVI time-series data set based on the savitzky– golay filter. *Remote sensing of Environment*, 91(3–4):332–344, 2004.
60. R. Stockli and P.L. Vidale. European plant phenology and climate as seen in a 20-year AVHRR land-surface parameter dataset. *International Journal of Remote Sensing*, 25(17):3303–3330, 2004.
61. R. Fensholt, K. Rasmussen, T.T. Nielsen, and C. Mbow. Evaluation of earth observation based long term vegetation trends - intercomparing ndvi time series trend analysis consistency of Sahel from AVHRR GIMMS, TERRA MODIS and SPOT VGT data. *Remote Sensing of Environment*, 113(9):1886–1898, 2009.
62. Michael A. White, Kirsten M. De Beurs, Kamel Didan, David W. Inouye, Andrew D. Richardson, Olaf P. Jensen, John O'Keefe, Gong Zhang, Ramakrishna R. Nemani, Willem J. D. Van Leeuwen, Jesslyn F. Brown, Allard De Wit, Michael Schaepman, Xioamao Lin, Michael Dettinger, Amey S. Bailey, John Kimball, Mark D. Schwartz, Dennis D. Baldocchi, John T. Lee, and William K. Lauenroth. Intercomparison, interpretation, and assessment of spring phenology in North America estimated from remote sensing for 1982 to 2006. *Global Change Biology*, 15(10):2335–2359, 2009.
63. B. Tan, J. T. Morisette, R. E. Wolfe, F. Gao, G. A. Ederer, J. Nightingale, and J. A. Pedelty. An enhanced TIMESAT algorithm for estimating vegetation phenology metrics from MODIS data. *IEEE Journal of Selected Topics in Applied Earth Observations and Remote Sensing*, 4(2):361–371, June 2011.

64. Mark Broich, Alfredo Huete, Matt Paget, Xuanlong Ma, Mirela Tulbure, Natalia Restrepo Coupe, Bradley Evans, Jason Beringer, Rakhesh Devadas, Kevin Davies, and Alex Held. A spatially explicit land surface phenology data product for science, monitoring and natural resources management applications. *Environmental Modelling & Software*, 64:191–204, 2015.
65. Elias Fernando Berra, Rachel Gaulton, and Stuart Barr. Assessing spring phenology of a temperate woodland: A multiscale comparison of ground, unmanned aerial vehicle and Landsat satellite observations. *Remote Sensing of Environment*, 223:229–242, 2019.
66. D.J. Mulla. Twenty five years of remote sensing in precision agriculture: Key advances and remaining knowledge gaps. *Biosystems Engineering*, 114(4):358–371, 2013.
67. T. Sakamoto, M. Yokozawa, H. Toritani, M. Shibayama, N. Ishitsuka, and H. Ohno. A crop phenology detection method using time-series MODIS data. *Remote Sensing of Environment*, 96(3–4):366–374, 2005.
68. J. Jönsson and L. Eklundh. TIMESAT - a program for analysing time-series of satellite sensor data. *Computers and Geosciences*, 30:833–845, 2004.
69. J.A. Sobrino and Y. Julien. Global trends in NDVI-derived parameters obtained from GIMMS data. *International Journal of Remote Sensing*, 32(15):4267–4279, 2011.
70. A.D. Richardson, T.F. Keenan, M. Migliavacca, Y. Ryu, O. Sonnentag, and M. Toomey. Climate change, phenology, and phenological control of vegetation feedbacks to the climate system. *Agricultural and Forest Meteorology*, 169:156–173, 2013.
71. C. Atzberger. Advances in remote sensing of agriculture: Context description, existing operational monitoring systems and major information needs. *Remote Sensing*, 5(2):949–981, 2013.
72. Daniel Lloyd. A phenological classification of terrestrial vegetation cover using shortwave vegetation index imagery. *International Journal of Remote Sensing*, 11(12):2269–2279, 1990.
73. Nicolas Delbart, Thuy Le Toan, Laurent Kergoat, and Violetta Fedotova. Remote sensing of spring phenology in boreal regions: A free of snow-effect method using NOAA-AVHRR and SPOT-VGT data (1982–2004). *Remote Sensing of Environment*, 101(1):52–62, 2006.
74. Michael A. White and Ramakrishna R. Nemani. Real-time monitoring and short-term forecasting of land surface phenology. *Remote Sensing of Environment*, 104(1):43–49, 2006.
75. Wen-bin Wu, Peng Yang, Hua-jun Tang, Qing-bo Zhou, Zhong-xin Chen, and Ryosuke Shibasaki. Characterizing spatial patterns of phenology in cropland of China based on remotely sensed data. *Agricultural Sciences in China*, 9(1):101–112, 2010.
76. Claire Boryan, Zhengwei Yang, Rick Mueller, and Mike Craig. Monitoring us agriculture: the US department of agriculture, national agricultural statistics service, cropland data layer program. *Geocarto International*, 26(5):341–358, 2011.
77. Noel Gorelick, Matt Hancher, Mike Dixon, Simon Ilyushchenko, David Thau, and Rebecca Moore. Google Earth Engine: Planetary-scale geospatial analysis for everyone. *Remote sensing of Environment*, 202:18–27, 2017.
78. USDA NASS. Usda-national agricultural statistics service, cropland data layer. *United States Department of Agriculture, National Agricultural Statistics Service, Marketing and Information Services Office, Washington, DC [Available at http//nassgeodata. gmu. edu/ Crop-Scape, Last accessed September 2012.]*, 2003.
79. Fran¸cois-Marie Bréon, Eric Vermote, Emilie Fedele Murphy, and Belen Franch. Measuring the directional variations of land surface reflectance from modis. *IEEE Transactions on Geoscience and Remote Sensing*, 53(8):4638–4649, 2015.
80. Eric Vermote, Christopher O Justice, and François-Marie Bréon. Towards a generalized approach for correction of the BRDF effect in MODIS directional reflectances. *IEEE Transactions on Geoscience and Remote Sensing*, 47(3):898–908, 2008.
81. Belen Franch, Eric F Vermote, José A Sobrino, and Yves Julien. Retrieval of surface albedo on a daily basis: Application to MODIS data. *IEEE Transactions on Geoscience and Remote Sensing*, 52(12):7549–7558, 2014.
82. Crystal B Schaaf, Feng Gao, Alan H Strahler, Wolfgang Lucht, Xiaowen Li, Trevor Tsang, Nicholas C Strugnell, Xiaoyang Zhang, Yufang Jin, Jan-Peter Muller, et al. First operational

BRDF, albedo nadir reflectance products from MODIS. *Remote sensing of Environment*, 83(1–2):135–148, 2002.

83. Fran¸cois-Marie Bréon, Eric Vermote, Emilie Fedele Murphy, and Belen Franch. Measuring the directional variations of land surface reflectance from modis. *IEEE Transactions on Geoscience and Remote Sensing*, 53(8):4638–4649, 2015.
84. Aisling J Daly, Jan M Baetens, and Bernard De Baets. Ecological diversity: measuring the unmeasurable. *Mathematics*, 6(7):119, 2018.
85. María Piles, Kaighin A McColl, Dara Entekhabi, Narendra Das, and Miriam Pablos. Sensitivity of aquarius active and passive measurements temporal covariability to land surface characteristics. *IEEE Transactions on Geoscience and Remote Sensing*, 53(8):4700–4711, 2015.
86. C. Bacour, F. Baret, D. Béal, M. Weiss, and K. Pavageau. Neural network estimation of LAI, fAPAR, fCover and LAI×Cab, from top of canopy MERIS reflectance data: Principles and validation. *Remote Sensing of Environment*, 105(4):313–325, 2006.
87. Belen Franch, Eric F Vermote, Sergii Skakun, Jean-Claude Roger, Inbal Becker-Reshef, Emilie Murphy, and C Justice. Remote sensing based yield monitoring: Application to winter wheat in United states and Ukraine. *International Journal of Applied Earth Observation and Geoinformation*, 76:112–127, 2019.
88. H. Jiang, H. Hu, R. Zhong, J. Xu, J. Xu, J. Huang, S. Wang, Y. Ying, and T. Lin. A deep learning approach to conflating heterogeneous geospatial data for corn yield estimation: A case study of the US corn belt at the county level. *Global Change Biology*, 26(3):1754–1766, 2020.
89. Jeremy Isaac Fisher, John F Mustard, and Matthew A Vadeboncoeur. Green leaf phenology at Landsat resolution: Scaling from the field to the satellite. *Remote sensing of environment*, 100(2):265– 279, 2006.
90. Katharine White, Jennifer Pontius, and Paul Schaberg. Remote sensing of spring phenology in northeastern forests: A comparison of methods, field metrics and sources of uncertainty. *Remote Sensing of Environment*, 148:97–107, 2014.
91. J.R.G. Townshend and C.O. Justice. Towards operational monitoring of terrestrial systems by moderate-resolution remote sensing. *Remote Sensing of Environment*, 83(1–2):351–359, 2002.
92. M. Berger, J. Moreno, J.A. Johannessen, P.F. Levelt, and R.F. Hanssen. ESA's sentinel missions in support of Earth system science. *Remote Sensing of Environment*, 120:84–90, 2012.
93. Hyungsuk Kimm, Kaiyu Guan, Chongya Jiang, Bin Peng, Laura F Gentry, Scott C Wilkin, Sibo Wang, Yaping Cai, Carl J Bernacchi, Jian Peng, et al. Deriving high-spatiotemporal-resolution leaf area index for agroecosystems in the us corn belt using planet labs cubesat and stair fusion data. *Remote Sensing of Environment*, 239:111615, 2020.
94. Clement Atzberger and Paul HC Eilers. A time series for monitoring vegetation activity and phenology at 10-daily time steps covering large parts of South America. *International Journal of Digital Earth*, 4(5):365–386, 2011.
95. Santiago Belda, Luca Pipia, Pablo Morcillo-Pallarés, and Jochem Verrelst. Optimizing gaussian process regression for image time series gap-filling and crop monitoring. *Agronomy*, 10(5):618, 2020.
96. JJ Walker, KM De Beurs, and RH Wynne. Dryland vegetation phenology across an elevation gradient in Arizona, USA, investigated with fused MODIS and Landsat data. *Remote Sensing of Environment*, 144:85–97, 2014.
97. Rodney D McKellip, Kenton W Ross, Joseph P Spruce, James C Smoot, Robert E Ryan, Gerald E Gasser, Donald L Prados, and Ronald D Vaughan. Phenological parameters estimation tool. 2010.
98. Bin Tan, Jeffrey T Morisette, Robert E Wolfe, Feng Gao, Gregory A Ederer, Joanne Nightingale, and Jeffrey A Pedelty. An enhanced timesat algorithm for estimating vegetation phenology metrics from MODIS data. *IEEE Journal of Selected Topics in Applied Earth Observations and Remote Sensing*, 4(2):361–371, 2010.
99. Thomas Udelhoven. Timestats: A software tool for the retrieval of temporal patterns from global satellite archives. *IEEE Journal of Selected Topics in Applied Earth Observations and Remote Sensing*, 4(2):310–317, 2010.

100. Arlete Rodrigues, Andre RS Marcal, and Mario Cunha. Phenology parameter extraction from timeseries of satellite vegetation index data using phenosat. In *2012 IEEE International Geoscience and Remote Sensing Symposium*, pages 4926–4929. IEEE, 2012.
101. Jie Zhou, Li Jia, Guangcheng Hu, and Massimo Menenti. Reconstruction of global MODIS NDVI time series: Performance of harmonic analysis of time series (HANTS). *Remote Sensing of Environment*, 163:217–228, 2015.
102. Sofanit Araya, Bertram Ostendorf, Gregory Lyle, and Megan Lewis. Cropphenology: An R package for extracting crop phenology from time series remotely sensed vegetation index imagery. *Ecological Informatics*, 46:45–56, 2018.
103. Lia Duarte, Ana C Teodoro, Ant´onio T Monteiro, Mário Cunha, and Hernâni Gonçalves. Qphenometrics: An open source software application to assess vegetation phenology metrics. *Computers and Electronics in Agriculture*, 148:82–94, 2018

Drone Imagery in Support of Orchards Trees Vegetation Assessment Based on Spectral Indices and Deep Learning

Ionuț Șandric, Radu Irimia, George P. Petropoulos, Dimitrios Stateras, Dionissios Kalivas, and Alin Pleșoianu

1 Introduction

The increase in the availability of drone technology, coupled with the latest discoveries in image segmentation and classification, has to lead to the emerging of new opportunities for researches from environmental sciences. Ultra-high spatial resolution imagery offers excellent advantages over the classical very-high spatial resolution satellite imagery. Some of its key advantages include its ability for rapid and repetitive collections over small areas without almost any influence from the cloud-cover distribution. Very important for drone imagery collection is to plan and apply an optimised flight plan that will minimise the shadow effect, will offer similar environmental conditions for repetitive flights and will have the best speed over image sharpness ratio [1, 2].

Over the years, the detection and classification of crown trees raised much interest for the scientists from the forest and environmental sciences, due to their essential role for landscape ecology and forestry management [1, 3–5]. Approximately a decade ago, detection of individual trees was focused on Lidar coupled with

I. Șandric (✉)
Faculty of Geography, University of Bucharest, Bucharest, Romania

Esri Romania, Bucharest, Romania
e-mail: ionut.sandric@geo.unibuc.ro

R. Irimia · A. Pleșoianu
Faculty of Geography, University of Bucharest, Bucharest, Romania

G. P. Petropoulos
Department of Geography, Harokopio University of Athens, Athens, Greece

D. Stateras · D. Kalivas
Department of Land Resources and Agricultural Engineering, Agricultural University of Athens, Athens, Greece

D. D. Bochtis et al. (eds.), *Information and Communication Technologies for Agriculture—Theme I: Sensors*, Springer Optimization and Its Applications 182,
https://doi.org/10.1007/978-3-030-84144-7_9

high-resolution ortho imagery or hyperspectral images [3, 5, 6], more recent advances in Unmanned Aerial Vehicles determined a graduated shift towards Structure from Motion application [7, 8]. The mapping techniques applied to individual trees detection also made the same technological switch from pixel classification [6] to object-based image analysis and lately to convolutional neural networks (CNN or deep learning methods) [9–18]. For forestry management, particular interest was also shown in the advantages offered by the high and ultra-high spatial resolution images coupled with Lidar data for mapping the morphometrical characteristics of each tree [19].

Besides crown trees delineation and trees classification, an essential part of the trees health assessment is the extraction and estimation of vegetation indices (VI). These VI are exploiting the differences between the visible spectrum (RGB) and the near-infrared spectrum (NIR) [20–22]. Due to this relationship between green, red and near-infrared spectrum, the state of vegetation health facilitates estimations and comparisons between individual trees with similar morphometrical characteristics. Even though the absence of information from the near-infrared spectrum can have an impact on the vegetation indices, it has been proven that it is possible to estimate reliable VI only from visible spectrum [20] and further the vegetation heath per tree.

Having a spatial distribution of the health of the trees gives the orchard managers valuable insight information. Trees health vegetation is based on detecting the tree crown, classify the tree according to the fruit species, extract the VI for each tree and use spatial statistics to identify the trees which are under development.

Building on the advantages of drone technology and the latest deep learning algorithms, the present study aims at assessing the combined use of ML techniques with spectral VIs derived from visible cameras mounted on drones, to be used as a proxy to characterise vegetation health of individual trees in an orchard field. To achieve this, several UAV flights were performed over two sites located in Greece and Romania with different environmental characteristics. For consistence and comparability, the same methodology was applied for both sites. Samples for each tree species were collected by manual digitizing the crown of each tree. Further, object instance segmentation techniques were used for training and detecting individual crown trees. The crown trees were crossintersected with the vegetation indices and based on their mean values, the z-score was applied to estimate the health state of each tree.

2 Methodology

To accomplish the study objectives, several image processing methods were implemented to the acquired drone data. The tree's crown was extracted by applying a deep learning object instance segmentation method. For mapping, the vegetation health, VI from visible spectrum (Red, Green and Blue) were used. These indices were chosen due to the non-constraint of using a multispectral camera, less affordable at the time the study was done. Further, to map each tree's health state several statistical methods were applied, to identify the under-developed and over-developed trees.

2.1 Tree Crown Detection and Classification

Each crown tree was spatially delineated by applying deep learning methods. Due to the aim of identifying the height and width of each tree, the Mask R-CNN algorithm [23, 24] was used, a widely used method for object instance segmentation. This method combines object detection with pixel classification methodologies to accurately locate an object and to accurately separate the object from the background. Thus, the tree is not only delimited by a bounding box, but it is also separate by other objects by drawing a contour that matches the crown. By applying Mask R-CNN, we achieved the goal of spatially delineate the crown tree without any impact or negative influence from the ground features. Similar approaches, but using a Single Shot Detector (SSD) were applied for the detection and classification of individual trees [25].

For the estimation of the height and the width of each tree, the polygons obtained in the crown tree delineation step were intersected with the normalised digital surface model (n-DSM), and making the difference between the maximum and minimum n-DSM pixel values.

2.2 Vegetation Indices (VIs)

Vegetation indices for assessing the state of trees health have been widely used in forestry and ecological studies [21]. Vegetation indices are designed to exploit the significant different behaviour of vegetation in the spectrum range from blue to near-infrared. For the current study, we focused on the use of only vegetation indices that are created with spectral intervals collected in the visible spectrum, respectively red, green and blue. The reason for this choice is related to the availability of infrared sensors mounted of drone, which, at the time this chapter was written, there were not widely available for consumer graded drones. Even though these indices do not take into consideration the infrared spectrum, excellent results were obtained in the past [20], which led to their extensive use in the crop studies [26, 27].

2.2.1 VARI – Visible Atmospherically Resistant Index

The VARI index (see Eq. 1), based on the atmospherically resistant vegetation index [28], makes use of the blue channel to account for the water vapours presents in the air. Thus, VARI index is less sensitive to atmospheric effects and suitable for assessing the state of the vegetation health.

$$\text{VARI} = \frac{\text{Green} - \text{Red}}{\text{Green} + \text{Red} - \text{Blue}} \tag{1}$$

2.2.2 GLI – Green Leaf Index

The GLI index (see Eq. 2) [29], in comparison with the VARI index, was specially designed to work with imagery obtained from an RGB camera, and it is focused on giving more weight to the Green spectrum.

$$\text{GLI} = \frac{(\text{Green} - \text{Red}) + (\text{Green} - \text{Blue})}{(2 * \text{Green}) + \text{Red} + \text{Blue}} \tag{2}$$

2.3 Tree Health Assessment

To assess the vegetation health of each tree, the standard score or also known as z-score was used, using the mean values for both GLI and VARI vegetation indices. All the trees that had standard scores below and over 1.96 standard deviation points were considered as having a low vegetation health state or a very high vegetation health state

$$\text{Standard score} = \frac{\text{Value} - \text{Mean}}{\text{Std dev}} \tag{3}$$

3 Study sites

To satisfy the study objectives, two sites of entirely different environments, one located in Romania and one in Greece were chosen to assess the suitability of VI indices obtained from drone's imagery for trees vegetation health estimation. Each site was surveyed with a different drone. For the Romanian site, a DJI Phantom 4 with an RGB with a resolution of 12MP was used. For the Greek site, a DJI Matrice 100 with Parrot Sequoia drone with an RGB camera with 16 MP resolution was used.

3.1 Romanian Study Site (No 1)

The study area is located in the centre of the Romanian Plain, northeast of Bucharest, Romania (coordinates 44°30′07′N and 26°15′45′E – Fig. 1). The research was carried out at the Belciugatele Didactic Resort/Moara Domnească Horticultural Farm, which is owned by the University of Agronomic Sciences and Veterinary Medicine in Bucharest. The orchard surface is structured on native and foreign fruit species (walnut, plum, apricot, apple), where the evolution of the trees is monitored according to the climatic, pedological, hydrological characteristics. This region is characterised by a temperate climate, with western, southern Mediterranean and eastern arid influences. Summers are hot, with average monthly air temperatures of 20–22 °C [30], with maximum temperatures that can exceed 39–40 °C. More than half of the amount of precipitation falls in the warm season (375–400 mm), and about 200 mm fall in the cold season.

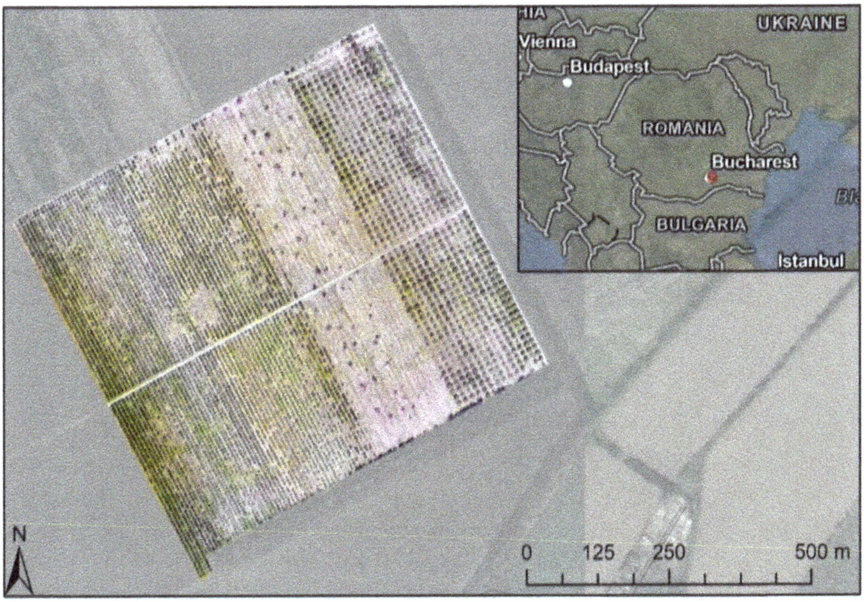

Fig. 1 Mora Domneasca orchard location (site no 1)

An area of about 19 ha was surveyed, covered with three different fruit species: plums, apricots and walnuts. The trees are planted in rows with northwest-southeast orientation. The rows density is different, with a distance between 4 m in a row and 5 m between rows for plums and apricots and 8 m per row and 10 m between rows for walnuts.

- Plum trees (7 ha) predominantly with Stanley, Centenar and Anna Spath. Plum trees have an early flowering, from late March to early April and, depending on the variety planted and its particularities, harvesting can be done from July to the end of September.
- Apricot trees (6 ha) with predominant Dacia and Tudor varieties, having an early flowering (in March), and harvesting takes place in late June and early July.
- Walnut trees (6 ha) with the Romanian varieties Vâlcea and Jupâneşti; flowering takes place from April to May, and harvesting takes place at the end of September.

3.2 Greek Study Site (No 2)

The study site of this research is Lygourio, which is a small Greek town located in the centre of Argolis prefecture (coordinates: 37°36′N 23°02′E – Fig. 2). It is built under the shadow of Arachnaion Mountain, near the Ancient Theatre of Epidaurus. Lygourio area is famous for the production of extra virgin olive oil, having been certified as Protected Designation of Origin (PDO). "Lygourio-Asklipiou". The climate is characterised as the typical Mediterranean, with hot, dry summers and mild, wet winters, while the average annual temperature of the region is 18.6 °C and the mean annual rainfall is 460.9 mm. Olive trees within this region flower during April and establish final fruit set in June. Harvesting occurs from October to December, followed by pruning (late January to March), in order to maximise light interception and maintain access to all tree canopies. The selected orchard extends into a hilly area (about 310 meters above sea level), resulting in high inclination percentage of the terrain. Thus, olive tree planting arrangement on grades facilitates the cultivation management and reduces the risk of erosion and desertification.

3.3 Drone Images Acquisition

For the Romanian study site, the aerial images were acquired with a DJI Phantom 4 quadcopter. Several flights with images having different spatial resolutions, ranging from 2 to 10 cm, were flown in the spring and summer seasons, close to the harvesting period. The flights during the spring period captured the degree of the flowering of the trees in different stages of flowering and, at the same time, measurements were made of the width of the crown (we extracted the tree height and crown width), distance and height of the trees. Only the summer images were used for the current study.

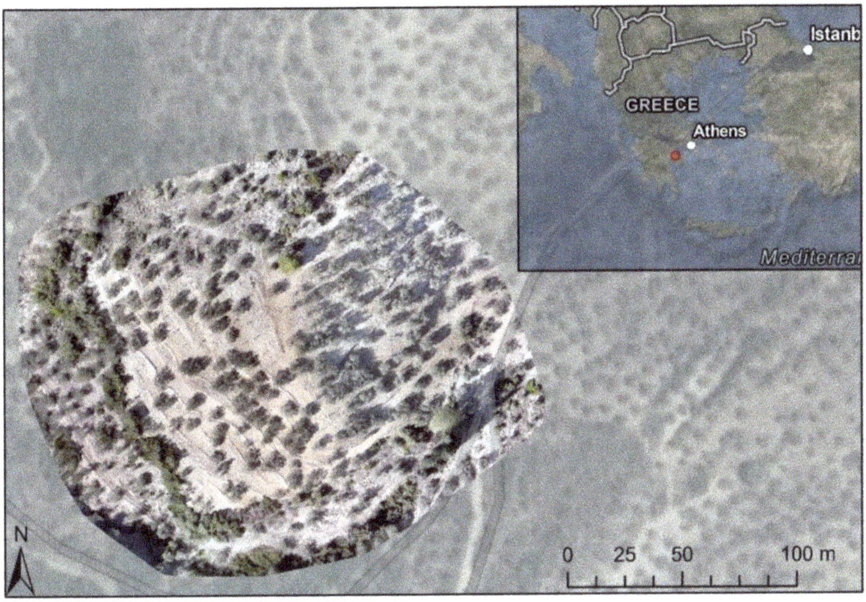

Fig. 2 Lygourio orchard location (site no 2)

For the Greek study site, the drone data acquisition occurred over a 1.4 ha olive orchard, which included 210 productive olive trees (over 35 years old) and 50 newly planted ones (about 3 years old), according to a non-linear plantation. Data were collected in early October 2017, when olive fruit colour had reached an optimum hue, before the commencement of harvesting.

4 Results and Discussion

4.1 Trees Detection Using Deep Learning

Mask R-CNN algorithm was used for crown trees detection and delineation, implemented in ArcGIS Python API using Pytorch and Fast.ai. The training and validation datasets were produced by manually digitising all the trees from both study sites (see Table 1). Each dataset was split into two datasets using 70% of the trees for training and the rest of 30% for validation datasets. This was done automatically for each epoch, hence with each epoch randomly selected trees were used for validation and training. In total, for each site, a number of 50 epochs were used to train the algorithm. For tree crown detection, we used ArcGIS Pro version 2.5, where the detection tool makes use of the trained CNN. Each tree was detected on an image chip with a weight and height of 128 pixels. The final datasets were obtained by

Table 1 The number of trees available for each orchard and for each tree specie

Site location	Total trees	Specie	Training	Validation
Romania	799	Walnut	560	239
Romania	2209	Apricot	1546	663
Romania	3231	Plum	2262	969
Greece	205	Olive	144	61

applying a dissolve algorithm based on the geometry and on the specie of the detected trees. All the trees that shared a boundary and had the same species were dissolved into one tree. This algorithm allowed us to obtain excellent crown detection per tree (see Fig. 3).

For site no 1 located in Romania, an accuracy of 0.95 for tree crown detection (Fig. 3a), 0.83 for the plum tree, 0.84 for apricot tree and 0.85 for the nuts tree was reached. For the Greek site no 2, the accuracy for tree crown detection was a bit lower of 0.73 (Fig. 3b). Because for the site no 2 there was only one tree species, there was no need to calculate per specie accuracy.

In some areas, trees were detected as several close polygons that share a boundary (see the middle of Fig. 3 for both a and b parts), and they were not joined into one single polygon. This was caused by the detection algorithm that identifies parts of one tree as being other tree specie. This issue can be solved by modifying the dissolve algorithm to take into account the species with the highest overlap between the overlapping polygons. Another option would be to have an operator to decide in which species the tree belongs, but in this case, the entire procedure will not be completely automated. Even though it can have an impact on the overall accuracy of the tree vegetation health assessment, this impact is expected to be very low because the confusion between species is made for apricot and plum trees, which have very similar characteristics.

Some trees, during the dissolving step, were joined together into one single big tree. This issue appears where the trees are closely located, and the algorithm is not able to separate the crowns of each individual tree. Even though the result has an impact on the overall accuracy for tree crown detection as a number, it does not play an essential role in the vegetation health status assessment. This argument is supported by the high similarities between close trees, part of the same species.

For olive trees located in the site no 2, the low accuracy for tree crown detection is mostly caused by the low number of samples available for the present study. Even though the number of trees was not high, by applying image transformation techniques common for deep learning studies, we managed to increase the number of samples to a few thousands of image chips. The same algorithm was applied to site no 1 located in Romania. The specific structure of the olives trees, having small leaves and large spaces between branches, played an important influence in the crown detection process. This made that some bushes to be confused as small olive trees and mapped by the algorithm as young olive trees. Similar, olive trees located at the border of the study area were partially mapped or mapped from several polygons closely located and sharing a common border. These issues were reduced by

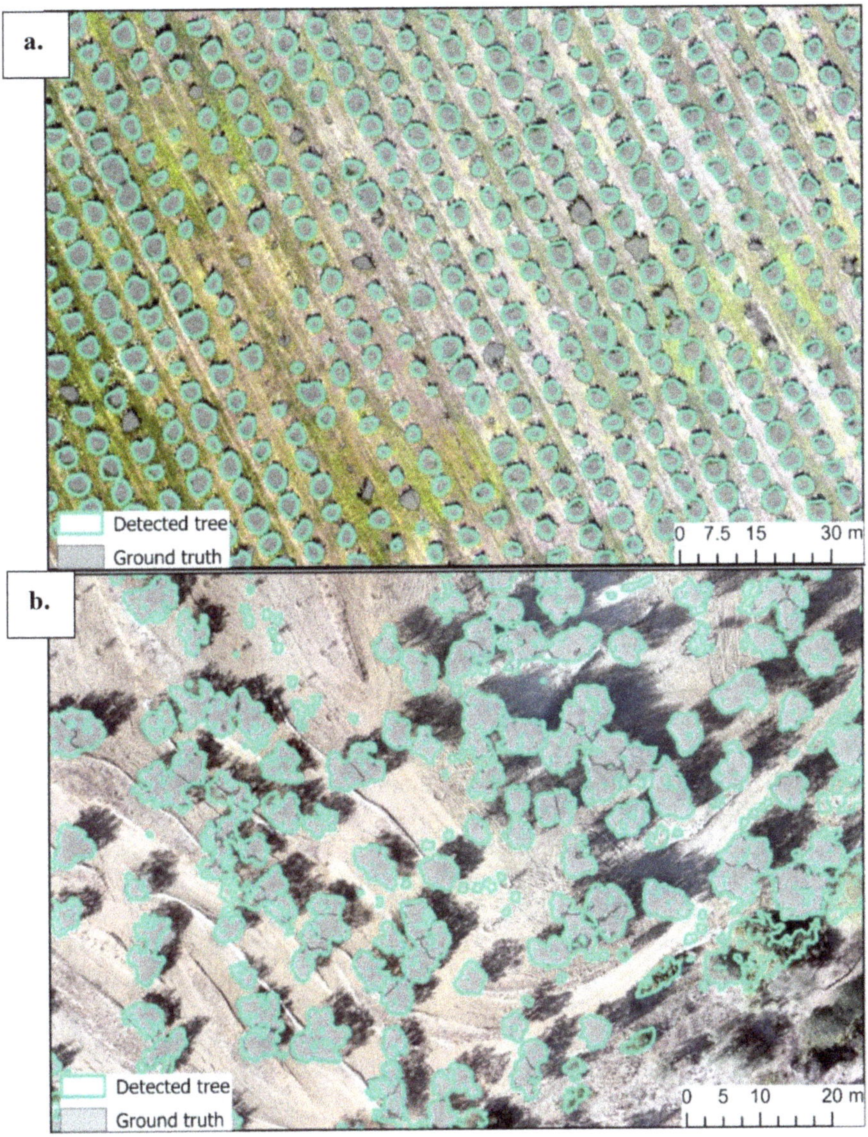

Fig. 3 Trees detection using Mask R-CNN deep learning algorithm. (**a**) site no 1 (plum and apricot trees); (**b**) site no 2 (olive trees)

applying the same dissolve algorithm used in the case of the site no 1. The significant difference between the dissolve made for site no 1 and site no 2 was the absence of the species, hence for site no 2 the polygons were dissolved with a common border only by their geometry. As expected, the dissolving step created artificial big olive trees where the branches of the closed tree were touching.

For both sites, the algorithm managed very well to distinguish between the shadows generated by the trees and the exact position of the crown tree. This was possible to achieve due to the high number of samples that were generated during the training process. Some trees were partially detected, mostly in the Greek site because of the high similarity between shrubs and olive trees.

4.2 Trees Vegetation Health Assessment

Although both vegetation indices were processed, after the poor results obtained for VARI index, we decided to drop the analysis for this index. For each tree, the spatial cross-intersection between the tree crown and the VARI and GLI rasters made possible the extraction of the following statistics per tree: mean, maximum, minimum and standard deviation values for the GLI index. The z-score was calculated for the mean values, as the mean values provide a more accurate description of the spatial distribution of the pixel values inside a crown tree. Also, the mean GLI values per tree were less influenced by the size of the trees. The standard deviation for each tree was bellowed 1e-3, meaning that the mean values are representative from a statistical point of view.

The vegetation health state was assessed for three statistical confidence thresholds: 90%, 95% and 99%. Under the normal distribution assumption, we considered that all the trees with a z-score below a value of −1.65 have poor vegetation health and those with a z-score value above 1.65 have good vegetation health for a p-value lower or equal to 0.1 (at 90% statistical confidence). Furthermore, all trees with values for z-score bellow −1.96 and above 1.96 were considered with poor or good vegetation health for a p-value lower or equal to 0.05. The 99% statistical confidence for the vegetation health per tree was considered for z-score values which were bellow −2.57 and above 2.57.

In accordance with Table 2, the analysis presents a very good correspondence between the number of trees mapped and the number of trees detected by Mask R-CNN.

In the case of the Greek study site, approximately 92% of the mapped trees were considered with average vegetation health, a percentage very similar with the one (approximately 94%) calculated for the detected trees (Fig. 4). A good agreement is also for the number of trees identified with good vegetation health, for statistical confidence equal or above 90%, where the mapped trees have 3%, and the detected trees have 4%. A slightly bigger difference is recorded for the trees having a poor vegetation health status, for statistical confidence equal or above 90%. In this case, 5% of the mapped trees were identified with poor vegetation health, and only 2% of the detected trees were considered with poor vegetation health. The difference can be explained by the fact that olive trees have larger spaces between branches and these spaces were delineated only for the detected trees dataset and not for the mapped trees dataset. Overall, a very good agreement between the assessment of the vegetation health for the mapped and detected trees, is considered.

Table 2 Tree health assessment for both study sites. The table presents the number of trees and their vegetation health for both detected and mapped trees

	Site location	Specie	99% poor	95% poor	90% poor	Average	90% good	95% good	99% good	Total
Mapped	**Greece**	**Olive**		3	7	189	1	4	1	205
	Romania	**Walnut**		143	137	507				799
		Apricot		5	9	2068	91	36		2209
		Plum		3	18	2982	99	96	32	
Detected	**Greece**	**Olive**		1	3	253		6	5	268
	Romania	**Walnut**	3		18	727	22	20	10	800
		Apricot		69	60	1932	51	28		2152
		Plum		2	21	3257	144	127	14	3565

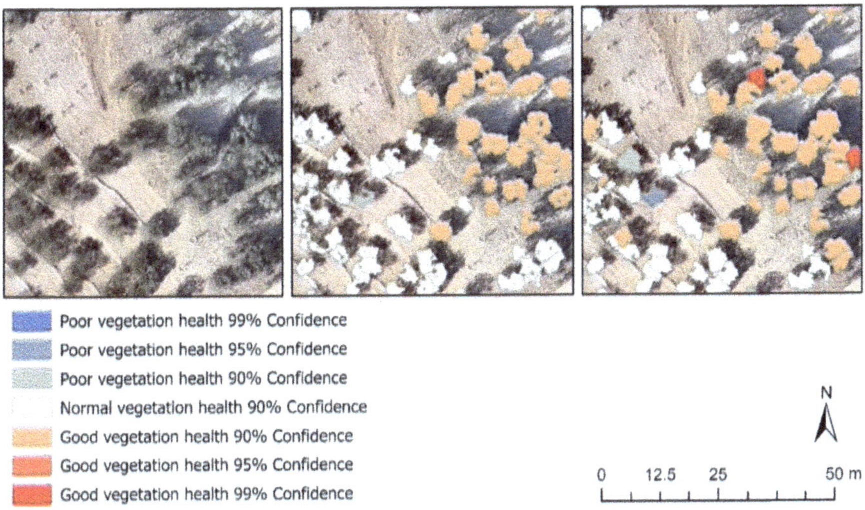

Fig. 4 Trees selected with a good or a low health state of the vegetation for the site no 2. The trees with good vegetation health are marked in red, and the ones with low vegetation health are marked in blue. From left to right, the images present the orthoimage with the spatial distribution of the olive trees; the olive trees detected by Mask R-CNN and mapped by their vegetation health status; the olive trees manually mapped and their vegetation health status

In the case of the study site no 1 (Romania), the overall agreement between both datasets, mapped and detected trees, remains similar with the one recorded for the site no 2 (Greece). This good agreement is proof that the methodology is stable, and it can be extrapolated to different environments.

The biggest difference, in terms of percentage of the trees identified with an average vegetation health status, is recorded for the walnut trees (Fig. 5). Here, approximate 64% of the mapped trees were identified with normal vegetation health in comparison with approximate 90% of the detected trees mapped with the same vegetation health status. We believe that this remarkable difference is mostly caused

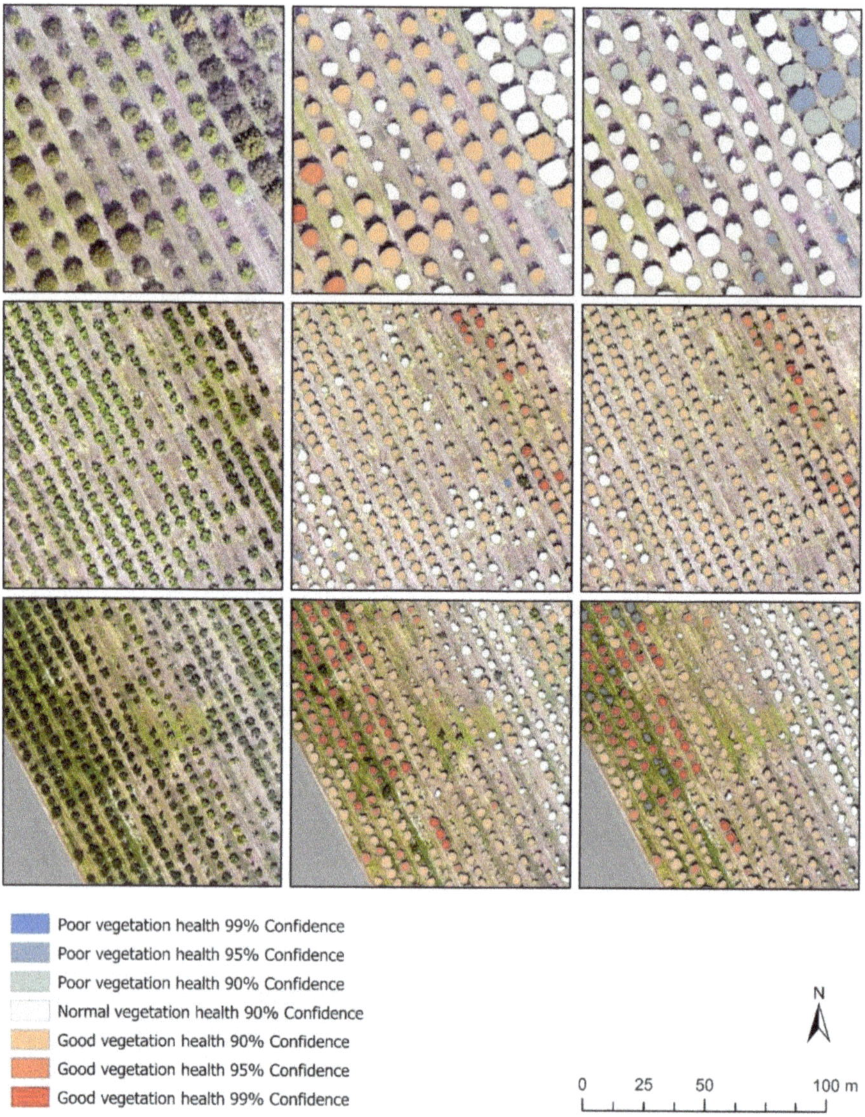

Fig. 5 Trees selected with a good or a low health state of the vegetation for site no 1. The trees with good vegetation health are marked in red, and the ones with low vegetation health are marked in blue. From left to right, the images present: the orthoimage with he spatial distribution if the trees; the trees detected by Mask R-CNN and mapped by their vegetation health status; the trees manually mapped and their vegetation health status. From bottom to top, the images present: plum trees, apricot trees and walnut trees

by the structure of the walnut's crown, that made the algorithm detect one big tree instead of several trees. This issue caused by the crown trees identification influenced the mean GLI value, hence leading to an overestimate of the number of trees with an average vegetation health status. For the mapped trees dataset no trees were identified having a good vegetation health status in comparison to the detected trees, where the percentage is approximate 7%. The same noticeable difference as in the normal vegetation health is recorded for the trees identified with a poor vegetation health status. In the case of the mapped trees dataset, approximate 35% of the trees were identified having poor vegetation health in comparison to approximately 2% identified by the detected trees having similar vegetation health status. These values are calculated for statistical confidence equal to or above 90%.x.

Very similar results are also recorded for the apricot and plum trees. These species have a similar spectral response, with apricot having a slightly smaller reflectance in the green spectrum. For both species, the percentage of the trees identified with a normal vegetation health status is close to 90% with differences of about 2%. The apricot trees, for the detected dataset, have approximately 6% of the trees identified with a poor vegetation heath in comparison with the mapped dataset were the percentage is 1%.

5 Conclusions

In this study, the combination of deep learning for tree crown delineation and vegetation indices was used for assessing the tree vegetation health. Using deep learning for tree crown delineation and tree detection gave outstanding results and proved to be reliable for future studies. The use of vegetation indices extracted from an RGB camera mounted on drones, also proved to return reliable results for mapping and to assess the trees health status.

Very good results were obtained in the case of the plum, apricot and walnut trees, mostly because these trees have the leaves oriented towards the camera and the spaces between leaves and branches are much smaller in comparison with the olives trees.

Less reliable results were obtained for olive trees crown delineation because of its specific texture with small leaves and large spaces between branches. The signal received from the ground had an essential influence in the assessment of the vegetation health status by increasing the GLI index mean values with a small fraction.

The similar agreement for the mapped and detected trees for both case studies, located in different environmental conditions, proves that the methodology presented in the current study is robust and can be applied with good results in areas with different environmental conditions. Computational time is acceptable, with less than 24 h necessary to compute the images and calculate the z-score values per tree, having a good perspective to be considered for future developments as an operational product.

We anticipate having an outstanding improvement in future studies if a multispectral camera, collecting at least one band in the near-infrared spectrum, will be used for producing the vegetation indices used in the assessment of tree vegetation health status. The VI indices produced with a multispectral camera will provide more accurate information in the green and near-infrared spectrum, related to the content of chlorophyll and leaf coverage, and hence a better estimate of the vegetation health status.

Overall, the study demonstrates the real potential of drone applications and deep learning methods for spatial and temporal rapid assessment of trees vegetation heath In comparison with the acquisition of high-resolution satellite imagery, the use of drone technology is cost-effective and more suitable for small to medium orchards. The increase of accessibility to this technology from the cloud and personal smartphones will make this approach less computationally intensive than it is now, and it is expected that the farmers will have the tools for optimal management of the orchards.

Acknowledgments Part of the datasets used for this study was obtained from the collaboration of the Faculty of Geography from the University of Bucharest with the Faculty of Horticulture from University of Agronomic Sciences and Veterinary Medicine of Bucharest (Prof. Adrian Peticila).

References

1. Tu, Yu-Hsuan, Stuart Phinn, Kasper Johansen, Andrew Robson, and Dan Wu. 2020. "Optimising Drone Flight Planning for Measuring Horticultural Tree Crop Structure." *ISPRS Journal of Photogrammetry and Remote Sensing* 160 (February): 83–96. https://doi.org/10.1016/j.isprsjprs.2019.12.006.
2. Xiong, Juntao, Zhen Liu, Shumian Chen, Bolin Liu, Zhenhui Zheng, Zhuo Zhong, Zhengang Yang, and Hongxing Peng. 2020. "Visual Detection of Green Mangoes by an Unmanned Aerial Vehicle in Orchards Based on a Deep Learning Method." *Biosystems Engineering* 194 (June): 261–72. https://doi.org/10.1016/j.biosystemseng.2020.04.006.
3. Ke, Yinghai, and Lindi J. Quackenbush. 2011. "A Review of Methods for Automatic Individual Tree-Crown Detection and Delineation from Passive Remote Sensing." *International Journal of Remote Sensing* 32 (17): 4725–47. https://doi.org/10.1080/01431161.2010.494184.
4. Kang, Hanwen, and Chao Chen. 2020b. "Fruit Detection, Segmentation and 3D Visualisation of Environments in Apple Orchards." *Computers and Electronics in Agriculture* 171 (April): 105302. https://doi.org/10.1016/j.compag.2020.105302.
5. Zhen, Zhen, Lindi Quackenbush, and Lianjun Zhang. 2016. "Trends in Automatic Individual Tree Crown Detection and Delineation—Evolution of LiDAR Data." *Remote Sensing* 8 (4): 333. https://doi.org/10.3390/rs8040333.
6. Larsen, Morten, Mats Eriksson, Xavier Descombes, Guillaume Perrin, Tomas Brandtberg, and François A. Gougeon. 2011. "Comparison of Six Individual Tree Crown Detection Algorithms Evaluated under Varying Forest Conditions." *International Journal of Remote Sensing* 32 (20): 5827–52. https://doi.org/10.1080/01431161.2010.507790.
7. Iglhaut, Jakob, Carlos Cabo, Stefano Puliti, Livia Piermattei, James O'Connor, and Jacqueline Rosette. 2019. "Structure from Motion Photogrammetry in Forestry: A Review." *Current Forestry Reports* 5 (3): 155–68. https://doi.org/10.1007/s40725-019-00094-3.
8. Kim, Wan-Soo, Dae-Hyun Lee, Yong-Joo Kim, Taehyeong Kim, Rok-Yeun Hwang, and Hyo-Jai Lee. 2020. "Path Detection for Autonomous Traveling in Orchards Using Patch-

Based CNN." *Computers and Electronics in Agriculture* 175 (August): 105620. https://doi.org/10.1016/j.compag.2020.105620.

9. Kang, Hanwen, and Chao Chen. 2020a. "Fast Implementation of Real-Time Fruit Detection in Apple Orchards Using Deep Learning." *Computers and Electronics in Agriculture* 168 (January): 105108. https://doi.org/10.1016/j.compag.2019.105108.
10. Li, Weijia, Haohuan Fu, Le Yu, and Arthur Cracknell. 2016. "Deep Learning Based Oil Palm Tree Detection and Counting for High-Resolution Remote Sensing Images." *Remote Sensing* 9 (1): 22. https://doi.org/10.3390/rs9010022.
11. Ma, Lei, Yu Liu, Xueliang Zhang, Yuanxin Ye, Gaofei Yin, and Brian Alan Johnson. 2019. "Deep Learning in Remote Sensing Applications: A Meta-Analysis and Review." *ISPRS Journal of Photogrammetry and Remote Sensing* 152 (June): 166–77. https://doi.org/10.1016/j.isprsjprs.2019.04.015.
12. Majeed, Yaqoob, Jing Zhang, Xin Zhang, Longsheng Fu, Manoj Karkee, Qin Zhang, and Matthew D. Whiting. 2020. "Deep Learning Based Segmentation for Automated Training of Apple Trees on Trellis Wires." *Computers and Electronics in Agriculture* 170 (March): 105277. https://doi.org/10.1016/j.compag.2020.105277.
13. Nezami, Somayeh, Ehsan Khoramshahi, Olli Nevalainen, Ilkka Pölönen, and Eija Honkavaara. 2020. "Tree Species Classification of Drone Hyperspectral and RGB Imagery with Deep Learning Convolutional Neural Networks." *Remote Sensing* 12 (7): 1070. https://doi.org/10.3390/rs12071070.
14. Tianyang, Dong, Zhang Jian, Gao Sibin, Shen Ying, and Fan Jing. 2018. "Single-Tree Detection in High-Resolution Remote-Sensing Images Based on a Cascade Neural Network." *ISPRS International Journal of Geo-Information* 7 (9): 367. https://doi.org/10.3390/ijgi7090367.
15. Weinstein, Ben G., Sergio Marconi, Stephanie A. Bohlman, Alina Zare, and Ethan P. White. 2020. "Cross-Site Learning in Deep Learning RGB Tree Crown Detection." *Ecological Informatics* 56 (March): 101061. https://doi.org/10.1016/j.ecoinf.2020.101061.
16. Wu, Jintao, Guijun Yang, Hao Yang, Yaohui Zhu, Zhenhai Li, Lei Lei, and Chunjiang Zhao. 2020. "Extracting Apple Tree Crown Information from Remote Imagery Using Deep Learning." *Computers and Electronics in Agriculture* 174 (July): 105504. https://doi.org/10.1016/j.compag.2020.105504.
17. Zhou, Zhongxian, Zhenzhen Song, Longsheng Fu, Fangfang Gao, Rui Li, and Yongjie Cui. 2020. "Real-Time Kiwifruit Detection in Orchard Using Deep Learning on Android™ Smartphones for Yield Estimation." *Computers and Electronics in Agriculture* 179 (December): 105856. https://doi.org/10.1016/j.compag.2020.105856.
18. Zou, Xinhuai, Ming Cheng, Cheng Wang, Yan Xia, and Jonathan Li. 2017. "Tree Classification in Complex Forest Point Clouds Based on Deep Learning." *IEEE Geoscience and Remote Sensing Letters* 14 (12): 2360–64. https://doi.org/10.1109/LGRS.2017.2764938.
19. Iizuka, Kotaro, Taichiro Yonehara, Masayuki Itoh, and Yoshiko Kosugi. 2017. "Estimating Tree Height and Diameter at Breast Height (DBH) from Digital Surface Models and Orthophotos Obtained with an Unmanned Aerial System for a Japanese Cypress (Chamaecyparis Obtusa) Forest." *Remote Sensing* 10 (2): 13. https://doi.org/10.3390/rs10010013.
20. Gitelson, Anatoly A., Yoram J. Kaufman, Robert Stark, and Don Rundquist. 2002. "Novel Algorithms for Remote Estimation of Vegetation Fraction." *Remote Sensing of Environment* 80 (1): 76–87. https://doi.org/10.1016/S0034-4257(01)00289-9.
21. Cârlan, Irina, Dagmar Haase, André Große-Stoltenberg, and Ionut Sandric. 2020. "Mapping Heat and Traffic Stress of Urban Park Vegetation Based on Satellite Imagery – A Comparison of Bucharest, Romania and Leipzig, Germany." *Urban Ecosystems* 23 (2): 363–77. https://doi.org/10.1007/s11252-019-00916-z.
22. Zha, Y., J. Gao, and S. Ni. 2003. "Use of Normalized Difference Built-up Index in Automatically Mapping Urban Areas from TM Imagery." *International Journal of Remote Sensing* 24 (3): 583–94. https://doi.org/10.1080/01431160304987.
23. Abdulla, Waleed. 2017. "Mask R-CNN for Object Detection and Instance Segmentation on Keras and TensorFlow." *GitHub Repository*. Github.

24. He, Kaiming, Georgia Gkioxari, Piotr Dollár, and Ross Girshick. 2017. "Mask R-CNN," March. http://arxiv.org/abs/1703.06870.
25. Pleşoianu, Alin-Ionuţ, Mihai-Sorin Stupariu, Ionuţ Şandric, Ileana Pătru-Stupariu, and Lucian Drăguţ. 2020. "Individual Tree-Crown Detection and Species Classification in Very High-Resolution Remote Sensing Imagery Using a Deep Learning Ensemble Model." *Remote Sensing 2020, Vol. 12, Page 2426* 12 (15): 2426. https://doi.org/10.3390/RS12152426.
26. Ballester, Carlos, John Hornbuckle, James Brinkhoff, John Smith, and Wendy Quayle. 2017. "Assessment of In-Season Cotton Nitrogen Status and Lint Yield Prediction from Unmanned Aerial System Imagery." *Remote Sensing* 9 (11): 1149. https://doi.org/10.3390/rs9111149.
27. Prabhakara, Kusuma, W. Dean Hively, and Gregory W. McCarty. 2015. "Evaluating the Relationship between Biomass, Percent Groundcover and Remote Sensing Indices across Six Winter Cover Crop Fields in Maryland, United States." *International Journal of Applied Earth Observation and Geoinformation* 39 (July): 88–102. https://doi.org/10.1016/j.jag.2015.03.002.
28. Kaufman, Y.J., and D. Tanre. 1992. "Atmospherically Resistant Vegetation Index (ARVI) for EOS-MODIS." *IEEE Transactions on Geoscience and Remote Sensing* 30 (2): 261–70. https://doi.org/10.1109/36.134076.
29. Louhaichi, Mounir, Michael M. Borman, and Douglas E. Johnson. 2001. "Spatially Located Platform and Aerial Photography for Documentation of Grazing Impacts on Wheat." *Geocarto International* 16 (1): 65–70. https://doi.org/10.1080/10106040108542184.
30. Sandu, I., V. I. Pescaru, I. Poiană, A. Geicu, I. Cândea, and D. Ţâştea. 2008. *Clima României*. Bucureşti: Editura Academiei Române.

Part IV
Proximal Sensing Applications

What Does the NDVI Really Tell Us About Crops? Insight from Proximal Spectral Field Sensors

Jon Atherton, Chao Zhang, Jaakko Oivukkamäki, Liisa Kulmala, Shan Xu, Teemu Hakala, Eija Honkavaara, Alasdair MacArthur, and Albert Porcar-Castell

1 Introducing the Normalized Difference Vegetation Index (NDVI)

Crops sit at the base of food chains, absorbing sunlight to fuel the photosynthetic reactions. The sunlight that plants absorb is referred to as Photosynthetically Active Radiation (PAR) and is measured as irradiance, in units of light quanta, integrated over the wavelength interval of 400 to 700 nm [1]. At wavelengths greater than about 750 nm incident light is preferentially scattered by leaves. This scattering is the reason that plants appear as bright objects in infrared photography. The transition between absorption and scattering causes the sharp jump in reflectance between the visible and near-infrared region of the reflectance spectrum known as the red edge (Fig. 1).

J. Atherton (✉) · C. Zhang · J. Oivukkamäki · L. Kulmala · A. Porcar-Castell
Optics of Photosynthesis Laboratory, Institute for Atmospheric and Earth System Research/Forest Sciences, University of Helsinki, Helsinki, Finland
e-mail: jon.atherton@helsinki.fi

S. Xu
Optics of Photosynthesis Laboratory, Institute for Atmospheric and Earth System Research/Forest Sciences, University of Helsinki, Helsinki, Finland

Faculty of Geographical Science, Beijing Normal University, Beijing, China

T. Hakala · E. Honkavaara
Department of Remote Sensing and Photogrammetry, Finnish Geospatial Research Institute, National Land Survey of Finland, Masala, Finland

A. MacArthur
School of Geosciences, University of Edinburgh, Edinburgh, UK

D. D. Bochtis et al. (eds.), *Information and Communication Technologies for Agriculture—Theme I: Sensors*, Springer Optimization and Its Applications 182,
https://doi.org/10.1007/978-3-030-84144-7_10

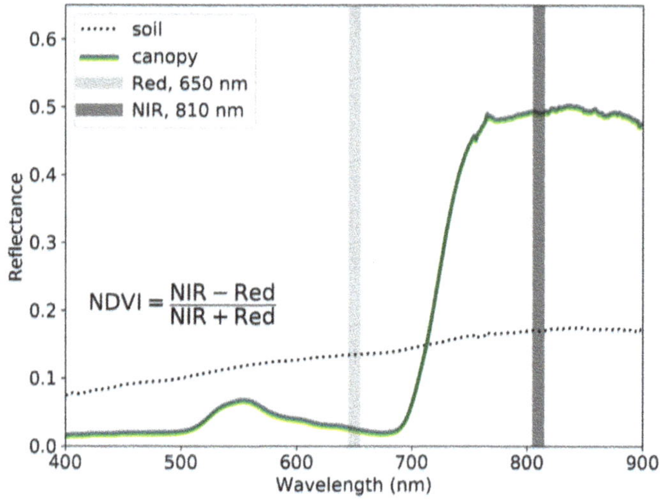

Fig. 1 Reflectance of potato canopy and soil. Canopy spectra present low reflectance in the visible due to absorption by pigments, and increased reflectance in the NIR due to scattering. The NDVI quantifies this difference. The soil spectrum is relatively constant across the shown interval, hence results in a much smaller NDVI relative to the potato. The Red and NIR band centres correspond to the METER NDVI band locations

The passive remote sensing of crops has a long and storied history which is entwined with the development of Earth Observing spectral radiometers in the latter half of the twentieth century (see [2] for a recent perspective). Early instruments measured radiance in a few multispectral bands spanning the visible and near-infrared region of the spectrum. It was evident that algebraically combining band data from these radiometers, to form so called vegetation indices and band ratios, increased the signal to noise content of the data greatly enhancing the observation of vegetation from space [3]. Studies in the 1970s and 1980s demonstrated how such data could be used to estimate vegetation parameters from space, including the leaf area index (LAI), defined as the ratio of the one-sided surface area of leaves to surface area of ground [4]. Global maps of vegetation phenology soon followed [5].

Note that in this early work there was a distinction between simple band ratios which are the division of bands, and vegetation indices (VIs) which are band functions that feature differencing as the main operation [6]. Myneni et al.'s [7] theoretical work showed a functional relationship between VIs and the first derivative of the reflectance spectrum. The advantage of using the first derivative of a spectrum rather than the measured zero order values of a given wavelength band is that confounding variation due to constant offsets between observations (e.g., in time) of single bands, which could occur due to instrumental or target related factors, cancel out in differences [7, 8].

The best-known vegetation index is the normalized difference vegetation index (NDVI), which quantifies the strong contrast between photosynthetic light absorption in the visible region of the spectrum and scattering in the near-infrared (NIR)

region [9, 10]. In simple terms, NDVI puts a single number on the red edge jump in the canopy spectrum, and is typically calculated using NIR and Red reflectance values as:

$$NDVI = \frac{NIR - Red}{NIR + Red} \tag{1}$$

The difference to which NDVI gives its name is in the numerator of Eq. 1, the denominator normalises difference values between −1 and 1. The normalisation makes NDVI potentially easier to interpret compared to the unbounded range possible with, for example, the simple ratio NIR/Red. The normalisation has also been suggested to reduce the effect of sensor degradation [3]. Crippen [11] suggested an interesting reason for the relative success of NDVI, he proposed that the NDVI was "self-perpetuating" as the established standard, having gathered sufficient user inertia.

The main limitation of the NDVI, which was already apparent in early applications [10], is referred to as the saturation effect, and denotes a non-linear asymptotic flattening, or loss of sensitivity, of the curve between NDVI and LAI (or biomass). The loss of sensitivity typically starts at LAI values ranging from 2–4 depending on the crop [12]. Further shortcomings of the NDVI were gradually revealed in the 1980s using physically based methods.

Seller's [6, 13] adapted the two stream (dual direction) radiative transfer formalism used in atmospheric science to model maize NDVI as a function of vegetation structural and optical parameters such as the LAI. His work explored the non-linear relationship between NDVI and LAI, finding that non-linearity was amplified by bare ground in the sensor field of view. However, and unlike for LAI, Seller's [6] went on to demonstrate a linear relationship between the fraction of absorbed PAR (fAPAR) and NDVI. This result makes intuitive sense as the fraction of light absorbed by a canopy will also saturate at a given leaf area. Taken together his results can be interpreted as suggesting that NDVI is useful as a measure of near instantaneous productivity, which depends on fAPAR, but of limited use for LAI, or total biomass, in most green crops or forests due to the saturation effect.

It is tempting to assume that if NDVI is linearly related to fAPAR, then NDVI should also be related to foliar chlorophyll, which is the main light absorbing molecule, or nitrogen content which is used to build leaf proteins.[1] However, as with LAI, the relationship between (red band) NDVI and chlorophyll content saturates at low chlorophyll values [14]. As Seller's [13] work showed, it is more likely that NDVI is influenced by canopy structural factors such as the leaf angle distribution, especially at low LAI values. The dissociation between pigments and NDVI, was also confirmed in latter studies [15] who found no relationships of merit between pigment content and NDVI.

[1] The use of chlorophyll to infer nitrogen is complicated by the fact that the ratio of total nitrogen to chlorophyll nitrogen varies substantially within a plant. More specifically sun leaves have less nitrogen allocated to chlorophyll than shade leaves.

The limitations described above, which are further confounded by atmospheric effects, fractional cover sensitivity and variance in soil colour and brightness, motivated the development of new and more complicated, in terms of mathematical formulae, VIs, designed to address these shortcomings [3, 12, 16]. These include the perpendicular and orthogonal vegetation indices where reflectance values in NIR-Red space are projected onto the so-called soil line, with greater distances (projections) from the line representing increased vegetation fractions [3, 17].

An additional advance that was used as the foundation of NASA's MODIS productivity algorithm, was the Enhanced Vegetation Index (EVI) that corrects for soil effects and is less sensitive to saturation than the NDVI [18]. More recent techniques for tracking productivity from space include Sun-induced Fluorescence (SIF) [19], and the NIRv index which is an adjusted form of NDVI whose derivation is rooted in Seller's [6] theory and which aims to minimise the effects of background variation in the signal [20].

A further development related to vegetation indices is the use of multi-angular observations to characterise the anisotropy (directional dependency) of observations. For a sensor above a sunlit scene, the observable radiance is a function of the inherent optical properties of the objects within the scene, their structure and also the view and solar geometry, and the ratio of diffuse to direct radiation. The Bidirectional Reflectance Distribution Function (BRDF) is a theoretical concept that formalizes the directional dependency as a function of view and direct beam incident light angles. Most measurements of NDVI are therefore subject to directional artefacts, and care must be taken to compare data observed with differing geometrical configurations. However, as well as being a possible hindrance, reflectance anisotropy can also be utilised to retrieve structural information, on for example canopy clumping, from multi-angle reflectance data [21].

Although no longer at the cutting edge of satellite remote sensing, the use of NDVI persists and may even be growing [22, 23]. New remote and proximal (close to canopy) NDVI platforms and sensors differ from the relatively coarse resolution satellites of old, and include high resolution uncrewed aerial vehicles (UAVs) [24, 25], field based robots [26], close contact spectral sensors mounted on mobile phenotyping platforms [27], active NDVI field sensing [22], and fleets of Earth Observing CubeSats [28]. Applications are also migrating from the traditional global photosynthesis prediction and change detection [2] to the rapidly evolving field scale commercial crop analytics, nutrient, and yield prediction [22] and phenotyping [27] disciplines which include a significant commercial element. A parallel and integrated development, is the uptake of data driven analytical modelling methods, referred to as machine learning, to relate NDVI or other optical data to crop parameters [29]. Such methods are useful as they can handle the vast amounts of data generated by high resolution imaging spectroscopy sensors to selectively arrive at accurate predictive models for e.g., chlorophyll content or LAI retrieval [30].

The emergence of the NDVI in new applications is probably due to the relative simplicity of the formula and the ease of measurement, requiring only an NIR and visible sensitive instrument. Obviously from the discussion above, the interpretation of NDVI is far from straightforward. Hence there is a gap that requires bridging between decades of physically based knowledge derived from satellite remote sensing and the state of the art in field and plot scale data, which tends to a data driven focus. This chapter digs into the issue of relating the historical satellite-derived theory [6] to new crop applications and sensors. This is achieved by analysing proximal data acquired at high temporal and spectral resolution in two agricultural sites in Finland, to attempt to answer the question: what does NDVI and spectral reflectance data really tell us about crops in the field? For the sake of simplicity, the chapter is focused on point based spectral reflectance sensing. However, the lessons learned here are also applicable to well calibrated imaging spectroscopy data.

2 Methods

2.1 *Sites, Sensors and Supporting Observations*

Data were acquired from spectral reflectance sensors using proximal field and UAV platforms at two agricultural experimental sites in Finland shown in Fig. 2. The sensors were designed to capture temporal and spectral variation, and the main sensor and site characteristics are listed in Table 1. The temporal dimension was investigated using data collected at the Qvidja research site, in South Western Finland in 2019 and the spectral dimension was investigated using data collected at the Viikki experimental field site in Helsinki in 2018, using a UAV platform.

The Qvidja estate site is an experimental grass site located in southwestern Finland. At this site, mainly timothy and meadow fescue grasses (*Phleum pratense* L. and *Festuca pratensis* Huds. respectively) grown in small 4 by 4 m plots underwent differing fertilization treatments classified into four groupings: no

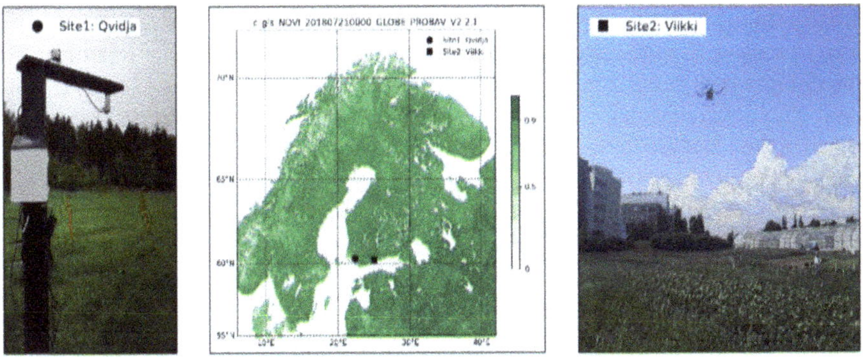

Fig. 2 NDVI instrumentation and site locations. Left frame shows proximal METER NDVI sensor above grasses at Qvidja site. The central frame shows site locations in southern Finland overlaid on a PROBA-V satellite NDVI retrieval from July 2018. The right frame shows the UAV carrying the Piccolo-Doppio payload flying over the potato crop at Viikki, Helsinski

Table 1 Site and NDVI sensor characteristics. Note that there were many additional instruments operating at both sites, as both sets of observations were collected during heavily instrumented field campaigns

Site	Crop	Experimental treatment	Dimension of interest	Sensor name and characteristics	Platform
1. Qvidja, south western Finland 60.2963, 22.3945	Grasses (timothy and meadow fescue)	Fertilization	Temporal	METER (formerly Decagon) NDVI Spectral Reflectance Sensor	Small field supports
2. Viikki, Helsinki, 60.2268, 25.0180	Potatoes (Lady Felicia)	Drought and nutrient	Hyperspectral	Piccolo Doppio hyperspectral DFOV spectrometer	Multi-rotor UAV

fertilization, 50% of optimal mineral fertilization, optimal mineral fertilization, and organic fertilization (For brevity, these categories are referred to as: No fertilization, 50%, Optimised and Organic fertilization.). The plots were harvested at multiple times of year. METER (Pullman, WA, USA), formerly Decagon, NDVI Spectral Reflectance Sensors were deployed at the site during summer 2019. One sensor was placed over each treatment using a wood support structure with a view zenith angle of 45 ° which resulted in a ground instantaneous field of view (IFOV) major axis length of 2.9 m. These devices use the GSM network to transfer data from the field to a cloud service, which is subsequently visualised in a web browser and downloaded as text files for further processing. According to the METER specifications, the red and NIR bands are centred at 650 and 810 nm, with 10 nm FWHM. Data were collected at a 5-minute interval, and daily averaged between 10:00 and 17:00. Supporting LAI measurements and SPAD chlorophyll readings were also taken and are briefly mentioned in the results.

The second experimental site was located within the University of Helsinki's Viikki campus in Helsinki, Finland. Here the potato variety Lady Felicia (*Solanum tuberosum* L.) was grown as row crops from seed in 2018. Two differing treatments were imposed: a paired drought treatment and a multi-level nutrient treatment. In the drought treatment, there were 10 plots of 6 m by 6 m, including a 1 m buffer, in total with a paired (treatment and control) sampling design where one of each pair was irrigated and the corresponding pair was under drought treatment. In the nutrient plots, which were the same dimensions, there were four levels of nutrient addition. The four nutrient addition levels varied between two different levels of nitrogen fertilizer (YaraBela Suomensalpietari, Yara International, Norway) and two different levels of general macronutrient fertilizer (Yara Mila Hevi3, Yara International, Norway) and were replicated four times. Due to the exceptionally good weather in Summer 2018, no rain exclusion was necessary to achieve the required drought effect.

For chlorophyll concentration ([Chl.]) sampling, leaves were picked during the measurement period and frozen until analysis. The frozen leaves were mixed with dimethyl sulfoxide (DMSO), which has been shown to be an effective solvent for chlorophyll and pigment analysis [31]. The samples were then homogenized and extracted in an oven for 4 h at 50 °C before analysis with a Shimadzu UV-1800 spectrometer (Shimadzu Corporation, Japan). [Chl.] was estimated as total chlorophyll *a* and *b* on a leaf area basis.

UAV flights were conducted by Finnish Geospatial Research Institute's (FGI) drone laboratory using a custom-built UAV based on a Gryphon Dynamics quadcopter frame, a Pixhawk autopilot and Applanix APX-15-EI UAV positioning system. The main payload of the UAV was a Piccolo Doppio (PD) Dual Field of View (DFOV) spectrometer system which uses the Cos-conical approach. This system is principally designed to retrieve SIF but also measures visible and near infrared radiance and irradiance, which is the purpose to which it was deployed here. The PD is based around two Ocean Optics spectrometers (Ocean Optics/Insight, Dunedin, FL, USA), a Flame and QE Pro and a bifurcated fibre optic assembly manufactured by Alker Ltd. (Alker Fibre Optic Specialists Ltd., Surrey, UK). The PD DFOV system collects incident irradiance through a cosine corrected diffuser fore-optic attached to one fibre optic leg and upwelling radiance through a bare tipped optical fibre. In the present study both the irradiance and radiance fibre optic cables were mounted to a stabilising gimbal (Photohigher, Wellington, New Zealand) and upwelling radiance was collected in the nadir view. Here data is presented from the Flame spectrometer, which has a usable spectral range of 400 to 950 nm, spectral sampling interval of 0.4 nm and a sampling band width (FWHM) of 1.3 nm. More details concerning the PD can be found in MacArthur et al. [32] and Atherton et al. [33].

The PD UAV flights were conducted on 25th July 2018 by hovering approximately 9 m above ground level for each of the plots. A 9 m height above canopy results in a top of canopy field of view diameter of 4 m, given the bare fiber IFOV angle of 25°. Repeats of 25 spectral radiance and irradiance samples were collected above each plot and subsequently averaged to estimate per plot reflectance. Per flight median optimised integration times were between 7–8 milliseconds for downwelling and were between 12–14 milliseconds for upwelling.

Hemispherical-conical reflectance factors (HCRF) were estimated as the ratio of upwelling radiance spectra, multiplied by π, to downwelling irradiance spectra [34]. Note that when measuring reflectance in the field, the observation depends on the atmospheric conditions, and in particular the diffuse to direct ratio of irradiance which influences the sampling of the BRDF [35, 36]. As such, measuring irradiance and radiance a small distance above the canopy target affects the resulting spectra due to atmospheric scattering and absorption in the path to the target. These points should be kept in mind when comparing data across sites and instruments.

The UAV was also used to collect RGB imagery which was processed to point clouds using AgiSoft Professional software (AgiSoft LLC, St. Petersburg, Russia). The imagery was collected using two Sony A7R II digital cameras with Sony FE 35 mm f/2.8 ZA Carl Zeiss Sonnar T* lens. They were mounted at +15° and − 15° oblique angles in a stabilized rack. The flights were carried out at a flight height of 50 m that resulted in a ground sample distance of 0.64 cm. A double grid setup with six north-south flight lines and nine east-west flight lines were used to obtain a minimum of nine overlapping images over the entire area of interest. Protocols describing the photogrammetric processing chain developed by FGI can be found in Viljanen et al. [37].

2.2 *Data Processing and Analysis*

Data collected at site 1 (Qvidja) required little extra processing except for daily statistics and quality control which were conducted in R 3.5.3 programming language [38]. A field cross calibration was conducted between radiance and irradiance sensors using a near lambertian Spectralon panel on June 5th and these coefficients used to calculate reflectance. At site 2 (Viikki), NDVI was calculated by first interpolating reflectance spectra to a 1 nm wavelength scale and then applying Eq. 1 to PD data using the reflectance values at wavelengths 650 nm and 810 nm.

Leaf angle distributions were estimated from UAV retrieved photogrammetric point cloud normal vectors. The Average Leaf Inclination Angle (ALA) per plot, referenced to the upwards (zenith) pointing vector, was computed as a summary statistic from these angular distributions. A description of this new approach can be found in Xu et al. [39]. In this chapter, the Singular Value Decomposition method was not used, rather the pre-computed vectors from Agisoft software as a higher correlation was found between leaf angles and spectral data using the latter approach. Data below a single reference height (4.4 m above coordinate system zero level) were excluded due to shadow and soil contamination. Similar results were obtained when no height threshold was applied, however the resulting correlations with spectral data were not as strong as when using the height filtered data. Linear correlations between spectral data and other variables (e.g., ALA) were calculated using Pearson's correlation coefficient.

3 Results and Discussion

3.1 *Temporal Variability at Site 1*

Figure 3 shows the time course of NDVI measurements during summer 2019 at the Qvidja site. There were two harvest periods which are marked on the figure at the start of June and mid-way through August. Post-harvest NDVI values were around 0.6–0.7 rising to maxima of around 0.8 for the organic and optimized fertilized sites a few weeks after the harvest period. The fertilized sites show elevated NDVI for most of the time course. At end of the season, variance in each of the time-series is increased which could be due to the lower light levels at the end of the year.

In these grasses, NDVI clearly tracks growth post-harvest and differentiates between fertilized and untreated plots. As corroborating evidence, LAI increases were measured over the experimental period with highest values in the fertilized plots; data from a SPAD chlorophyll meter showed no increasing trends over the sampling period (data not shown). It is therefore probable that changes in NDVI related to an associated increase in the fraction and total amount of absorbed PAR by photosynthetic elements which was determined by the fractional coverage and also related, probably asymptotically, to LAI [6].

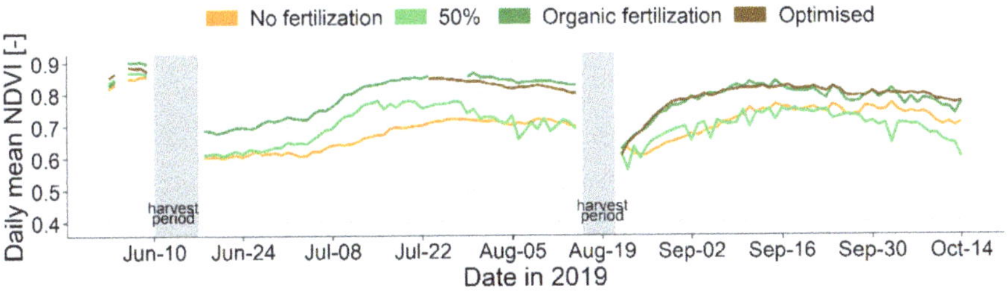

Fig. 3 Time-series of proximally sensed NDVI of grasses undergoing nutrient treatments at Qvidja estate site, Southwestern Finland. Data are daily averaged

Variation in the NDVI is driven by canopy scattering which occurs principally in the NIR, and soil reflectance which occurs across all wavelengths [6, 13]. To explore these issues further, the focus shifts to the second site where variability across space using a UAV-based hyperspectral instrument was observed.

3.2 *Spectral-Spatial Variability at Site 2*

When retrieved proximally from a UAV platform, there was a negative relationship between Average Leaf inclination Angle (ALA) and NDVI in the potato crops at site 2 (Pearson's $r = -0.77$, Fig. 4). Further, those plots under greatest water deficit presented the largest ALA and smallest NDVI values. This results chimes with previous research where the link between canopy structure and NDVI under water limited conditions has been highlighted, and where NDVI is referred to as a structural vegetation index [40].

Water limitation causes a lack of leaf turgidity, commonly referred to as wilting, resulting in greater leaf inclination angles and hence larger plot-wise ALA. This shift towards an erectophile distribution influences the NDVI due to the increased canopy gap fraction both within and between crop rows. Increasing the gap fraction exposes a greater proportion of soil, increasing the contribution of soil reflectance to the total signal. An increase in gap fraction also increases the distance between scattering and absorbing leaf elements and changes the canopy shadowing properties. It is not possible to separate between all these factors using the NDVI alone, here hyperspectral data is instructional.

ALA is correlated with canopy scale reflectance (Fig. 5) across large parts of the visible and NIR spectrum. However, both the strength and sign of the correlation (Fig. 5 lower panel) is dependent on wavelength. Visible reflectance is positively correlated with ALA, whereas in the NIR the sign switches to a negative correlation of increased magnitude. This can be explained by the interacting effects of light scattering by soil and vegetation. In sections of the visible region, soil is of higher reflectivity than vegetation (Fig. 1). Increasing the ALA means that the sensor FOV contains a higher proportion of soil background, resulting in the positive

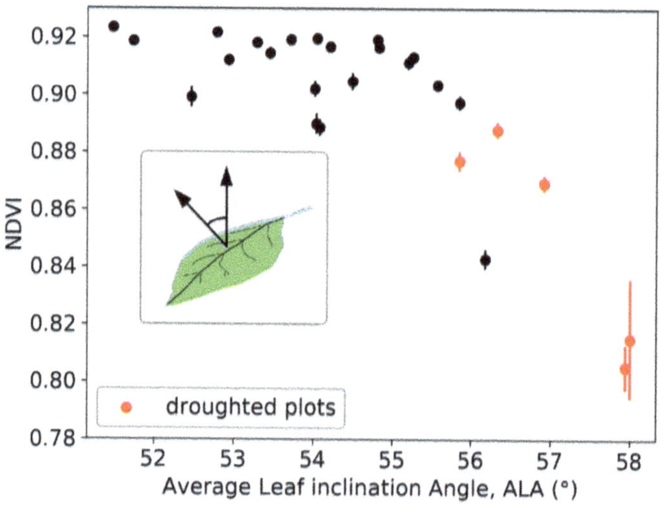

Fig. 4 Average Leaf inclination Angle (ALA) per plot Vs NDVI for a combination of water and nutrient treatment potato plots. Drought treatment plots with wilted leaves tend to the bottom right of the plot, with relatively high ALA and low NDVI values. NDVI error bars are one sample standard deviation of batch means

correlation. In contrast, in the NIR region scattering in the sensor FOV is dominated by canopy leaves [6]. An increase in ALA reduces the probability that a photon is scattered in the direction of the nadir pointing sensor, resulting in a negative correlation at those wavelengths greater than approximately 700 nm. These results combine to decrease reflectance under water stressed conditions in the NIR, due to lessening leaf scattering, but increase reflectance in much of the visible region due to increased soil scattering. This result is in contradiction to Moran et al. [41] who found water related scattering decreases in the visible and NIR regions.

Returning to the NDVI, then it appears that variation in the index is caused by the interplay of soil and canopy scattering. Although these mechanisms have opposite signs of correlation with the structural variable ALA, they both decrease the NDVI relative to increasing ALA (loss of turgidity) due to the mathematical formulation of the index. This is because NDVI can be reduced due to either an increase in red reflectance, caused by soil scattering, or by a decrease in NIR reflectance, caused by a reduction in leaf scattering. However, just because NDVI responds to ALA does not mean NDVI is an optimal remote estimator of ALA. Note that NIR reflectance has slightly higher correlation with ALA relative to NDVI (Fig. 5 lower panel). Further, Zou and Mõttus [42] found that although NDVI was related to ALA in crops, other indices demonstrated stronger correlations.

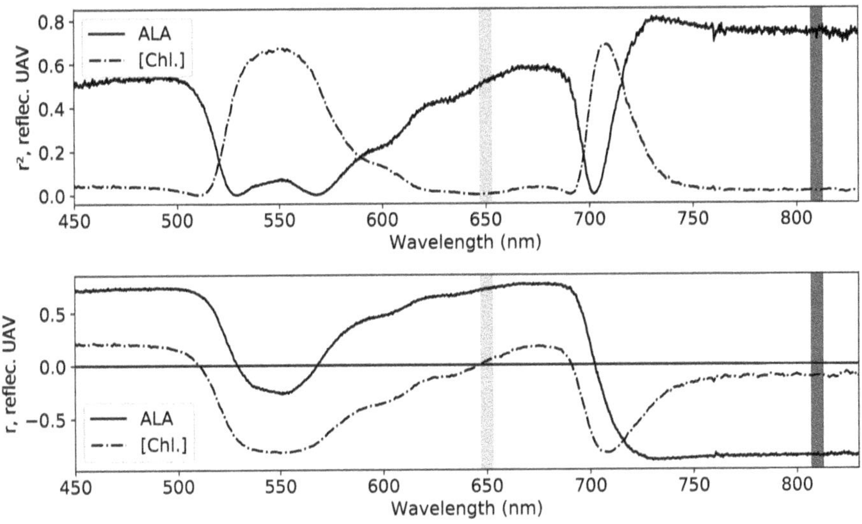

Fig. 5 Spectral correlation with structural (ALA) and leaf level ([Chl.]) variables. Top panel shows the square of Pearson's correlation coefficient between spectral reflectance (HCRF) and variables, and bottom panel shows Pearson's correlation between spectral reflectance (HCRF) and variables. Vertical lines mark the location of NDVI bands

The two mechanisms discussed above, which control the NDVI in the potato crop, do not directly[2] relate to photosynthetic light absorption at the leaf scale. In this case variation in leaf optical properties, and by extension chlorophyll, is largely immaterial to NDVI which is under the control of variance in canopy scale topology. The breakdown of the relationship between pigment content and NDVI under water limited conditions has previously been noted and alternative indices applied for chlorophyll and nitrogen estimation such as the MERIS terrestrial chlorophyll index that focus specifically on the red edge region [15]. Note that, in addition to the relevance to field data discussed in the current chapter, retrieval of chlorophyll from space is currently a major goal in satellite remote sensing [43].

Clearly broadband NDVI cannot be used to estimate leaf properties such as chlorophyll content in the water limited case. However, the results shown in Fig. 5 suggest that a sensor with differing band location could potentially be used to infer pigment content. Figure 5 shows a clear correlation between canopy reflectance and foliar chlorophyll concentration in the red edge close to 700 nm, and in the green region where correlation with ALA is minimal. In contrast, the NDVI bands fall squarely in the soil-structure dominated spectral regions (Fig. 5 top panel). As an interesting aside, the empirical results shown in the top panel of Fig. 5 closely resemble the model predictions found in Fig. 1 of Zou and Mõttus [44].

[2]As an indirect effect, wilting does reduce absorbed PAR, but this mechanism is a canopy rather than leaf scale process.

To summarise the above, NDVI does respond to water deficit. However, this response is due to structural changes in the canopy, not variance in pigments or leaf optical properties. These results suggest that, whilst care must be taken in interpretation, NDVI can indeed be used as an indicator of water stress capturing water related variation in soil and canopy scattering properties in the red and NIR bands respectively.

4 Conclusions and Outlook

At both sites, the NDVI conveyed information relating to canopy structure. At the first site, the NDVI followed post-harvest grass development, differentiating between nutrient treatments (Fig. 3). At the second site, NDVI responded to water limitation in potatoes via variation in canopy scattering but was unrelated to pigment content (Figs. 4 and 5). Therefore, reflectance in the red-edge or green region is required to relate canopy observations to leaf pigments, especially under conditions of structural variance caused by water limitation.

At the two study sites, area integrated point data were used to investigate variance in canopy reflectance in the temporal and spatial dimensions. Such data are useful as a robust empirical benchmark and are potential candidates for the validation of CubeSat-based NDVI and LAI retrievals. However, the future of field spectroscopy lies in high spatial resolution imagery [24]. Once properly calibrated, such systems can probe variance in ecosystem processes at the individual plant scale across the field. So how best to capitalize on our current knowledge going forward in this direction?

The application of mature radiative transfer codes to estimate biophysical parameters (e.g., pigment content) from multi-angular UAV data is an avenue under current investigation [45]. However, it is worth emphasising that the spatial scales that data is now generated at are unprecedented and call for the development of new physically based tools and models. Verrelst [30] discusses options for hybrid imaging spectroscopy approaches combining physically based remote sensing with data driven methods which is an interesting avenue of future research. The method of radiosity [46, 47], which models the canopy as a collection of planar canopy elements throughout 3D space, is a worthwhile avenue of research to revisit in this direction. Following from this, perhaps the most interesting direction of research is the extension of imaging spectroscopy into the 3rd spatial dimension and the opportunities that this affords [48].

Acknowledgments Laura Heimsch is thanked for her help in the field and LK acknowledges Business Finland's (project number 6905/31/2018) funding of the Qvidja study. Niko Viljanen is thanked for processing point clouds. Anu Riikonen is thanked for her help with the Viikki experimental plot. The measurements at Viikki were funded by Academy of Finland project decision number 304097. Regarding satellite data shown in Fig. 2: "The products were generated by the Global Land Service of Copernicus, the Earth Observation programme of the European Commission. The research leading to the current version of the product has received funding from various European Commission Research and Technical Development programs. The product is based on PROBA-V data provided by ESA and distributed by VITO NV".

References

1. McCree, K.J., 1971. The action spectrum, absorptance and quantum yield of photosynthesis in crop plants. Agricultural Meteorology, 9, pp.191-216.
2. Ryu, Y., Berry, J.A. and Baldocchi, D.D., 2019. What is global photosynthesis? History, uncertainties and opportunities. Remote sensing of environment, 223, pp.95-114.
3. Bannari, A., Morin, D., Bonn, F. and Huete, A.R., 1995. A review of vegetation indices. Remote sensing reviews, 13(1-2), pp.95-120.
4. Wiegand, C.L., Richardson, A.J. and Kanemasu, E.T., 1979. Leaf Area Index Estimates for Wheat from LANDSAT and Their Implications for Evapotranspiration and Crop Modeling 1. Agronomy Journal, 71(2), pp.336-342.
5. Justice, C.O., Townshend, J.R.G., Holben, B.N. and Tucker, E.C., 1985. Analysis of the phenology of global vegetation using meteorological satellite data. International Journal of Remote Sensing, 6(8), pp.1271-1318.
6. Sellers, P.J., 1985. Canopy reflectance, photosynthesis and transpiration. International journal of remote sensing, 6(8), pp.1335-1372.
7. Myneni, R.B., Hall, F.G., Sellers, P.J. and Marshak, A.L., 1995. The interpretation of spectral vegetation indexes. IEEE Transactions on Geoscience and Remote Sensing, 33(2), pp.481-486.
8. Curran, P.J., Dungan, J.L. and Gholz, H.L., 1990. Exploring the relationship between reflectance red edge and chlorophyll content in slash pine. Tree physiology, 7(1-2-3-4), pp.33-48.
9. Rouse, J. W., Haas, R. W., Schell, J. A., Deering, D. W. and Harlan, J. C., 1974. Monitoring the vernal advancement and retrogradation (Greenwave effect) of natural vegetation NASA/ GSFCT Type III Final Report, Greenbelt, MD, USA.
10. Tucker, C.J., 1979. Red and photographic infrared linear combinations for monitoring vegetation. Remote Sensing of Environment, 8(2), pp.127-150.
11. Crippen, R.E., 1990. Calculating the vegetation index faster. Remote sensing of Environment, 34(1), pp.71-73.
12. Carlson, T.N. and Ripley, D.A., 1997. On the relation between NDVI, fractional vegetation cover, and leaf area index. Remote sensing of Environment, 62(3), pp.241-252.
13. Sellers, P.J., 1987. Canopy reflectance, photosynthesis, and transpiration, II. The role of biophysics in the linearity of their interdependence. Remote sensing of Environment, 21(2), pp.143-183.
14. Gitelson, A.A., Kaufman, Y.J. and Merzlyak, M.N., 1996. Use of a green channel in remote sensing of global vegetation from EOS-MODIS. Remote sensing of Environment, 58(3), pp.289-298.
15. Eitel, J.U.H., Long, D.S., Gessler, P.E. and Hunt, E.R., 2008. Combined spectral index to improve ground-based estimates of nitrogen status in dryland wheat. Agronomy journal, 100(6), pp.1694-1702.
16. Rondeaux, G., Steven, M. and Baret, F., 1996. Optimization of soil-adjusted vegetation indices. Remote sensing of environment, 55(2), pp.95-107.
17. Richardson, A.J. and Wiegand, C.L., 1977. Distinguishing vegetation from soil background information. Photogrammetric engineering and remote sensing, 43(12), pp.1541-1552.
18. Huete, A., Didan, K., Miura, T., Rodriguez, E.P., Gao, X. and Ferreira, L.G., 2002. Overview of the radiometric and biophysical performance of the MODIS vegetation indices. Remote sensing of environment, 83(1-2), pp.195-213.
19. Mohammed, G.H., Colombo, R., Middleton, E.M., Rascher, U., van der Tol, C., Nedbal, L., Goulas, Y., Pérez-Priego, O., Damm, A., Meroni, M. and Joiner, J., Cogliati, S., Verhoef, W., Malenovský, Z., Gastellu-Etchegorry, J.P., Miller, J.R., Guanter, L., Moreno, J. Moya, I., Berry, J.A., Frankenberg, C. and Zarco-Tejada, P.J. 2019. Remote sensing of solar-induced chlorophyll fluorescence (SIF) in vegetation: 50 years of progress. Remote sensing of environment, 231, p.111177.
20. Badgley, G., Field, C.B. and Berry, J.A., 2017. Canopy near-infrared reflectance and terrestrial photosynthesis. Science advances, 3(3), p.e1602244.

21. Chen, J.M., Liu, J., Leblanc, S.G., Lacaze, R. and Roujean, J.L., 2003. Multi-angular optical remote sensing for assessing vegetation structure and carbon absorption. Remote Sensing of Environment, 84(4), pp.516-525.
22. Franzen, D., Kitchen, N., Holland, K., Schepers, J. and Raun, W., 2016. Algorithms for in-season nutrient management in cereals. Agronomy Journal, 108(5), pp.1775-1781.
23. Xue, J. and Su, B., 2017. Significant remote sensing vegetation indices: A review of developments and applications. Journal of Sensors,.
24. Aasen, H., Honkavaara, E., Lucieer, A. and Zarco-Tejada, P. J., 2018. Quantitative remote sensing at ultra-high resolution with UAV spectroscopy: a review of sensor technology, measurement procedures, and data correction workflows. Remote Sensing, 10(7), p.1091
25. Manfreda, S., McCabe, M.F., Miller, P.E., Lucas, R., Pajuelo Madrigal, V., Mallinis, G., Ben Dor, E., Helman, D., Estes, L., Ciraolo, G. and Müllerová, J., Tauro, F., De Lima, M.I., De Lima, J.L.M.P., Maltese, A., Frances, F., Caylor, K., Kohv, M., Perks, M., Ruiz-Pérez, G., Su, Z., Vico, G. and Toth, B., 2018. On the use of unmanned aerial systems for environmental monitoring. Remote sensing, 10(4), p.641.
26. Bai, G., Ge, Y., Scoby, D., Leavitt, B., Stoerger, V., Kirchgessner, N., Irmak, S., Graef, G., Schnable, J. and Awada, T., 2019. NU-Spidercam: A large-scale, cable-driven, integrated sensing and robotic system for advanced phenotyping, remote sensing, and agronomic research. Computers and Electronics in Agriculture, 160, pp.71-81.
27. Enciso, J., Maeda, M., Landivar, J., Jung, J. and Chang, A., 2017. A ground based platform for high throughput phenotyping. Computers and Electronics in Agriculture, 141, pp.286-291.
28. Houborg, R. and McCabe, M.F., 2018. Daily Retrieval of NDVI and LAI at 3 m Resolution via the Fusion of CubeSat, Landsat, and MODIS Data. Remote Sensing, 10(6), p.890.
29. Condorelli, G.E., Maccaferri, M., Newcomb, M., Andrade-Sanchez, P., White, J.W., French, A.N., Sciara, G., Ward, R. and Tuberosa, R., 2018. Comparative aerial and ground based high throughput phenotyping for the genetic dissection of NDVI as a proxy for drought adaptive traits in durum wheat. Frontiers in plant science, 9, p.893.
30. Verrelst, J., Malenovský, Z., Van der Tol, C., Camps-Valls, G., Gastellu-Etchegorry, J.P., Lewis, P., North, P. and Moreno, J., 2019. Quantifying vegetation biophysical variables from imaging spectroscopy data: a review on retrieval methods. Surveys in Geophysics, 40(3), pp.589-629.
31. Wellburn, R.W., 1994. The spectral determination of chlorophylls a and b, as well as total carotenoids, using various solvents with spectrophotometers of different resolution. Journal of plant physiology, 144(3), pp.307-313.
32. MacArthur, A, Robinson, I, Rossini, M, Davis, N and MacDonald, K, 2014. A dual-field-of-view spectrometer system for reflectance and fluorescence measurements (Piccolo Doppio) and correction of etaloning. Proceedings of the Fifth International Workshop on Remote Sensing of Vegetation Fluorescence. European Space Agency, Fifth International Workshop on Remote Sensing of Vegetation Fluorescence, Paris, United Kingdom, 22-24 April 2014.
33. Atherton, J., MacArthur, A., Hakala, T., Maseyk, K., Robinson, I., Liu, W., Honkavaara, E. and Porcar-Castell, A., 2018, July. Drone Measurements of Solar-Induced Chlorophyll Fluorescence Acquired with a Low-Weight DFOV Spectrometer System. In IGARSS 2018-2018 IEEE International Geoscience and Remote Sensing Symposium, pp.8834-8836.
34. Hakala, T., Markelin, L., Honkavaara, E., Scott, B., Theocharous, T., Nevalainen, O., Näsi, R., Suomalainen, J., Viljanen, N., Greenwell, C. and Fox, N., 2018. Direct reflectance measurements from drones: sensor absolute radiometric calibration and system tests for forest reflectance characterization. Sensors, 18(5), p.1417.
35. Schaepman-Strub, G., Schaepman, M.E., Painter, T.H., Dangel, S. and Martonchik, J.V., 2006. Reflectance quantities in optical remote sensing—Definitions and case studies. Remote sensing of environment, 103(1), pp.27-42.
36. Stuckens, J., Somers, B., Verstraeten, W.W., Swennen, R. and Coppin, P., 2009. Evaluation and normalization of cloud obscuration related BRDF effects in field spectroscopy. Remote Sensing, 1(3), pp.496-518.

37. Viljanen, N., Honkavaara, E., Näsi, R., Hakala, T., Niemeläinen, O. and Kaivosoja, J., 2018. A novel machine learning method for estimating biomass of grass swards using a photogrammetric canopy height model, images and vegetation indices captured by a drone. Agriculture, 8(5), p.70.
38. R Core Team, 2019. R: A language and environment for statistical computing. R Foundation for Statistical Computing, Vienna, Austria. URL http://www.R-project.org/
39. Xu, S., Zaidan, M.A., Honkavaara, E., Hakala, T., Viljanen, N., Porcar-Castell, A., Liu, Z. and Atherton, J., 2020. On the Estimation of the Leaf Angle Distribution from Drone Based Photogrammetry. In IGARSS 2020-2020 IEEE International Geoscience and Remote Sensing Symposium, pp.4379-4382.
40. Ihuoma, S.O. and Madramootoo, C.A., 2017. Recent advances in crop water stress detection. Computers and Electronics in Agriculture, 141, pp.267-275.
41. Moran, M.S., Pinter Jr, P.J., Clothier, B.E. and Allen, S.G., 1989. Effect of water stress on the canopy architecture and spectral indices of irrigated alfalfa. Remote sensing of environment, 29(3), pp.251-261.
42. Zou, X. and Mõttus, M., 2017. Sensitivity of common vegetation indices to the canopy structure of field crops. Remote Sensing, 9(10), p.994.
43. Croft, H., Chen, J.M., Wang, R., Mo, G., Luo, S., Luo, X., He, L., Gonsamo, A., Arabian, J., Zhang, Y. and Simic-Milas, A., Noland, T.L., He, Y., Homolová, L., Malenovský, Z., Yi, Q. Beringer, J. Amiri, R. Hutley, L., Arellano, P., Stahl, C., Bonal, D., 2020. The global distribution of leaf chlorophyll content. Remote Sensing of Environment, 236, p.111479.
44. Zou, X. and Mõttus, M., 2015. Retrieving crop leaf tilt angle from imaging spectroscopy data. Agricultural and Forest Meteorology, 205, pp.73-82.
45. Roosjen, P.P., Brede, B., Suomalainen, J.M., Bartholomeus, H.M., Kooistra, L. and Clevers, J.G., 2018. Improved estimation of leaf area index and leaf chlorophyll content of a potato crop using multi-angle spectral data–potential of unmanned aerial vehicle imagery. International journal of applied earth observation and geoinformation, 66, pp.14-26.
46. Borel, C.C., Gerstl, S.A. and Powers, B.J., 1991. The radiosity method in optical remote sensing of structured 3-D surfaces. Remote Sensing of Environment, 36(1), pp.13-44.
47. Qin, W. and Gerstl, S.A., 2000. 3-D scene modeling of semidesert vegetation cover and its radiation regime. Remote Sensing of Environment, 74(1), pp.145-162.
48. Aasen, H., Burkart, A., Bolten, A. and Bareth, G., 2015. Generating 3D hyperspectral information with lightweight UAV snapshot cameras for vegetation monitoring: From camera calibration to quality assurance. ISPRS Journal of Photogrammetry and Remote Sensing, 108, pp.245-259.

Geophysical Sensors for Mapping Soil Layers – A Comparative Case Study Using Different Electrical and Electromagnetic Sensors

Erika Lück, Julien Guillemoteau, Jens Tronicke, Jana Klose, and Benjamin Trost

1 Introduction

In smart farming and precision agriculture, site-specific management is based, among other things, on information about field heterogeneity, which can be obtained from plant observations, and on soil variability resulting, for example, from data of geophysical measurements. Electrical conductivity data provide reliable information that can be used to image lateral field heterogeneities fast and cost-effectively] [1]. In many studies, apparent electrical conductivity (ECa) data are correlated with parameters such as salinity [2, 3], soil moisture [4–6], and soil texture [7–9]. Because soil fertility and yield depend on these parameters, there are also studies linking ECa to yield [8, 10–13]. Particularly, Vitharana et al. Vitharana et al. [14] and Van Meirvenne et al. [15] concluded in their studies that ECa can be one of the key features for delineating agricultural management zones.

In a first step, ECa data are usually used to image lateral soil heterogeneity over an integrated depth interval of approximately 1–2 m below Earth's surface. Today, ECa maps are increasingly used to optimize the location of soil sampling points [16, 17], to establish management zones [18–20] and to explain the variability of biomass and yield [21–23]. Additional information regarding vertical heterogeneity (e.g., soil layering including depths of layers) helps, for example, to understand in more detail soil-moisture relationships, the interaction between soil and plants, to select plant types with appropriate root depth, and to optimize the operation of agricultural machines.

E. Lück (✉) · J. Guillemoteau · J. Tronicke · J. Klose
University of Potsdam, Institute of Geosciences, Potsdam, Germany
e-mail: elueck@geo.uni-potsdam.de

B. Trost
Leibniz Institute for Agricultural Engineering and Bioeconomy (ATB), Potsdam, Germany

D. D. Bochtis et al. (eds.), *Information and Communication Technologies for Agriculture—Theme I: Sensors*, Springer Optimization and Its Applications 182,
https://doi.org/10.1007/978-3-030-84144-7_11

To close the gap between two- and three- dimensional (2D and 3D) soil data, several novel techniques have been evaluated [24–26] including new instruments and inversion routines. Today, various multi-channel electromagnetic (EM) systems (such as DUALEM, Dualem.com, Canada; CMD-Explorer, GF-instruments, Czech Republic and Topsoil Mapper, Geoprospectors, Austria) as well as multi-frequency EM-systems (such as Profiler EMP-400, GSSI, USA; GEM-2 Geophex Ltd., USA) are available to measure simultaneously with different depth sensitivities. Furthermore, different systems with galvanically coupled electrodes are used to obtain 3D soil information [27–30]. Further recent developments include 1D inversion routines to reconstruct conductivity-depth-functions from EM and direct current (DC) data [31–38], as well as quasi-2D/3D using 1D inversion with lateral constraints [39–41] and full 3D inversion routines [42] to generate 2D and 3D subsurface conductivity models.

Data quality and consistency are critical when performing an inversion of measured ECa data into conductivity models. Sufficient data quality can be achieved by suppressing or eliminating noisy data and performing shift corrections [43, 44]. In the past, drift in electromagnetic data was often considered as a major problem and the need for stable instruments has been formulated [45]. Many modern instruments (including the DUALEM system used in this study) are equipped with a temperature stabilization. Studies investigating drift-related phenomena for the DUALEM sensor have been carried out by Hampe [46]. He found a minor drift of ~4 mS m^{-1} only at the beginning of his measurements (20 min after strong temperature changes) for one of the four channels of this system. Moreover, persistent minor temperature drifts can be significantly reduced using a non-conductive thermal insulator [47]. Consistent data means that all channels of an instrument are calibrated and the relation between several channels is only influenced by their individual depth-sensitivity function and soil stratigraphy. To evaluate different sensors and interpretations, the resulting conductivity-depth models are typically compared with:

1. Direct-push measurements [8, 42, 48];
2. Results from boreholes soil sampling [30];
3. Conductivity models obtained using other sensors [49–51].

In practice, a particular challenge is to map soils with only minor conductivity contrasts. This corresponds to a typical situation in northern Germany, where often sandy and loamy soils with small-scale heterogeneities characterize the landscape. The thickness of the sand cover varies and thus also the water holding capacity within the rooting zone of the plant. In addition to small contrasts, low conductivities with average values between 10 and 20 mS m^{-1} make the application of the electromagnetic induction method (EMI) particularly difficult in such environments. In the case study presented here, the DUALEM-21 system (Dualem, Inc., Canada) is used on a sandy soil with small variability, to evaluate the potential of this sensor for such environments. The study examines to what extent the DUALEM-21 is a suitable tool for mapping lateral and vertical changes in soil properties at the field scale. For comparison and evaluation, DC geoelectrical measurements with fixed static electrodes along three transects and direct soil sampling including texture analysis are used. Furthermore, the results of ECa mapping are compared with the results obtained using the multi-array direct current (DC) system Geophilus [28]. To complete the frequency range of electrical (static) and electromagnetic (10 kHz) geophysical methods, the results of a 250 MHz GPR (ground-penetrating radar) survey performed at our field site are also considered.

2 Materials and Methods

2.1 *Site Characteristics*

The field site is a 0.85-hectare test field in Marquardt (52°27′50″ N, 12°57′31″ E) located about 30 km west of Berlin, Germany (Fig. 1a). At a height of 42 meters above sea level, the field shows a flat terrain with a slight topographic increase of 3 m from north-west to south-east. The glacial embossed site is located in the Havelland region – about 150 m east of the river Wublitz, a branch of the river Havel (Fig. 1b). The topsoil is a sandy loam, and the bedrock consists of glacial till. Because the soil was formed by glacial and postglacial deposits, it is characterized by small-scale variabilities with only minor variations in soil texture. The data were collected in April 2016. At this time, the volumetric water content in the topsoil, as measured with a TDR probe, varied between 12 and 18%.

Today, the field site (Fig. 1c) is part of a research site operated by the Leibniz Institute for Agricultural Engineering and Bioeconomy (ATB) to develop sensor-based approaches for precision agriculture and horticulture. In the years before, the

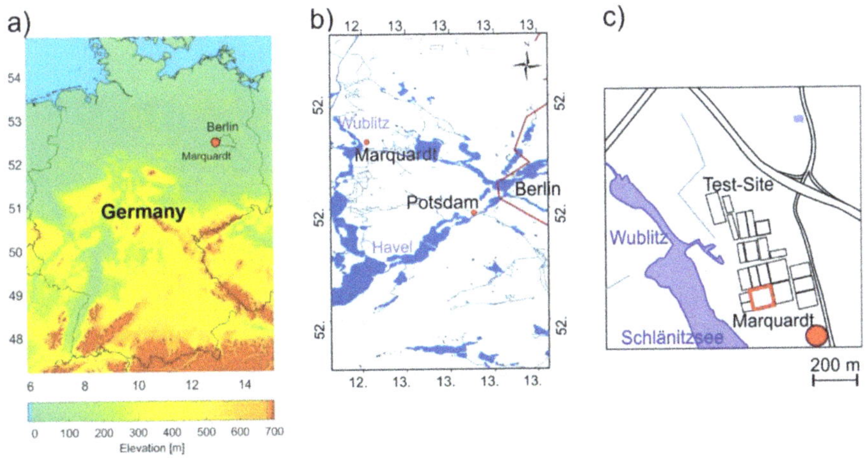

Fig. 1 Location of the study site. (**a**) about 30 km west of Berlin in north-east-Germany (data source: GLOBE1 [52]), (**b**) near the river Wublitz (data source: OpenStreetMap [53]) and (**c**) in the Fieldlab for Digital Agriculture operated by the ATB. The test field is highlighted by a red rectangle

field served as an orchard test site of the Federal Plant Variety Office of Germany. To ensure that natural soil heterogeneity is depicted instead of a former experimental design, the test field was uniformly managed since 2015.

2.2 Data Acquisition

2.2.1 ECa Mapping with DUALEM-21

The DUALEM-21 is a Slingram-based electromagnetic induction sensor that operates at a source frequency of 9 kHz. With four receiver coils in horizontal coplanar (HCP) and perpendicular (PERP) orientation (Fig. 2a), four different depth sensitivities are realized [54]. While the out-of-phase data are mainly influenced by the

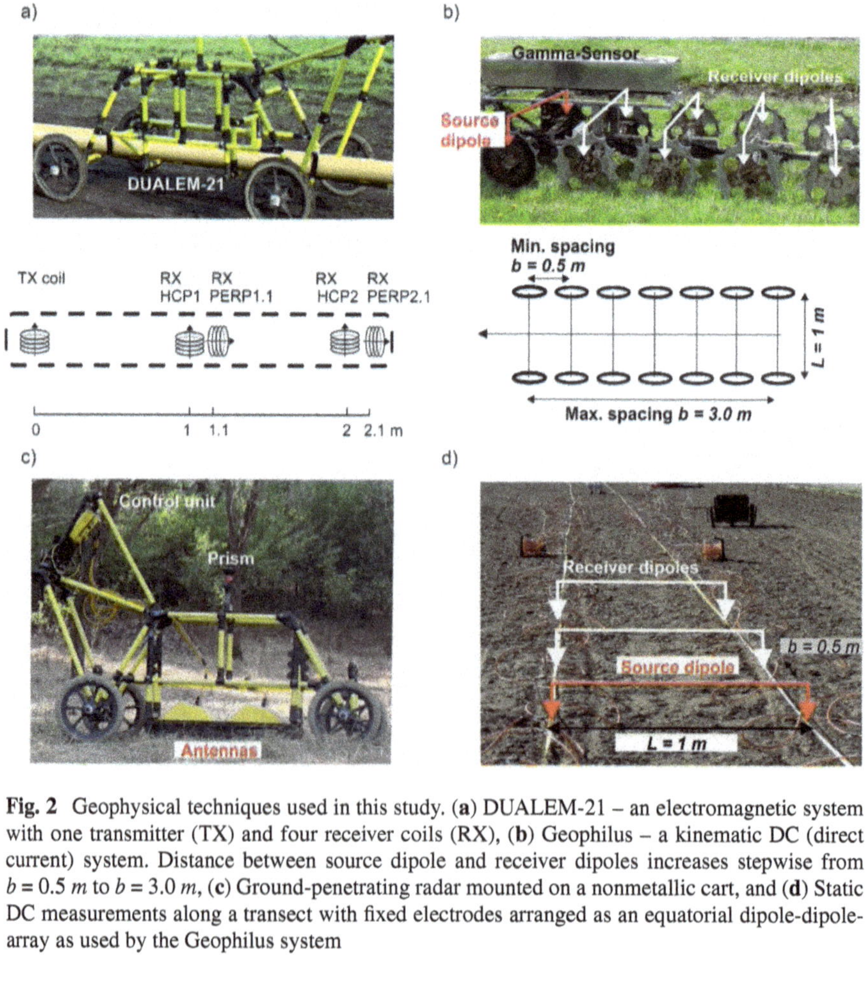

Fig. 2 Geophysical techniques used in this study. (**a**) DUALEM-21 – an electromagnetic system with one transmitter (TX) and four receiver coils (RX), (**b**) Geophilus – a kinematic DC (direct current) system. Distance between source dipole and receiver dipoles increases stepwise from $b = 0.5\ m$ to $b = 3.0\ m$, (**c**) Ground-penetrating radar mounted on a nonmetallic cart, and (**d**) Static DC measurements along a transect with fixed electrodes arranged as an equatorial dipole-dipole-array as used by the Geophilus system

Table 1 Field parameters of the geophysical surveys performed at the field site

	DUALEM	Geophilus	GPR
Crossline spacing	~5 m	~5 m	~1 m
Inline spacing	0.06–0.10 m	1.43 m	0.06 m
Number of channels	4	6	1 Time series

subsurface electrical conductivity [55], information about the magnetic soil properties is contained in the in-phase component [56–60]. In this study, only out-of-phase data are considered.

During field measurements, the DUALEM instrument was mounted on a metal-free cart (Fig. 2a) and moved manually across the field with a line spacing of ~5 m. Therefore, the spatial point distribution of the resulting DUALEM dataset shows a higher inline sample point density than in crossline direction. A sampling interval of 0.1 s resulted in an inline point spacing of ~10 cm (Table 1). Although the DUALEM is equipped with an internal non-differential global positioning system (GPS), a more precise positioning (in the order of several cm) was realized by a total station with an automatic tracking option and a prism mounted on the cart [61]. As a by-product, this strategy also provides a detailed digital terrain model (DTM) of the surveyed area.

The data were collected using a so-called zigzag mode. Because the spatial sensitivity for perpendicular coil orientations of the DUALEM is not symmetrical in the inline direction, it would be better to collect the data in parallel mode. Due to the relatively large profile spacing compared to the footprint of the DUALEM, this is not critical for the presented data.

2.2.2 ECa Mapping with Geophilus

Geophilus [28] is a DC sensor and works with rolling electrodes arranged as an equatorial dipole-dipole array (Fig. 2b) pulled by a cross country vehicle. All seven dipoles (one source dipole and six receiver dipoles termed channel 1 to channel 6) show the same dipole length of 1 m and a spacing to the neighbouring dipoles of b = 0.5 m. In contrast to previous published Geophilus specifications [28], a modified Geophilus sensor was used in this study. The electronics were replaced by an industrial instrument known as Expert Vibro (Delphin Technology, Germany), and the electrical coupling of the electrodes has been modified by the Geophilus company (Geophilus GmbH, Germany). An additional gamma sensor, which measures the total counts of gamma radiation, is added to the system to limit the ambiguity in the interpretation of the measured electrical conductivity data. Although gamma radiation data have also been measured, only the electrical conductivity data are considered in this study.

For the data set measured on our field site, the distance between the tracks was ~5 m (i.e., comparable to our DUALEM dataset). The inline data point spacing depends on the driving velocity of the vehicle and the used sampling frequency. With a velocity of ~5.5 km h^{-1} and a sampling frequency of 1 Hz, an inline point spacing of ~1.5 m was achieved (Table 1, Fig. 3b). In order to georeference the measured data, a differential GPS with an accuracy of ~0.5 m was used.

2.2.3 DC Measurements with Static Electrodes Along Reference Transects

Typically, the data quality of kinematic surveys is lower than that of static measurements. In order to evaluate the results of the mobile sensors used, small-scale measurements were carried out along three selected transects using conventional static geoelectrics with fixed electrodes (Fig. 2d) and an inline point spacing of 0.5 m. A GeoTom system (GeoLog, Germany) was used, and 100 electrodes were arranged as an equatorial dipole-dipole array with identical geometry as used for the Geophilus survey. In Fig. 3a, the position and the orientation of the transects 1, 2, and 3 are shown together with the data-point locations of the DUALEM survey, where the direction of these transects is indicated by arrows.

2.2.4 Ground-Penetrating Radar

In order to complete the frequency spectrum of electromagnetic methods, the field was also mapped with ground-penetrating radar (GPR). GPR is useful to image the topography of internal soil horizons as well as artificial objects (e.g., drainage pipes and other utilities) if there is a contrast in dielectric permittivity. A pulseEKKO system (Sensors & Software, Canada) with 250 MHz shielded antennas was used. The antennas were mounted with an offset of 0.38 m on the same cart used for DUALEM surveying. Thus, also the GPR measurements have been performed manually and georeferenced by the positioning data as measured by an auto-tracking

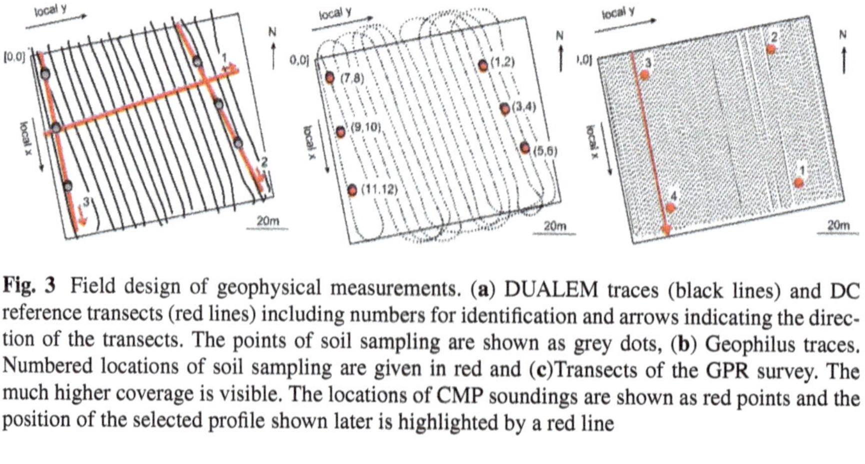

Fig. 3 Field design of geophysical measurements. (**a**) DUALEM traces (black lines) and DC reference transects (red lines) including numbers for identification and arrows indicating the direction of the transects. The points of soil sampling are shown as grey dots, (**b**) Geophilus traces. Numbered locations of soil sampling are given in red and (**c**)Transects of the GPR survey. The much higher coverage is visible. The locations of CMP soundings are shown as red points and the position of the selected profile shown later is highlighted by a red line

total station. The GPR data were recorded at a much higher spatial density compared to our DUALEM and Geophilus data (Table 1) and, thus, the GPR positioning data were used to generate an elevation model of our study site. The data represent relative elevation values in relation to the fixed position of the total station. Four additional CMP (common mid-point) measurements were carried out (red points in Fig. 3c) to derive subsurface velocity information as needed, for example, to convert travel times into depths.

2.3 *Data Processing*

Due to the different positioning methods, the original coordinate systems are different for both mobile electrical sensors. By means of a coordinate transformation, the positions of the Geophilus measurements (WGS coordinates) are transformed into the local system used for DUALEM surveying and for all other datasets. In order to identify the test field, for example on an aerial photograph of the research site, the local coordinate system was later rotated so that the y-axis points towards geographic north. For all sensors, the offset between the GPS receiver or prism and the individual electrodes, coils or antennas has been corrected and individual latency corrections have been applied.

For data denoising different strategies have been used. The evaluation of DUALEM data quality is mainly based on drift analysis, on comparing readings at points of crossing lines, and on the resulting conductivity-depth models. To estimate noise amplitudes, the original raw measurements are compared to slightly smoothed time series after removing a strong anomaly caused by a water pipe in the North of the field. From this, the short-period noise of DUALEM is estimated to be in the order of 0.1–0.2 mS m^{-1}. This is in the range of the value published by the producer (± 0.25) and also in accordance with the results presented by Hampe [46].

For the Geophilus data, useful criteria for noise reduction and detecting outliers are, for example, the signal frequency at the potential electrodes, the measured phase angle and the standard deviation at each point. Because for DC measurements the measured voltage decreases with increasing electrode spacing, the Geophilus near-surface data (channels 1 and 2) are less noisy than the larger offset channels sensitive to greater depths (channels 3 to 6). Thus, in our study the number of outliers increases from 8.8% for channel 1 and 11.5% for channel 2 to 28.7% for channel 4 and 39.6% for channel 6. Because there was a problem with the data recorded by channel 5, this channel was not considered for further analysis. To ease the comparison of all sensor data, the resistivity values determined with DC geoelectrics (Geophilus data and data measured with static electrodes) were converted into conductivities.

While only minor processing was necessary for the DUALEM data, the raw Geophilus data stored many outliers which had to be removed. After removing isolated spikes caused by small metallic objects in the DUALEM data and erroneous Geophilus data, a zero-phase second order polynomial smoothing filter was applied to the time series of the mobile ECa sensor data. The filter window size was adjusted in such a way that smoothing was carried out for both data sets over a similar distance of ~8 m. The ECa data measured with the mobile sensors were gridded on a regular grid of 0.5 × 0.5 m and smoothed with a 2D-Gaussian filter.

The DUALEM data along the reference transects 2 and 3 were extracted from these ECa maps. Their inversion is performed by an in-house nonlinear 1D inversion program [34] and, later, the results are compared with the conductivity depth models resulting from the conventional DC surveys with fixed electrodes. These DC data were inverted using both 1D and 2D assumptions. An in-house program was used for the nonlinear point-by-point 1D inversion [35]. The 2D inversion was carried out with the commercial software package RES2DINV (Geotomo software; [62]).

The focus of this study is on methods for mapping electrical properties. In addition, the potential of GPR was studied to image lateral soil heterogeneity and compare the imaged patterns to the conductivity maps. Before the GPR data were combined into a 3D cube, profile-wise standard processing (DC removal, zero-time correction, dewow, frequency filtering and spherical divergence correction) was applied. The 3D data set was gridded on a regular grid of 0.25 × 0.25 m covering an area of 83 × 100 m. The third coordinate of the data cube is the time axis. For interpretation, selected time slices of amplitude and energy were used.

2.4 Soil Sampling and Soil Texture Analysis

The various electrical methods provide information regarding subsurface architecture and parameter distributions (such as electrical conductivity maps), but do not allow to directly infer soil type and soil texture. Ground-truth data are typically needed for such a more detailed interpretation of geophysical parameter models. To gather such ground-truth data, 12 bore holes (red points in Fig. 3b) were drilled along the transects 2 and 3 (red lines in Fig. 3a). Two sampling points were always realized close to each other at 1 m. Soil samples were taken up to 1 m depth. The cores were divided into samples depending on visible soil discontinuities. Due to the partly weak contrasts in the material, the subdivision into depth intervals was not always easy and, thus, bears the potential of subjectivity. The number of samples varies between two and four for each sampling location. The soil samples were analyzed in the laboratory in terms of particle size distribution using sieve analysis and a laser diffractometer for clay and silt content.

Finally, the conductivity-depth models are interpreted and evaluated by comparing these models with the horizons visible in the drill cores (up to 1 m depth) and with the results of soil texture analysis. The soil is classified in terms of particle size distribution using the DIN standard 4022 (slightly modified Atterberg [63] scheme) to distinguish eight classes with a logarithmic grain size scale between 2 μm and 2 mm.

3 Results

First, the maps and cross sections of our measured data are presented. Then, all conductivity depth models obtained after data inversion are compared to our soil data.

3.1 *Lateral Soil Heterogeneity*

The ECa maps image mainly the lateral variability within the field (Fig. 4). While the eastern part of the field shows only small differentiation in ECa data, the western part is characterized by a more distinct anomaly with high ECa values. The anomalies within the field are associated with a variable thickness of the sand cover. In areas with higher ECa, the sand cover is thinner than in low-conductive areas. When comparing the patterns of the conductivity maps with the derived digital terrain model (DTM; Fig. 4g), most of the anomalies are not visible in the topography. However, it can be noticed that the general increase in terrain height towards the southeast correlates with an overall decrease in ECa values.

The main difference between EM- and DC-sensors is their different sensitivity to high conductive objects. The conductivity maps measured with the DUALEM (Fig. 4a–d) show a strong linear anomaly in the northern part of the site which is related to a metallic water pipe. In the vicinity of the pipe, it is difficult to see soil variability. South of the pipe, all conductivity maps show a similar anomaly pattern assumed to be related to soil heterogeneity. With exception of the linear anomaly

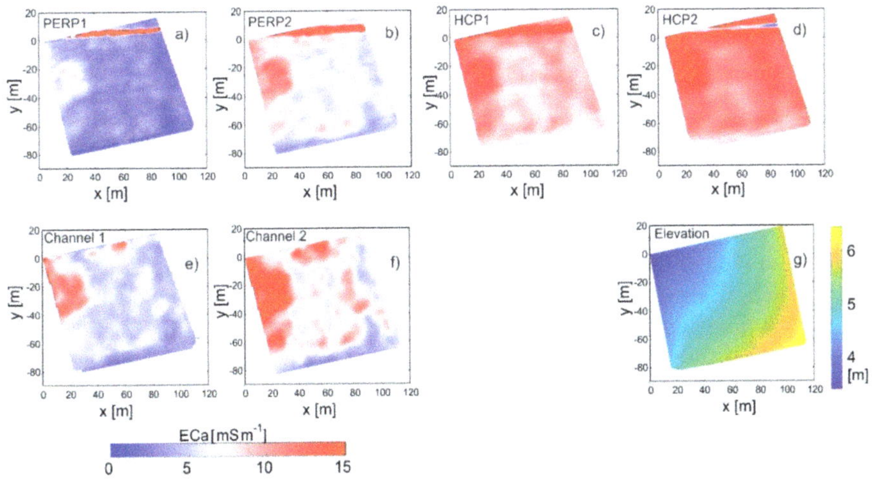

Fig. 4 Maps of ECa and DTM. (**a**), (**b**), (**c**), and (**d**) DUALEM data measured with the given coil orientation, (**e**) and (**f**) Geophilus data measured with $b = 0.5\ m$ (Channel 1) and $b = 1.0\ m$ (Channel 2), and (**g**) DTM derived from total station positioning data

Table 2 Correlation coefficients calculated between the ECa data measured with DUALEM and Geophilus. In the last column, the correlation of ECa data with the GPR energy slice (extracted from the GPR cube between 25 and 30 ns) is given

	PERP1	PERP2	HCP1	HCP2	Channel1	Channel2	GPR
PERP1	1	0.95	0.90	0.70	0.90	0.91	−0.30
PERP2	0.95	1	0.92	0.80	0.88	0.90	−0.39
HCP1	0.90	0.92	1	0.88	0.90	0.92	−0.29
HCP2	0.70	0.80	0.88	1	0.74	0.76	−0.24
Channel1	0.90	0.88	0.90	0.74	1	0.95	−0.27
Channel2	0.91	0.90	0.92	0.76	0.95	1	−0.29
GPR	−0.30	−0.39	−0.29	−0.24	−0.27	−0.29	1

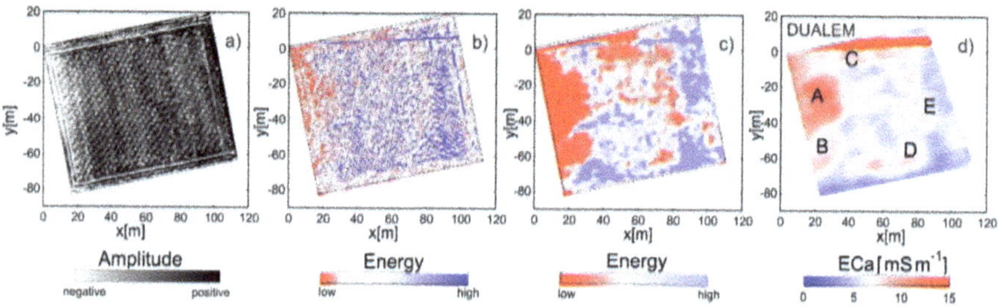

Fig. 5 GPR time slices of amplitudes and energy. (**a**) Amplitudes at 3 ns (ground wave) to map structures within the uppermost soil layer, (**b**) Energy for the time window between 15 and 25 ns, (**c**) Energy for the time window between 25 and 30, and (**d**) DUALEM – ECa map of the PERP2 configuration. The high conductive areas labelled with letters A to E correspond to areas of low GPR energy in Fig. 5c

related to the pipe, EM and DC maps show similar patterns. A simple cross-correlation was calculated to quantify this similarity (Table 2). Here, the region around the pipe was not considered to calculate the correlation coefficients.

The differences in the correlation coefficients can be explained by different vertical and horizontal sensitivities of the individual sensor configurations. For each sensor, the maps in Fig. 4 are arranged in row with increasing depth of sensitivity from left to right. All conductivity maps are plotted with the same color scale to provide an idea of the changes with the depth. The range is between 2 and 15 mS m^{-1}, which is typical for sandy soils in northern Germany. Obviously, at the Marquardt field site the conductivities generally increase with depth.

Figure 5 shows time-slices extracted from the GPR data cube and the ECa map of the DUALEM PERP2 configuration for comparison. Because GPR data are recorded at each point over a time window of 100 ns, this method is better suited to differentiate between different depths. Focusing on ground wave arrivals, Fig. 5a illustrates the near-surface structures within the uppermost soil layer. A regular diagonal pattern dominates the map which is not visible in the DTM. This pattern is related to differences in soil density (caused by the deep ploughing 1 year before our

Table 3 Statistics of ECa values for the considered mobile sensor configurations. The standard deviation (Std) is given as a measure of heterogeneity within a specific soil volume depending on the individual sensitivity function of the considered sensor configuration

	Mean [mS m^{-1}]	Median [mS m^{-1}]	Std [mS m^{-1}]
DUALEM PERP 1.1	3.9	3.7	1.1
DUALEM PERP 2.1	6.8	6.7	1.5
DUALEM HCP 1	9.2	8.9	1.7
DUALEM HCP 2	10.8	10.8	1.7
Geophilus 1	5.9	5.4	2.1
Geophilus 2	8.2	7.3	3.4

measurements), which are probably also related to soil moisture variations. The traces at the edges of the field are caused by topography effects (headland). Fig. 5b depicts the energy (i.e., squared amplitudes) summed in a time window between 15 and 25 ns. Using the velocity of 0.085 m ns^{-1} resulting from our CMP surveys, this corresponds to a depth range between 0.64 and 1.06 m. The regular pattern from the topsoil is still visible. The linear anomaly in the northern part of the field is caused by the metallic water pipe, which is also detected by the DUALEM (Fig. 5d). The energy map shown in Fig. 5c gives an idea about the depth of penetration of the GPR waves. The squared amplitudes are summed in a time window between 25 and 30 ns. Using a velocity of 0.085 m ns^{-1}, this corresponds to a depth range between 1.06 and 1.27 m. Because electrical conductivity is controlling GPR attenuation, the lateral heterogeneity of GPR energy correlates negatively with the conductivity map shown in Fig. 5d. For better illustration, the highly conductive areas of the DUALEM data are marked as zones A-E. These zones are visible in the GPR data as zones with lower energy. Due to the higher spatial resolution and the smaller footprint of the GPR data, the GPR image is not as smooth as the ECa map.

3.2 *Information About Soil Stratification*

As a first approximation, the mean values of the ECa maps (Fig. 4) can be used for a rough stratigraphic interpretation. As the values in northern part of the field are strongly affected by the pipe and not by the soil, the mean, median and standard deviation (Std) values were calculated only for the area south of the pipe (Table 3). With increasing depth of investigation, the apparent electrical conductivities increase from ~4 mS m^{-1} to ~11 mS m^{-1} as demonstrated by the mean and median values for the different configurations. The standard deviation mainly contains information about lateral variabilities within a certain depth range. An increase in standard deviation with depth indicates an increase of soil differentiation with depth and possibly also an increase of noise.

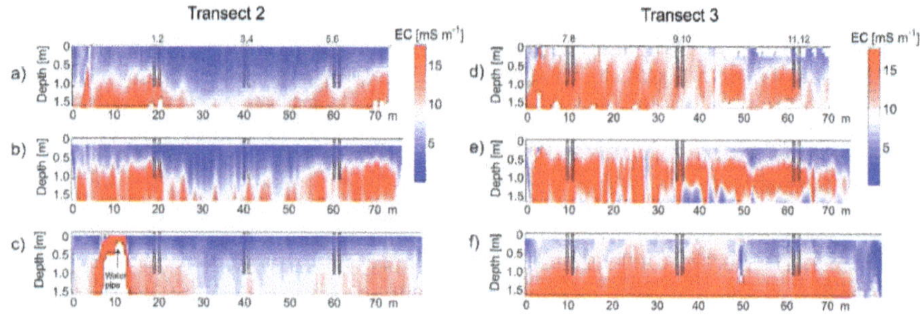

Fig. 6 Conductivity depth models along transects 2 (**a–c**) and 3 (**d–f**). (**a**) and (**d**) 1D (1 dimensional) inversion of equatorial dipole-dipole DC data acquired with fixed electrodes, (**b**) and (**e**) 2D inversion of static DC data using RES2DINV, and (**c**) and (**f**) point-by-point non-linear 1D inversion of DUALEM data. The positions of the soil sampling are shown by vertical bars. The numbers of the sampling points for texture analysis are shown in (**a**) and (**d**)

3.3 Two-Dimensional Conductivity Models Along Reference Transects

The conductivity models along the reference transects (Fig. 6) resulting from 1D and 2D inversion of static DC and DUALEM data confirm the trend of increasing conductivity with depth. Up to a depth of 1.5 m, all models indicate a two-layer case with varying thickness of the high-resistive top layer. Transect 2 shows a basin-like structure in the middle of the profile. In the western part of the field (transect 3), a high conductive layer can be found close to the surface. The thickness of the low conductive top layer increases towards the edges of the profile. These soil structures are imaged similarly by both methods.

One main difference between the DC and EM models is that the pipe related anomaly is not visible in the DC data. The EM data are strongly influenced by this anomaly at the beginning of transect 2. This is explained by the theory of the methods. If the sensitivity of DC is inversely proportional to the conductivity, the EM sensitivity is the opposite.

3.4 GPR Transects

Figure 7 is an example illustrating the quality of our GPR data. This profile was measured close to the position of two CMP surveys (Fig. 3c) running from NNW to SSE and is representative for the data collected at this site. The profile is dominated by the air- and ground-wave arrivals up to ~10 ns and by strong hyperbolic curves at times >10 ns (marked in Fig. 7b by red arrows). These hyperboles are caused by a 60 cm deep ploughing in 2014, more than 1 ½ year before our field work. No further sharp reflections are visible within the upper 1–2 m.

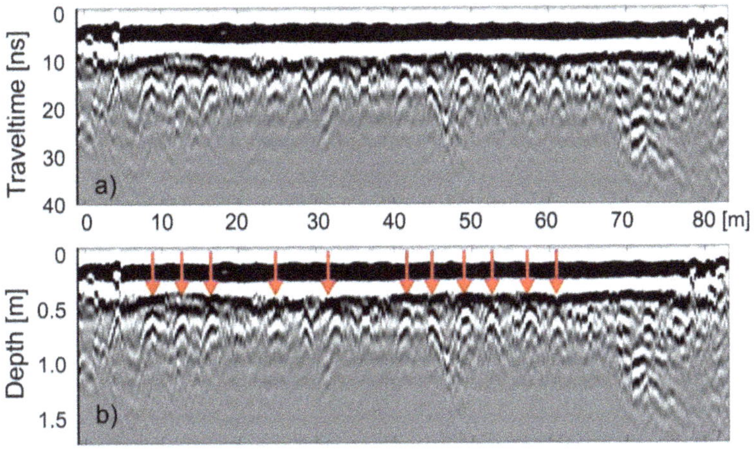

Fig. 7 GPR transect in the western part of the field close to the position of two CMP surveys. (**a**) Travel time section, and (**b**) Depth section calculated using a velocity of 0.085 m ns^{-1} as derived from CMP surveying. Prominent hyperboles are marked with red arrows

Table 4 Statistics of measured particle size fractions (all samples) of clay, silt, and sand up to 1 m depth

	Diameter [mm]	Minimum content [%]	Maximum content [%]	Mean content [%]	Std [%]
Coarse sand	0.6–2.0	2.9	6.9	4.4	0.8
Medium sand	0.2–0.6	27.4	52.3	36.3	5.5
Fine sand	0.06–0.2	22.4	38.9	33.3	3.6
Coarse silt	0.02–0.06	2.4	6.6	4.3	1.1
Medium silt	0.006–0.02	4.1	9.6	5.7	1.1
Fine silt	0.002–0.006	7.2	15.8	10.8	2.0
Clay	<0.002	1.7	6.8	3.5	1.3

3.5 Soil Texture as Ground-Truth Data

The results of our texture analysis are summarized in Table 4. Fine and medium sands are the main components of our soil samples (75%). The silt content is between 15 and 30%. Comparing these data with the German manual of soil mapping [64], the main soil types are Su2, Sl2, and Su3. 'S' is the symbol for sand, 'u' for silt and 'l' for loam. That means, sand dominates everywhere, and silt and loam content vary. Su2 is a medium silty sand with silt content between 10 and 25% and Su3 has a higher silt content (25–40%). The clay content of both is less than 5%.

In general, the grain size distributions for the individual soil samples are very similar (Fig. 8). The cumulative curves illustrate the differences between top and

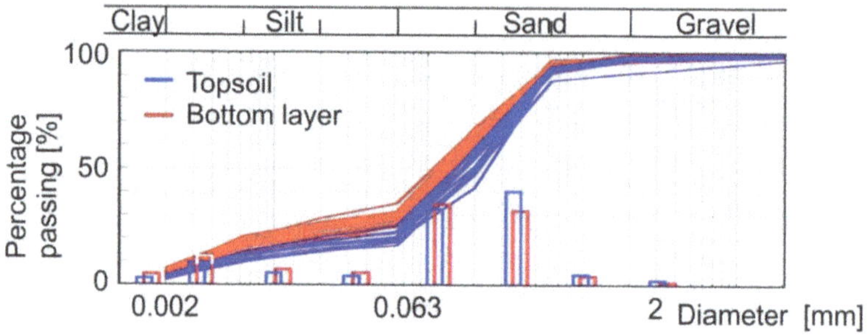

Fig. 8 Grain size distribution (absolute and cumulative) for all individual soil samples

bottom soil. While only one topsoil sample is available for each drilling point (from the surface up to a depth of ~30–50 cm), there are sometimes several soil samples for the sub-soil due to layered sequences. Clay and silt content increase with depth. The mean clay content varies between 2.5% for the topsoil and 4.5% for the bottom layer. The differences for fine silt (9.2% within the topsoil and 12.2% for greater depths) are more distinctive than for medium (4.9% and 6.5%) and coarse silt (3.7% and 5.0%).

The variability of the thickness of the top layer and also the spatial variability in soil texture is illustrated in Fig. 9a, where the sum of clay, fine and medium silt contents is plotted as a function of depth. The results of all 12 positions were arranged together along one line. While in the eastern part of the field (at points 3 and 4) the content of fine particles is low within the upper meter, the other depth profiles show a clear increase of fine particles with depth. Furthermore, an increase of soil variability with depth can be noticed (increasing Std values in Table 3). The differentiation in the topsoil is less clear than in the bottom soil (Table 5). This confirms the increasing variability in ECa data with depth, which was found for DUALEM and for Geophilus data.

A comparison of the texture data and the conductivity models shows that the stratigraphy recognized in the soil data is also visible in the conductivity models. The three conductivity depth models (Fig. 9b–d) are similar at the respective positions but differ slightly depending on the used instrument and the applied inversion routine. Overall, the site is characterized by increasing conductivities with depth, while the thickness of the low conductive top layer varies. In the western part of the field (point 7–12, transect 3), it shows a thickness of only a few decimeters, while along transect 2 (point 1–6) it is characterized by a thickness >0.5 m. The similarity of neighboring points with a distance of 1 m is obvious.

The conductivity models achieved from the 1D inversion of the DUALEM data and from the 2D inversion of the DC data show similar correlations with the texture data (Table 6). There is a negative correlation with the coarser soil components (sum of coarse silt, sand, and gravel) and a positive one with the fine particles.

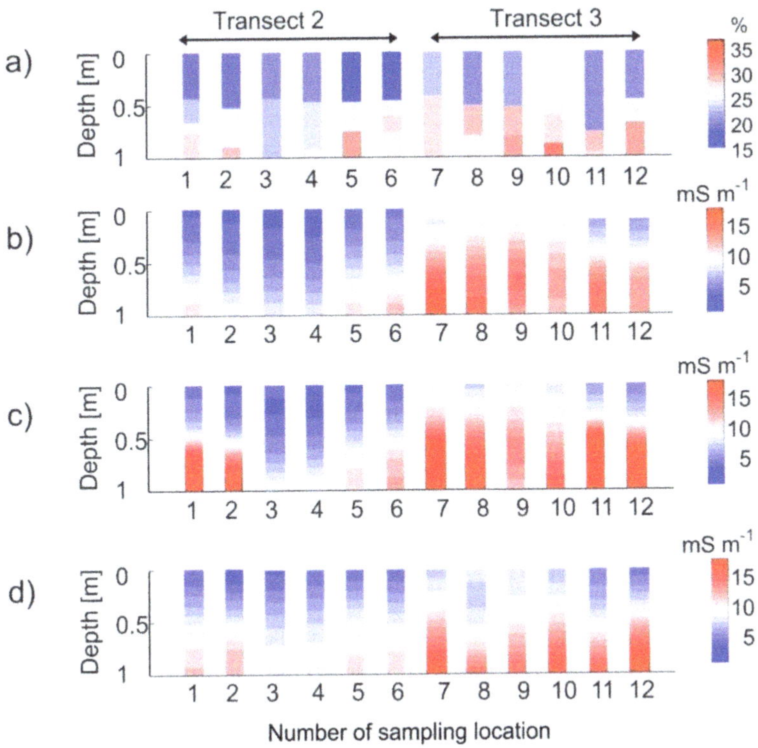

Fig. 9 Depth profiles at the sampling points 1–12 along the reference transects 2 and 3. The sampling positions are given in Fig. 3a. (**a**) Silt (fine and medium) and clay content from the soil texture analysis, (**b**) Conductivity-depth models from the 1D inversion of static DC data, (**c**) Conductivity-depth models from the 2D inversion of static DC data, and (**d**) Conductivity-depth models from the 1D inversion of DUALEM data

Table 5 Statistics of measured particle size fractions (all samples) of clay, fine silt, and medium silt differentiated for top and bottom soil

	Diameter [mm]	Depth [m]	Minimum content [%]	Maximum content [%]	Mean content [%]	Std [%]
Clay	<0.002	0–0.4	1.7	3.8	2.5	0.5
		0.6–1.0	2.6	6.8	4.5	1.1
Fine silt	0.002–0.006	0–0.4	7.2	10.8	9.1	1.1
		0.6–1.0	8.8	15.8	12.3	1.8
Medium silt	0.006–0.02	0–0.4	4.1	6.6	4.9	0.8
		0.6–1.0	4.9	9.6	6.3	0.9
Fine particles	<0.02	0–0.4	13.9	20.3	16.6	1.9
		0.6–1.0	17.4	28.6	23.2	2.9

Table 6 Correlation coefficients between soil texture (from all reference points) and conductivity-depth models (from different datasets and inversion routines)

Grain size	DC 1D inversion	DC 2D inversion	DUALEM 1D inversion
<0.002 mm	0.31	0.34	0.45
<0.006 mm	0.40	0.46	0.47
>0.006 mm	−0.42	−0.50	−0.56

4 Discussion

The main advantage of electromagnetic induction sensors is their ability to perform non-contact measurements and their overall good data quality. For DC geoelectrics, the galvanic coupling between the electrodes and the ground is influenced by the local ground conditions and, thus, data quality might vary from point to point. Especially for dry or hard topsoil, it becomes difficult to achieve sufficiently good data quality with mobile electrodes. Because the contact resistances are easier to control when measuring with static electrodes, measurements with static electrodes were carried out along individual transects to evaluate our DUALEM data and the resulting conductivity models.

Proximal multi-receiver systems, like DUALEM-21 and Geophilus, have a great potential for detailed imaging not only regarding lateral soil variability but also in view of soil stratification at the field scale. Here, the distances between coils or electrodes mainly determine the depth sensitivity of the data. When using more than one receiver coil or more than one pair of potential electrodes, information can be received about changes in electrical parameters even with the depth. How well the conductivity models reflect the reality, depends among other things, on the contrasts in physical parameters and on the quality and consistency of the measured data. A direct comparison of ECa data measured with different sensors is only of limited significance due to different spatial sensitivities, the integrated volumes, and the differences in physics between the methods and arrays used. DUALEM and DC data were inverted, and the resulting conductivity models were compared along the transects. For the test site in Marquardt, all electrical conductivity models, regardless of the used sensor and inversion algorithm, show a tendency of increasing values with depth. This increase in electrical conductivity can be explained by increasing content of fine particles with depth, as found with the texture analysis of soil samples. All of our data and models indicate a rather homogeneous top layer and an increasing differentiation with depth.

As with all geophysical methods, the quality of GPR data depends on the contrasts in the physical parameters and on the surveying parameters. The choice of the antenna frequency is always a compromise between depth penetration and resolution. With the 250 MHz antennas used, we could not depict any continuous reflector within the depth of investigation (approximately ~1.5 m at our site). However, the GPR data confirmed the patterns known from our ECa maps and also imaged traces of previous deep ploughing that were not visible with any of the other methods used. In addition, the depth of the water pipe in the north of the field could be determined.

5 Conclusion

A field study on the use of geophysical sensors to image lateral and vertical soil heterogeneity within the rooting zone of plants up to a depth of 1-2 m was presented. It was shown that partly similar and partly complementary information could be obtained with the instruments used. Even if the methods that have been used are suitable for mapping the electrical conductivity, the instruments differ both in their depth sensitivity and in their sensitivity to conductive or resistive zones. While the DC method is most sensitive to electrical resistivity (the inverse value of conductivity), the electromagnetic induction method is most sensitive to electrical conductivity, and GPR reflections can be observed when there are some contrasts in permittivity. The best choice of a sensor depends on the site-specific differentiation in soil parameters. Without prior knowledge of the field, a multi-sensor approach is recommended. Furthermore, it was shown that the DUALEM is a feasible tool to image soil heterogeneity and soil stratification also in sandy soils with low conductivity values.

Multichannel sensors like DUALEM-21 or Geophilus are powerful tools to depict three-dimensional soil structures. Inversion routines exist to transform multichannel ECa data into conductivity depth models. Consistent data with sufficient data quality are required to map layers and depths correctly. To ensure consistency of data, a regular check of the sensor calibration is necessary.

When geophysical sensors are used to map soil heterogeneity and to derive soil maps for smart farming, differences caused by previous cultivation (e.g., fertilization, irrigation or ploughing) might influence the outcome. The previous field setup of the farmer, which is not related to the natural soil properties, can hinder a meaningful interpretation of the measured physical parameters. A non-uniform agricultural management can lead to differences in plant growth and thus to differences in soil moisture and therefore also to differences in electrical conductivities.

At the end, for the test field in Marquardt, it can be concluded that the area is comparatively homogeneous and, therefore, is well suited for further agricultural experiments such as fertilizer tests.

Acknowledgments The authors would like to thank M. Dubnitzki for his support during the field campaign. Geophilus data were collected by J. Ruehlmann – thanks for providing these data. We also thank H. Kraudelt who supported the soil texture analysis.

References

1. Heil K, Schmidhalter U (2017) The application of EM38: determination of soil parameters, selection of soil sampling points and use in agriculture and archaeology. Sensors 17(11). https://doi.org/10.3390/s17112540
2. Corwin DL, Lesch SM (2003) Protocols and guidelines for field-scale measurement of soil salinity distribution with ECa-directed soil sampling. J Environ Eng Geoph 18(1):1-25. https://doi.org/10.2113/JEEG18.1.1

3. Lund EC, Christy CD, Drummond PE (2000) Using yield and soil electrical conductivity (EC) maps to derive crop production performance information. In: Proceeding of the 5th International Conference on Precision Agriculture, Bloomington, MN USA, 2000
4. Brevik EC, Fenton TE, Lazari A (2006) Soil electrical conductivity as a function of soil water content and implications for soil mapping. Precis Agric 7:393-404. doi:https://doi.org/10.1007/s11119-006-9021-x
5. Mele M, Inzoli S, Giudici M et al (2014) Relating electrical conduction of alluvial sediments to textural properties and pore-fluid conductivity. Geophys Prospect 62:631-645. https://doi.org/10.1111/1365-2478.12102
6. Saey T, De Smedt P, Meerschman E et al (2012) Electrical conductivity depth modelling with a multireceiver EMI sensor for prospecting archaeological features. Archaeol Prospect 19(1):21-30. doi/https://doi.org/10.1002/arp.425
7. Hampe T (2015) Evaluierung des Systems DUALEM-21 zur Leitfähigkeitskartierung. Master Thesis, University Potsdam, Not published.
8. Loke MH, Barker RD (1996) Rapid least-squares inversion of apparent resistivity pseudosections by a quasi-Newton method. Geophys Prospect 44:131-152. https://doi.org/10.1111/j.1365-2478.1996.tb00142.x
9. McNeill JD (1980) Electromagnetic terrain conductivity measurement at low induction numbers. Technical Note TN-6, Geonics Ltd, Mississauga Ontario, Canada.
10. Ezrin MH, Amin MSM, Anuar AR et al (2010) Relationship between rice yield and apparent electrical conductivity of paddy soils. Am J Applied Sciences 7(1):63-70. https://doi.org/10.3844/ajassp.2010.63.70
11. Lueck E, Ruehlmann J (2013) Resistivity mapping with GEOPHILUS ELECTRICUS – Information about lateral and vertical soil heterogeneity. Geoderma 199:2-11. https://doi.org/10.1016/j.geoderma.2012.11.009
12. Myers DB, Kitchen NR, Sudduth KA et al (2010) Combining proximal and penetrating soil electrical conductivity sensors for high-resolution digital soil mapping. In: Viscarra Rossel, McBratney and Minasny (eds) Proximal Soil Sensing, Progress in Soil Science, Springer, Dordrecht, pp. 233-243. https://doi.org/10.1007/978-90-481-8859-8_19
13. Sudduth KA, Kitchen NR, Drummond ST (2017) Inversion of soil electrical conductivity data to estimate layered soil properties. In: Taylor JA (ed) Proceedings of 11th ECPA, Edinburgh, UK, 2017, 8(2):433-438. https://doi.org/10.1017/52040470017001303
14. von Hebel C, Rudolph S, Mester A et al (2014) Three-dimensional imaging of subsurface structural patterns using quantitative large-scale multiconfiguration electromagnetic induction data. Water Resour Res 50(3):2732-2748. https://doi.org/10.1002/2013WR014864
15. Triantafilis J, Santos FAM (2010) Resolving the spatial distribution of the true electrical conductivitiy with depth using EM38 and EM31 signal data and a laterally constrained inversion model. Aust J Soil Res 28:434-446
16. Corwin DL, Lesch SM, Segal E et al (2010). Comparison of sampling strategies for characterizing spatial variability with apparent soil electrical conductivity directed soil sampling. J Environ Eng Geoph 15(3):147-162. https://doi.org/10.2113/JEEG15.3.147
17. Heil K, Schmidhalter U (2012) Characterisation of soil texture variability using the apparent soil electrical conductivity at a highly variable site. Comput Geosci-UK 39:98-110. https://doi.org/10.1016/j.cageo.2011.06.017
18. Bottega AL, Queiroz DM, Pinto F et al (2017) Precision agriculture applied to soybean: Part I – delineation of management zones. Aust J Crop Sci 11(05):573-579. https://doi.org/10.21475/ajcs.17.11.05.p381
19. Lavoué F, van der Kruk J, Rings J et al (2010) Electromagnetic induction calibration using apparent electrical conductivity modelling based on electrical resistivity tomography. Near Surf Geophys 8(6):553-561. https://doi.org/10.3997/1873-0604.2010037
20. Monteiro Santos FA (2004) 1-D laterally constrained inversion of EM34 profiling data. J Appl Geophys 56:123-134. https://doi.org/10.1016/j.jappgeo.2004.04.005
21. Eigenberg RA, Nienhaber JA, Woodbury BL et al (2008) Four-year summary of the use of soil conductivity as a measure of soil and crop status. In: Allred JJ et al (eds) Handbook of agricultural geophysics, CRC-Press, pp 273-280

22. Vitharana UWA, Van Meirvenne M, Simpson D et al (2008) Key soil and topographic properties to delineate potential management classes for precision agriculture in the European loess area. Geoderma 143(1):206-215. https://doi.org/10.1016/j.geoderma.2007.11.003
23. Boenecke E, Lueck E, Ruehlmann J et al (2018) Determining the within-field yield variability from seasonally changing soil conditions. Precis Agric 19:750-769. https://doi.org/10.1007/s11119-017-9556-z
24. Gebbers R, Lück E, Heil K (2007) Depth sounding with the EM38 – detection of soil layering by inversion of apparent electrical conductivity measurements. In: Stafford JV (ed) Precision agriculture'07. Proceeding of 6 ECPA, Skiathos Greece, Wageningen Academic Publisher, The Netherlands, pp 95-102
25. Pan L, Adamchuk VI, Prasher S et al (2014) Vertical soil profiling using a galvanic contact resistivity scanning approach. Sensors 14:13243-13255. https://doi.org/10.3390/2140713243
26. Sudduth KA, Drummond ST, Kitchen NR (2001) Accuracy issues in electromagnetic induction sensing of soil electrical conductivity for precision agriculture. Comput Electron Agr 31(3):239-264. https://doi.org/10.1016/S0168-1699(00)00185-X
27. Dabas M (2009) Theory and practice of the new fast electrical imaging system ARP©. In: Campana S, Piro S. (eds) Seeing the unseen, Geophysics and Landscape Archaeology, CRC Press, Taylor and Francis Group: Boca Raton, FL, USA, 2009, pp. 105–126
28. Lück E, Gebbers R, Ruehlmann J et al (2009) Electrical conductivity mapping for precision farming. Near Surf Geophys 7(1):15-25. https://doi.org/10.3997/1873-0604.2008031
29. OpenStreetMap contributors (2015) https://planet.openstreetmap.org. and https://download.geofabrik.de (latest access: 31.10.2020)
30. Sudduth KA, Myers DB, Kitchen NR et al (2013) Modeling soil electrical conductivity-depth relationships with data from proximal and penetrating Eca sensors. Geoderma 199:12-21. https://doi.org/10.1016/j.geoderma.2012.10.006
31. Moghadas D, Behroozmand AA, Christiansen AV (2020) Soil electrical conductivity imaging using a neural network-based forward solver: Applied to large-scale Bayesian electromagnetic inversion. J Appl Geophys 104012. doi. https://doi.org/10.1016/j.jappgeo.2020.104012
32. Buvat S, Julien T, Michelin J et al (2014) Mulit-depth electrical resistivity survey for mapping soil units within two 3 ha plots. Geoderma 232-234:317-327. https://doi.org/10.1016/j.geoderma.2014.04.034
33. Guillemoteau J, Sailhac P, Boulanger C et al (2015) Inversion of ground constant offset loop-loop electromagnetic data for a large range of induction number. Geophysics 80:E11-E21. https://doi.org/10.1190/geo2014-0005.1
34. Guillemoteau J, Simon FX, Lueck E et al (2016) 1D sequential inversion of portable multi-configuration electromagnetic induction data. *Near Surf Geophys* 14:411-420. https://doi.org/10.3997/1873-0604.2016029
35. Guillemoteau J, Lück E, Tronicke J (2017a) 1D inversion of direct current data acquired with a rolling electrode system. J Appl Geophys 146:167-177. https://doi.org/10.1016/j.jappgeo.2017.09.010
36. Klose T, Guillemoteau J, Simon FX et al (2018) Toward subsurface magnetic permeability imaging with electromagnetic induction sensors: Sensitivity computation and reconstruction of measured data. Geophysics 83(5):E335-E345. https://doi.org/10.1190/geo2017-0827.1
37. Misra RK, Padhi J (2014) Assessing field-scale soil water distribution with electromagnetic induction method. J Hydrol 516:200-209. doi.org/https://doi.org/10.1016/j.jhydrol.2014.02.049
38. Tite M, Mullins C (1970) Electromagnetic prospecting on the archaeological sites using a soil conductivity meter. Archaeometry 12(1):97–104. https://doi.org/10.1111/j.1475-4754.1970.tb00010.x
39. Auken E, Christiansen AV (2004) Layered and laterally constrained 2D inversion of resistivity data. Geophysics 69(3):752-761. https://doi.org/10.1190/1.1759461
40. Christiansen AV, Pedersen JB, Auken E et al (2016) Improved geoarchaeological mapping with electromagnetic induction instruments from dedicated processing and inversion. Remote Sens 8(12): 15

41. Wallor E, Kersebaum KC, Lorenz K et al (2017) Connecting crop models with highly resolved sensor observations to improve site-specific fertilisation. Advances in Animal Biosciences, 8(2), 689-693. https://doi.org/10.1017/S2040470017000358
42. Guillemoteau J, Christensen NB, Jacobsen BH et al (2017b) Fast 3D multi-channel deconvolution of electromagnetic induction loop-loop apparent conductivity data sets acquired at low induction number. Geophysics 82(6):1-46. https://doi.org/10.1190/geo2016-0518.1
43. Piikki K, Wetterlind J, Söderström M et al (2015) Three-dimensional digital soil mapping of agricultural fields by integration of multiple proximal sensor data obtained from different sensing methods. Precis Agric 16:29-45. https://doi.org/10.1007/s11119-014-9381-6
44. Shanahan PW, Binley A, Whalley WR et al (2015) The use of electromagnetic induction to monitor changes in soil moisture profiles beneath different wheat genotypes. Soil Sci Soc Am J 79(2):459–466. https://doi.org/10.2136/sssaj2014.09.0360
45. Su ASM., Adamchuk VI (2014) Evaluation of the temporal and operational stability of apparent soil electrical conductivity measurements. In: Proceedings of 12th ICPA, Sacramento, California, 2014, published on-line at http://www.ispag.org
46. Guillemoteau J, Simon FX, Hulin G et al (2019) 3-D imaging of subsurface magnetic permeability/susceptibility with portable frequency domain electromagnetic sensors for near surface exploration. Geophys J Int 219(3):1773-1785. https://doi.org/10.1093/gji/ggz382
47. Heil K, Schmidhalter U (2019) Theory and guidelines for the application of the geophysical sensor EM38. Sensors 19. 4293. doi.org/10.3390/s19194293
48. Moral FJ, Terrón JM, Da Silva JM (2010) Delineation of management zones using mobile measurements of soil apparent electrical conductivity and multivariate geostatistical techniques. Soil and Tillage Research 106(2):335-343. https://doi.org/10.1016/j.still.2009.12.002
49. Koganti T, Moral FJ, Rebollo FJ et al (2017) Mapping cation exchange capacity using a Veris-3100 instrument and invVERIS modelling software. Sci Total Environ 599-600:2156-2165. doi/https://doi.org/10.1016/j.scitotenv.2017.05.074
50. Van Meirvenne M, Islam MM, De Smedt P et al (2013) Key variables for the identification of soil management classes in the Aeolian landscapes of north-west Europe. Geoderma 199:99-105. https://doi.org/10.1016/j.geoderma.2012.07.017
51. Cavalcante Frage LH., Schamper C, Noel C et al (2019) Geometrical characterization of urban fill by integrating the multi-receiver electromagnetic induction method and electrical resistivity tomography: A case study in Poitiers, France. Eur J Soil Sci 70(5):1012-1024. https://doi.org/10.1111/ejss.12806
52. GLOBE tast team (1999) The Global Land One-Kilometer Base Elevation (GLOBE) Digital Elevation Model, Version 1.0. National Oceanic and Atmospheric Administration, National Geophysical Data Center. https://doi.org/10.7289/V5C8276M
53. Neudecker E, Schmidhalter U, Sperl C et al (2001) Site-specific soil mapping by electromagnetic induction. In: Grenier G, Blackmore S (eds) Proceedings of the 3rd ECPA-conference, Montpellier, 2001. pp 271-276
54. Robinson DA, Lebron I, Lesch SM et al (2004) Minimizing Drift in Electrical Conductivity Measurements in High Temperature Environments using the EM-38. Soil Sci Soc Am J 68(2):339–445. doi.org/10.2136/sssaj2004.3390
55. Malicki MA, Walczak RT (1999) Evaluating soil salinity status from bulk electrical conductivity and permittivity. Eur J Soil Sci 50(3):505–514. https://doi.org/10.1046/j.1365-2389.1999.00245.x
56. Arellano-Castro RF, Gómez-Treviño E, Méndez-Delgado S et al (2020) Feasibility of magnetic susceptibility tomography for shallow targets. J Appl Geophys 172:103868.
57. Huang J, Minasny B, Whelan BM et al (2017) Temperature-dependent hysteresis effects on EM induction instruments: An example of single-frequency multi-coil array instruments. Comput Electron Agr 132:76-85.
58. Sun W, Whelan BM, Minasny B et al (2012) Evaluation of a local regression kriging approach for mapping apparent electrical conductivtity of soil (Eca) at high resolution. J Plant Nutr Soil Sc 175(2):212-220. https://doi.org/10.1002/jpln.201100005

59. Tabbagh A (1986) Applications and advantages of the Slingram electromagnetic method for archaeological prospecting. Geophysics 51:576–584. https://doi.org/10.1190/1.1442112
60. Zhao D, Li N, Zare E et al (2020) Mapping cation exchange capacity using a quasi-3d joint inversion of EM38 and EM31 data. Soil and Tillage Research 200:104618. https://doi.org/10.1016/j.still.2020.104618
61. Böniger U, Tronicke J (2010) On the potential of kinematic GPR surveying using a self-tracking total station: evaluating system crosstalk and latency. IEEE Transactions on Geoscience and Remote Sensing 48(10):3792-3798. https://doi.org/10.1109/TGRS.2010.2048332
62. Li Y, Shi Z, Wu C et al (2008) Determination of potential management zones from soil electrical conductivity, yield and crop data. J Zhejiang Univ Sci B 9(1):68-76. https://doi.org/10.1631/jzus.B071379
63. Atterberg A (1916) Die Klassifikation der humusfreien und der humusarmen Mineralböden Schwedens nach den Konsistenzverhältnissen derselben. Int Mitt Bodenkd 6:27–37
64. Eckelmann W (ed) (2005) Bodenkundliche Kartieranleitung KA5– manual of soil mapping 5th Ed. Schweizerbart Science Publishers, Stuttgart, Germany

Geoinformation Technologies in Pest Management: Mapping Olive Fruit Fly Population in Olive Trees

Androniki Papafilippaki, George Stavroulakis, and George P. Petropoulos

1 Introduction

Agricultural production is highly dependent on climate and is adversely affected by anthropogenic climate change and increasing climate variability [1]. Losses in agricultural production are predicted to be intensified due to future climate changes, which may pose serious threat to food security and economic development of the human societies [2, 3]. An abiotically stressful environment in a changing climate is anticipated to affect negatively the diversity and abundance of insect pests and the damage caused in economically important agricultural crops [4]. Therefore, protection of agricultural yield from pest aggravation in the context of global climate change has been the focal point of ecological research [5]. Vulnerability of agriculture can be reduced through adaptation measures and tools aiming to increase climate variability [2]. In this respect, interest in forecasting models is radically growing among entomologists to predict the environmental suitability for new and invading agricultural insect pests. Early predictions on the future pest distribution and abundance through forecasting models may allow improving preparedness to combat episodes of serious insect pests by creating viable pest management strategies well in advance. In this respect, models are regarded as key priority to support a better understanding and prediction of both insect population dynamics and growth potential under diverse environmental conditions [6, 7]. Improved

A. Papafilippaki (✉) · G. Stavroulakis
Department of Environmental and Natural Resources Engineering, Hellenic Mediterranean University, Laboratory of Water and Soil Quality Control, Chania, Crete, Greece
e-mail: npapafilippaki@hmu.gr; gstav@chania.teicrete.gr; gstav@hmu.gr

G. P. Petropoulos (✉)
Department of Geography, Harokopio University of Athens, Athens, Greece
e-mail: gpetropoulos@hua.gr

D. D. Bochtis et al. (eds.), *Information and Communication Technologies for Agriculture—Theme I: Sensors*, Springer Optimization and Its Applications 182,
https://doi.org/10.1007/978-3-030-84144-7_12

knowledge on the basic physiological aspects of insects' growth and of their growth population, in relation to environmental factors, form a sound basis for successful pest management in varied agro-ecological conditions [8].

Olive (*Olea europaea* L) is a crop of great economic and social importance of all countries located around the Mediterranean basin. Olives and olive oil can be considered the most significant constituents of the Mediterranean diet due to their beneficial effects on human health which have been associated with the well-balanced fatty acid composition and the presence of vitamins and natural antioxidants [9]. Pest infestation consists one of the main threats for the olive production as it damages the olive-fruit [10]. *Bactrocera* or *Dacus oleae*, (Gmelin), (Diptera, Tephritidae), so called olive-fruit fly (Fig. 1) is the most serious threat of olive trees, as it can cause massive damage to olive production in all Mediterranean countries [11]. It is a monophagous species. In nature, the adult female which bears a distinct ovary, lays multiple eggs in olive fruits where the larvae are developed. Its biological cycle includes three to four generations a year in most areas, including Greece. Depending on the geographical region on which it grows, it overwinters as an adult in sheltered areas or as a nymph on the ground. When the fruit approaches its final size and becomes soft then the female opens the spawning hole with its ovary and inserts the egg into the mesocarp. Generally, it introduces one egg per fruit, but in cases of very dense population of olive trees or few fruits, more than one ovulation per fruit can be observed. The hatching larva then digs a hole in the mesocarp and is the one that causes the damage, devouring the mesocarp, making the fruit unsuitable for eating and at the same time reducing the quality of the olive oil. Most fruit flies are monophagous or stenophagous and breed in fruits or seedpods of their host plant [12, 13]. The olive fruit fly is considered one of the most serious pests of olives since it causes enormous damages both in quality and quantity of olive oil production. Particularly, the larvae of the third stage greatly reduce olive yield since they cause a removing of a significant proportion of the olive pulp. Also, when the olive fruit fly infestation is great, intense fruit fall is observed, negatively affecting the oil production. In addition, the holes dug by the larvae favor the establishment of olive anthracnose (*C. gloeosporioides*), an important phytosanitary problem of olive groves [9]. Furthermore, the abundance as well as the distribution of cold intolerant

Fig. 1 *Bactocera oleae* (olive fruit fly) and McPhail glass type trap [10]

pests such as olive fruit fly could be greatly affected by climate change and especially global warming [14].

To monitor olive-fruit fly population in the field, the classic approach is to use the McPhail glass type trap (Fig. 1), in which ammonia salt or hydrolyzed protein is used to attract the insects. This approach has been used successfully in practice for many years in order to monitor olive-fruit fly infestation and its population variations in an olive grove. The efficiency of this trap exceeds a radius of 20 m and is increasing under conditions of low relative humidity and high temperature. The main preventive measure implemented by the state agricultural division to reduce olive trees infestation from olive-fruit fly includes early bait spraying with specialized insecticides (organophosphates or pyrethrinoids) applied from the ground (in Greece), which usually takes place during the summer and early autumn period of each year. This precautionary measure aims at attracting and eventually destroying the olive fruit fly before it begins to lay eggs. The number and timing of bait spraying depends on the evolution of the population which is monitored with the traps [12, 15]. Although bait spraying is considered an environmentally friendly control practice against olive fruit fly, often causes surface and groundwater pollution or the detected residues in olives and olive oil may threaten human health [16].

The advent of geoinformation technology in agriculture, particularly of Geographic Information Systems (GIS), Remote Sensing and Global Positioning Systems (GPS), has opened up new paths for analysing spatial patterns in insect populations [8]. It allows analysis of georeferenced data on insects and has proved as an emerging technology in area-wide pest management because many dimensions of insect ecology have a spatial component [17]. Recent advantages in computational science have also made it easier to process in short time periods large amount of data in an efficient way and at variable spatial scales using advanced algorithms readily available in many software platforms managing such data. The recent developments in GIS and geostatistics have made it easier to analyse complex spatial patterns of insect species and also help in delineation of agri-ecological hotspots and future areas of pest risk [18].

In purview of the above, this study aims at demonstrating how the combined use of an olive-fruit fly trap network can be used synergistically with geoinformation - in particular the use of GIS and GPS - as a cost-effective and operational approach in providing information on insects' population in the field. For this purpose, as a case study the Keritis hydrological basin located in Crete, Greece, is used, an area representative of typical Mediterranean conditions cultivated primarily by olive trees.

2 Experimental Set Up

The study area was the hydrological Basin of Keritis (Fig. 2), which covers a total area of 176 km^2. It is a typical Mediterranean characteristics site, located in the northern central part of the province of Kydonia in the prefecture of Chania, with the direction of the North-South axis and at an average distance from the city of

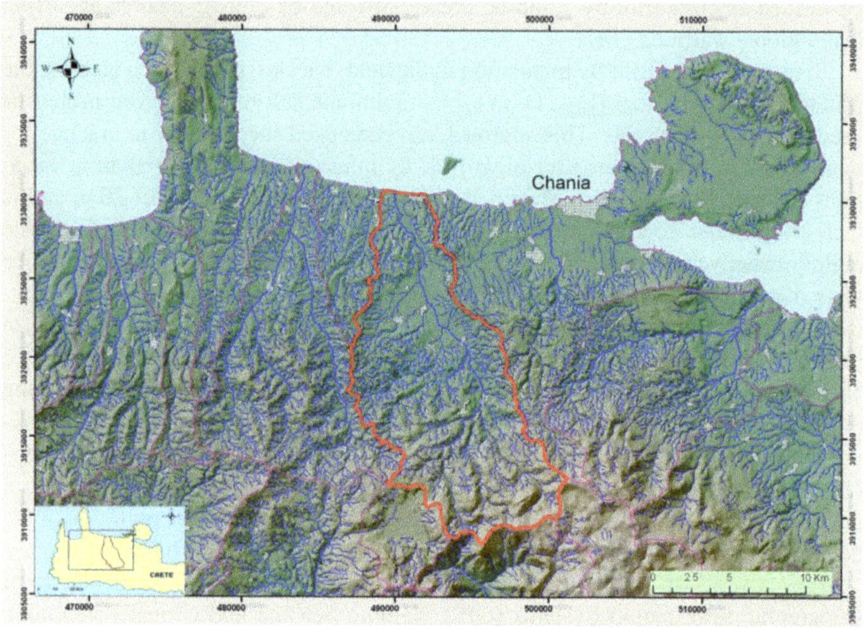

Fig. 2 Location of the Keritis hydrological basin in Chania, Crete

Chania, about 15 km. It is one of the most important hydrological basins in the prefecture of Chania and also of key economic value to the wider area due to olive oil production. The southern part of the catchment area is a part of the mountain range of Lefka Ori. The northern part is flat, with rich vegetation and aquifer. The area also includes the wetland of Agyia Lake and belongs to the European wetland protection network Natura 2000 (code: GR 4340006), hosting rare species of flora and fauna. The main crops in the basin area include olive and citrus trees.

In the area, an operational network of olive - fruit fly traps has been set up and operated by the local stuff of the Municipal Department. In our study, data from this operational network in August 2006 has been used to demonstrate the practical use of the technique presented next in mapping olive- fruit fly population.

3 Methods

Firstly, GPS devices were used to record the location of each trap which was part of the olive- fruit fly traps network. The location of these trap sites was recorded with the assistance of the local stuff of the Municipal Department who had also previously installed the traps network. The trap location data from the GPS was subsequently entered in a GIS environment. In this study, the ArcGIS software (ESRI) was used, but equally any other software with similar capabilities may be used. The traps location data were also enriched with auxiliary information which included, apart from the olive-fruit fly population data, some descriptive characteristics of each sampling location. Next, these files were converted to Google Earth-compatible format (i.e., KML) and subsequently the traps network was overlaid into the high-resolution Google Earth image available for our study area, as shown in Fig. 3 below. This map allowed obtaining an overview of the spatial distribution and density of the traps network over the studied area.

Following this step, the data about the number of insects per trap collected by the staff that had installed the traps in a frequency of 5 days was entered in the GIS environment (ArcGIS). In addition, the GIS database was populated with additional information that was available for the studied region from a previous research work conducted in the area. This included the following:

- The boundaries of the Keritis hydrological basin;
- The boundaries of the Municipal Districts of the wider area;
- The hydrographic network of the area;
- The elevation isolines available at a space of 20 m;
- Satellite photos of the area with a spatial resolution of 1 m (available from previous study in the area).

From the number of insects recorded in the traps, several maps were created at different time intervals that depicted the spatial-temporal change of the olive-fruit fly population [19]. These maps were derived from the point measurements (traps)

Fig. 3 Traps location in our experimental area shown in Google Earth

with the implementation of spatial interpolation using the Kriging technique and a cell size 20 × 20 m. The Kriging interpolation technique is possible to be integrated into a GIS model which can use data from an automated McPhail trap (number of insects) as inputs [10]. The automated McPhail trap subsystem could also be integrated with an automatic spraying subsystem in a united pest monitoring and optimized bait-sprays system [20].

In order to evaluate the olive-fruit fly population in areas of close proximity to water (e.g., wetlands, or river channels), the hydrographic network of the area was used to create a polygon file with a buffer zone of 200 m from the riverbed (Fig. 4). To control the population of the insects in areas with northern exposure, another raster file was created from the elevation data with a resolution also of 20 m (Fig. 5). From the intersection of the above thematic levels with the level of the traps, it was possible to identify which traps were located near the areas which were covered by water and which are in a northern exposure.

4 Results & Discussion

The olive-fruit fly population mapping which was obtained from the implementation of geoinformation technologies such as GIS and GPS in combination with the standard approach of traps network is a valuable tool to assist a more efficient pest control and allow also a conservative use of pesticides in the field. Agronomists involved in the implementation of pesticides against olive-fruit fly infestation as well as farmers can easily be informed by the GIS platform results (even online, since this platform can work as a simple web-GIS platform) about the pest population on time and make immediate decisions about whether, when and where to spray (as for example seen in Fig. 6). The collected data can be stored on a shared website from where anyone interested in this information can access it.

The map showing the olive fruit fly traps network, one of the early mapping products of the approach demonstrated herein, can offer valuable information on the insects' population interpretation maps. It can also inform procedures on pesticide implementation, as it shows whether the distribution of traps per municipal district is uniform or uneven. For example, in our study area, in regions such as Lakkoi (Fig. 7) the olive trees are almost the only crop that is being cultivated. On the other hand, in areas such as Fournes, the olive trees plantation does not cover the whole district. Thus, in the latter case, pesticides implementation can be easily controlled with the help of Google Earth or very high-resolution satellite or aerial images. Such information can also help in acquiring information about tree properties (e.g., tree distances, crown size), an approach that can potentially be supported by field visits as well, in order to enhance accuracy (as shown in Fig. 8). In areas where the distribution of traps is spatially uniform, a much clearer picture emerges on the spatio-temporal distribution of the pest population.

The spatio-temporal change of the population in the studied area in August 2006 is shown in Figs. 9a and 9b below, which is an example of olive fruit fly population

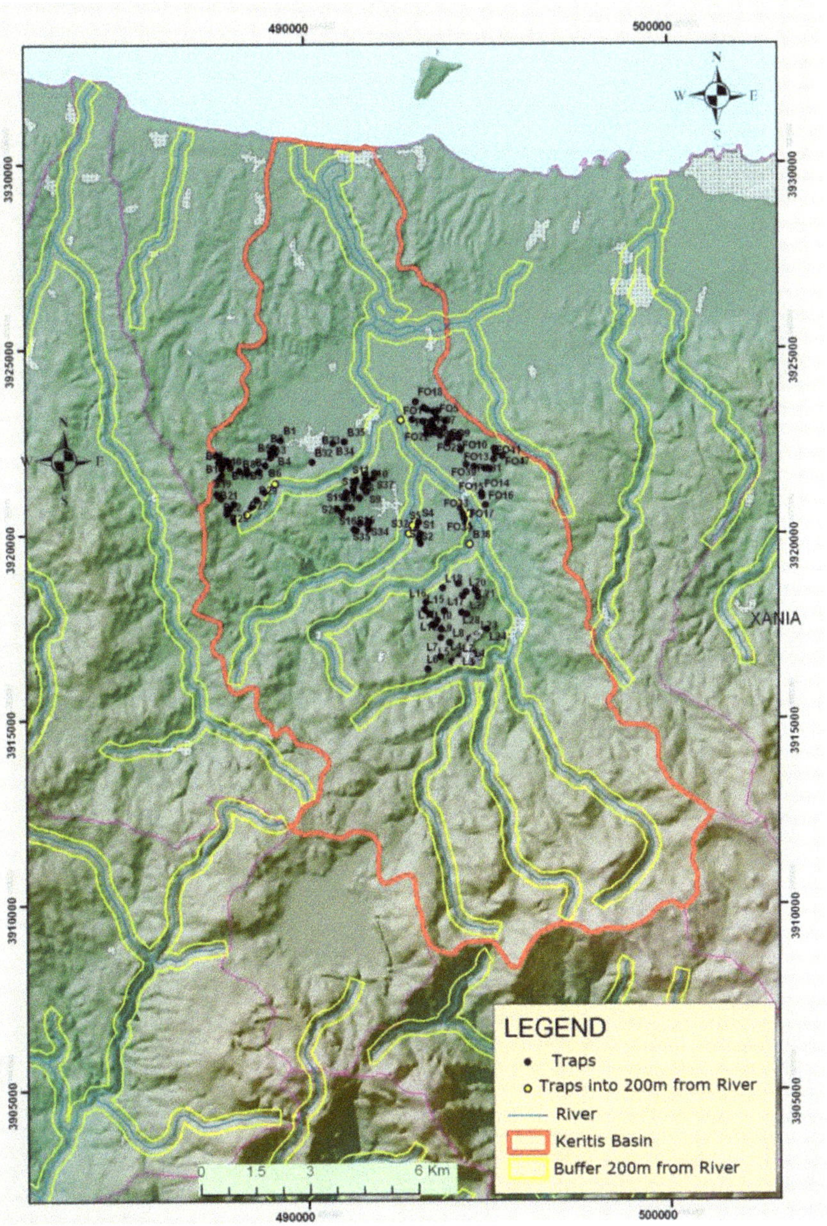

Fig. 4 Traps located in a distance of 200 m on either side of the river streams

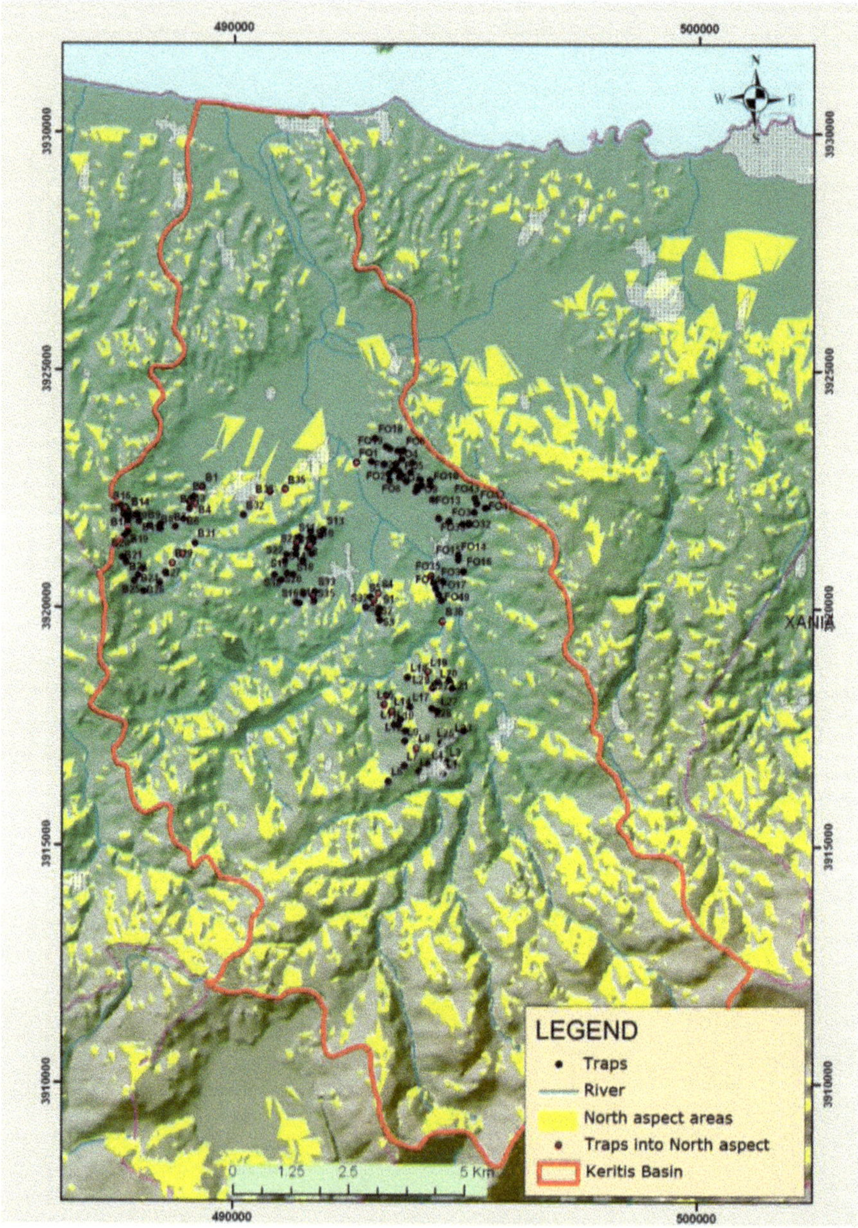

Fig. 5 Traps located in areas with northern exposure of the study area

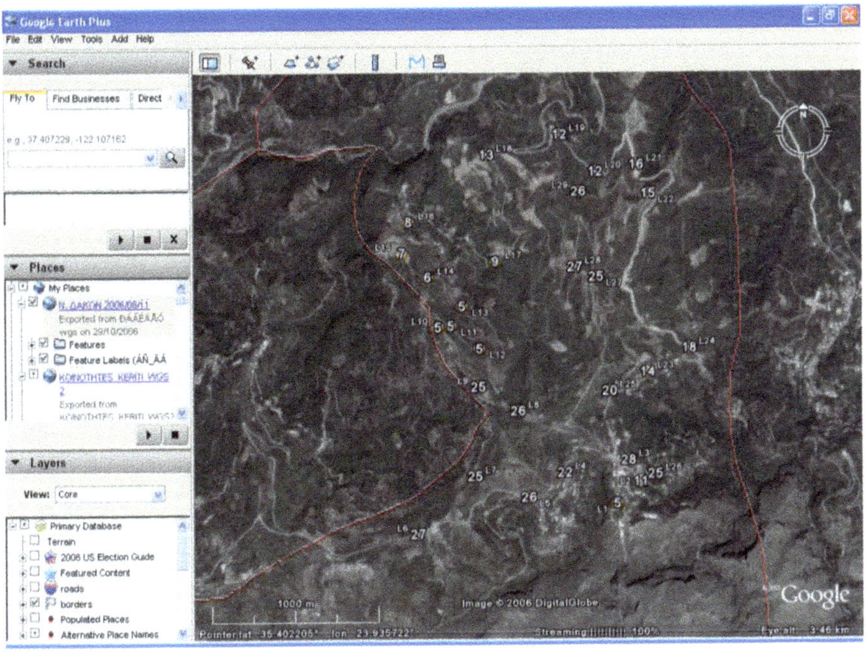

Fig. 6 Illustration of trap locations and the number of the insects in the easy-to-use environment of Google Earth

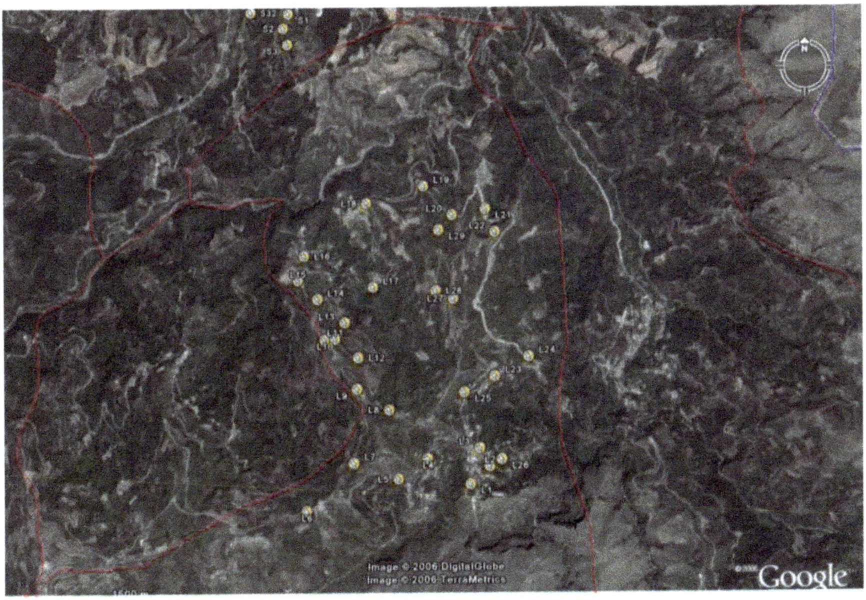

Fig. 7 Evaluation of the olive-fruit flydistribution in the borders of the Lakkoi local districts. As it can be observed there are no traps set up in the northern part of the district

Fig. 8 An illustration of the olive-fruit flytraps distribution in the area of Fournes. As it can be observed, there are no traps in some areas but those areas are cultivated with citrus which are characterised by a smaller crown and shorter planting distances

map produced for the studied area for two selected dates just before and after spraying. The areas shown on the map in green, are areas with low pest population, while areas in orange or red show a high population. As it can be observed, between Fig. 8a and b there is a decrease in the insect population due to spraying, while at the same time it seems that the spraying in areas of Fournes was inefficient since there was no reduction in the population. Also, the results illustrated in Figs. 9a and 9b suggest that the spraying can be applied locally in the areas where high population appears without taking into account only the average population per Municipal District as it is applied until present time.

It is well-known that the microclimatic conditions of a given area play a significant role on olive-fruit fly population growth, especially during summer. The insect growth is favoured in areas with mild temperature and high relative humidity such as areas with northern exposure and areas near rivers. Comparing the average number of traps in the areas with northern exposure and other traps shows that the traps in the northern exposure have an increased number of insects (map 4). Also, the traps located at 200 m on either side of the watercourse have an increased number of insect's captures (see Fig. 4). Furthermore, Table 1 shows the average populations of olive-fruit fly in traps located in a northern exposure, within a 200 m radius from rivers and in areas without special microclimatic conditions (such as high relative humidity or lower temperatures) before and after spraying. According to this table, the insect population was decreased by 57% in areas with north aspect, 50% in areas near rivers and 82% in the areas without special environmental conditions. The traps average number of olive-fruit flies confirm that the major limiting factor

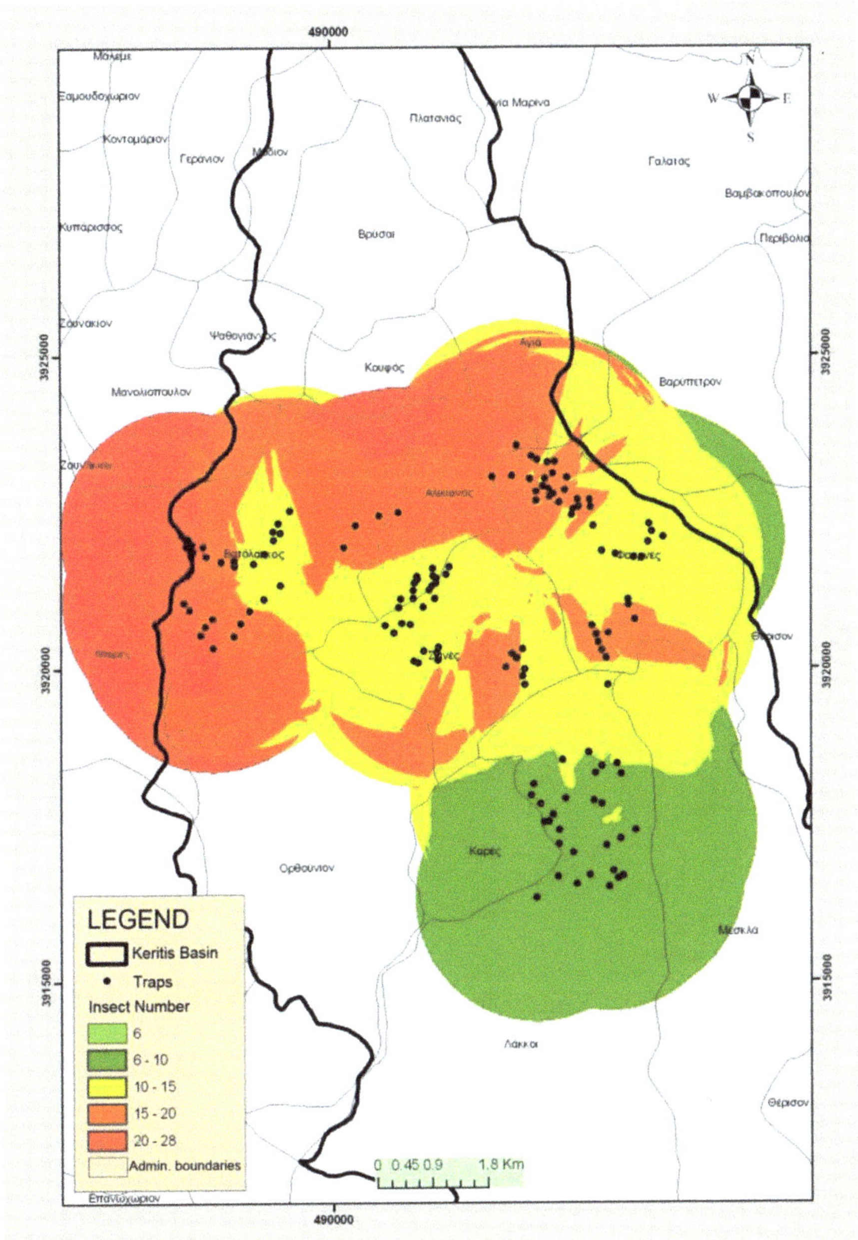

Fig. 9a Distribution of population before spraying (2006/8/21). The different colors illustrate the number of insects captured in the traps

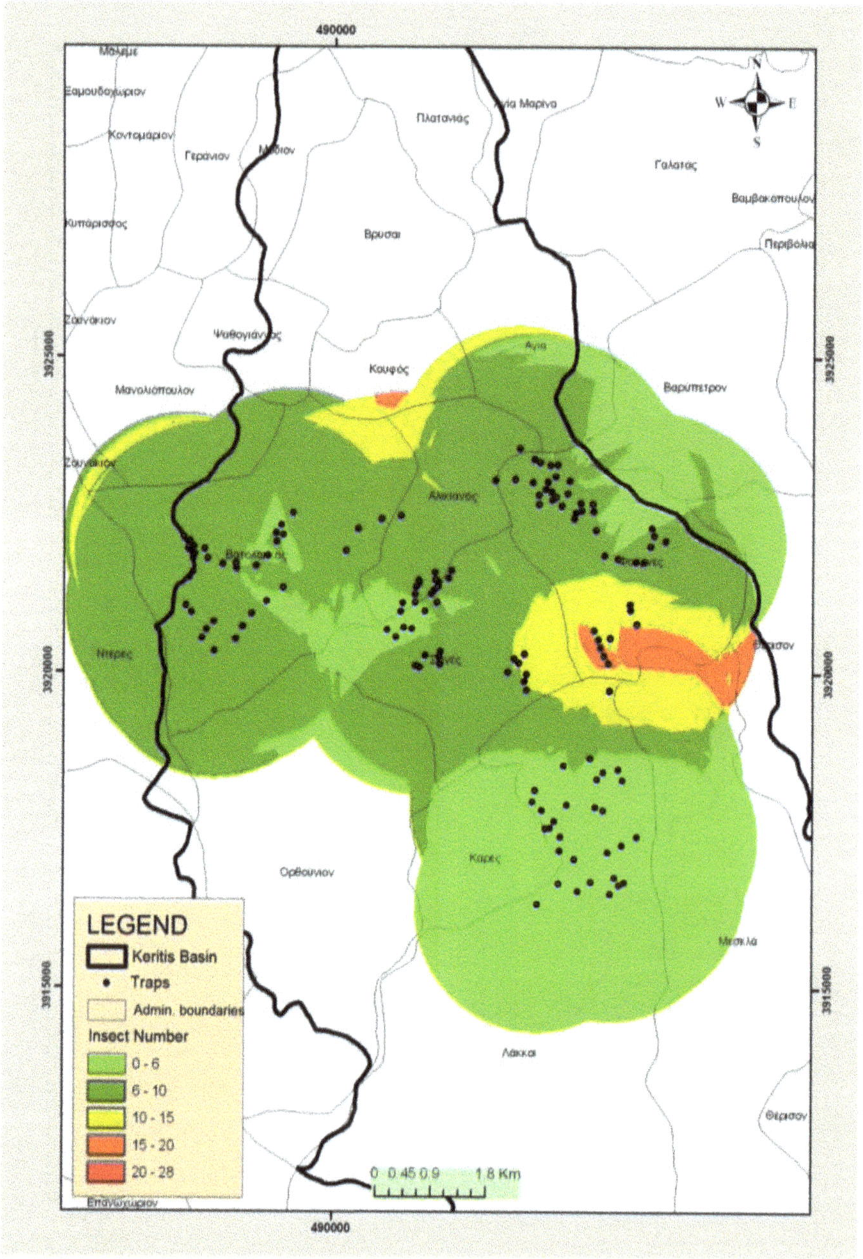

Fig. 9b Distribution of population after spraying (2006/8/26). The different colors illustrate the number of insects captured in the traps

Table 1 Average number of insects catches in traps before and after spraying located in areas without special microclimatic conditions (common areas) (1), in areas of northern exposure (2) and in areas within a radius of 200 m from rivers (3)

Average number of insects	Before spraying	After spraying
Common areas (1)	11	4
Areas of northern exposure (2)	21	9
Areas within a radius of 200 m from rivers (3)	22	11

for the population growth is the low relative humidity in combination to high temperatures, since the pest number per trap was the lowest both before and after spraying. These results suggest that the use of pesticides in this area should be implemented only when they are imperative, due to the high risk of surface water contamination. Dangerous zones can be converted within the GIS platform to Google Earth-compatible files and be subsequently displayed in it so that instantaneous satellite imagery guides the spraying from the ground teams. This can be an important strategy which can significantly help in protecting the water resources of the area.

All in all, the modelling framework presented herein clearly demonstrates the added value of geoinformation tools in pest management and in particular in the delineation of pest population hotspots. This approach could be of key practical value in assessing the effects of climate change on insect abundance, geographic distribution and severity of incidence under an array of environmental conditions. It has immense ecological applications to address various insect-related problems relevant to crops yield production. Furthermore, it can be transferable to the mapping of other entomological diseases to olive trees or other crops.

However, the presented approach, as any other modelling framework, cannot be used as a panacea to all the pest problems; rather it should be seen as a useful means for getting fist-hand knowledge about the intricacies of the issue and building up an approach to resolve it. In that respect, we note that our approached can also be further expanded in multiple directions. For example, the use of wireless sensing cameras, or wireless traps could be accommodated, in the future, in the framework, which would minimise or even eliminate the field visits to the network traps and thus the cost of operating such system. Furthermore, the collected data could be built in a web-GIS database installed in a cloud or a local server. The latter would allow a wide distribution via the internet of all the collected data and the quick communication of any analysis products to the end users provided directly to them (e.g., in their smartphones or via email).

Evidently, it is of urgent need today to develop tools and approaches that will allow us to respond to the increasing frequency of the diversity and abundance of insect pests and ultimately the extent of damage caused in economically important agricultural crops. In this context, the development of reliable and robust operational methodologies for quantifying their extent and for providing information on the risk of occurrence is essential. Hence, it is of vital importance to build a holistic approach for assessing the impacts of potentially increasing frequency and severity of extreme pest diseases events associated with global change to agriculture, including the development of robust economic models. The development of such products and tools constitutes a valuable and indispensable tool for tackling these problems at different scales, from local to global extents.

5 Conclusions

In this study a new methodology framework was presented for monitoring olive-fruit fly population and its changes based on the use of a locally installed pest trap network combined with geoinformation technologies, specifically GIS and GPS. The potential of the approach was demonstrated under real conditions in a typical Mediterranean olive groves field located in Crete, Greece.

As it was shown, the approach allows analysing the population data with all the technical tools and functionalities offered in a GIS environment. From the mapping of the trap network and the population mapping, valuable conclusions can be drawn for the uniformity of the distribution of traps in the studied area as well as for the improvement of the number and location of the traps. The cartography of the spatio-temporal change of the population for the entire period of the insecticide implementation allows the control of the effectiveness of the sprays per region. This offers us the ability to identify the specific zones requiring spraying around ecologically sensitive areas (rivers, lake, Natura 2000 area, organic crops), in which special care must be taken when spraying, thus assisting also environmental protection. The mapping of trap locations, the number of traps and ecologically sensitive zones in a form compatible with free online programs (Google Earth) and also the quick dissemination of these data to the public, provides everyone the opportunity to be involved in the study of the collected data in near real-time. At the same time, it supports the decision-making process on time even for the GIS non-expert.

In summary, this study advocates the promising potential of geoinformation in mapping the spatial and temporal change of olive-fruit fly population and also in evaluating the relationship between entomological and geographical parameters (distance from rivers, northern exposure). This technology has the potential to allow a targeted and conservative use of pesticides in the field, maximizing the economic benefits for both the government and the farmer and it also has a great potential of expandability with other contemporary technologies and not only.

Acknowledgments The present study was carried out in the framework of a sub-project of Natura 2000-Chania, funded by EPEAEK II ARCHIMEDES 75% from the European Social Fund and 25% from Greek Resources. The authors gratefully acknowledge the anonymous reviewers and the editor for their comments that resulted to improving the manuscript. Dr. George P. Petropoulos's participation to this study was funded under the European Union's Horizon 2020 Marie Skłodowska-Curie "ENVISION-EO" project (grant agreement No 752094) and the author gratefully acknowledges the financial support provided by the European Commission.

References

1. Brodt S. Six J. Feenstra G. Ingels C. & Campbell D. (2011) Sustainable Agriculture I Learn Science at Scitable
2. ICPP (2007) Climate Change. Cambridge University Press
3. Kastner T, Erb KH, Haberl H (2014) Rapid growth in agricultural trade: Effects on global area efficiency and the role of management. Environ Res Lett 9:. https://doi.org/10.1088/1748-9326/9/3/034015
4. Ogallo LA, Boulahya MS, Keane T (2000) Applications of seasonal to interannual climate prediction in agricultural planning and operations. Agric For Meteorol 103:159–166. https://doi.org/10.1016/S0168-1923(00)00109-X
5. Fand BB, Kamble AL, Kumar M (2012) Will climate change pose serious threat to crop pest management: A critical review? Int J Sci Res Publ 2:
6. Jarvis CH, Baker RHA (2001) Risk assessment for nonindigenous pests: 2. Accounting for interyear climate variability. Divers Distrib 7:237–248. https://doi.org/10.1046/j.1366-9516.2001.00114.x
7. Kroschel J, Sporleder M, Tonnang HEZ, et al (2013) Predicting climate-change-caused changes in global temperature on potato tuber moth Phthorimaea operculella (Zeller) distribution and abundance using phenology modeling and GIS mapping. Agric For Meteorol 170:228–241. https://doi.org/10.1016/j.agrformet.2012.06.017
8. Sporleder M, Simon R, Juarez H KJ (2008) Regional and seasonal forecasting of the potato tuber moth using a temperature-driven phenology model linked with geographic information systems. In: Kroschel J LL (ed) Integrated pest management for the potato Tuber Moth, Phthorimaea operculella Zeller – a potato pest of global importance. Tropical agriculture, advances in crop research. Margraf Publishers, Weikersheim, pp 15–30
9. Rodrigues MÂ, Coelho V, Arrobas M, et al (2019) The effect of nitrogen fertilization on the incidence of olive fruit fly, olive leaf spot and olive anthracnose in two olive cultivars grown in rainfed conditions. Sci Hortic (Amsterdam) 256:108658. https://doi.org/10.1016/j.scienta.2019.108658
10. Doitsidis L, Fouskitakis GN, Varikou KN, et al (2017) Remote monitoring of the Bactrocera oleae (Gmelin) (Diptera: Tephritidae) population using an automated McPhail trap. Comput Electron Agric 137:. https://doi.org/10.1016/j.compag.2017.03.014
11. Marchini D, Petacchi R, Marchi S (2017) Bactrocera oleae reproductive biology: new evidence on wintering wild populations in olive groves of Tuscany (Italy). Bull Insectology 70:121–128
12. Fletcher BS (1987) The Biology of Dacine Fruit Flies. Annu Rev Entomol 32:115–144. https://doi.org/10.1146/annurev.en.32.010187.000555
13. Schneider M, Wunder C, Reuss E, et al (2017) Evading plant defence: Infestation of poisonous milkweed fruits (Asclepiadaceae) by the fruit fly Dacus siliqualactis (Diptera: Tephritidae). Toxicon 139:13–19. https://doi.org/10.1016/j.toxicon.2017.09.011
14. Gutierrez AP, Ponti L, d'Oultremont T, Ellis CK (2007) Climate change effects on poikilotherm tritrophic interactions. Clim Change 87:167–192. https://doi.org/10.1007/s10584-007-9379-4
15. Fitt GP (1984) Oviposition behaviour of two tephritid fruit flies, Dacus tryoni and Dacus jarvisi, as influenced by the presence of larvae in the host fruit. Oecologia 62:37–46. https://doi.org/10.1007/BF00377370
16. Haniotakis GE (2005) Olive Pest Control Present Status and Prospects. In: Proc. IOBC/WPRS Conf. Integr. Prot. Olive Crop. Chania, 29-31 May 2003. - Ref. - Sci. Res. Publ.
17. Scheldeman X, van Zonneveld M (2010) Training manual on spatial analysis of plant diversity and distribution. Bioversity International, Rome
18. Yadav DS, Chander S, Selvaraj K (2010) Agro-ecological zoning of brown planthopper [Nilaparvata lugens (Stal)] incidence on rice (Oryza sativa L.)
19. Petacchi R., Ragaglini G. GD (2003) Spatial data analysis of bactrocera oleae infestation from large scale monitoring network: importance of this approach in IPM. In: 1st European meet-

ing of the IOBC/WPRS Study group “Integrated Control in Olives”, “Integrated Protection of Olive Crops”. Chania MAICH
20. Fouskitakis G, Doitsidis L, Varikou K, et al (2017) Integrated pest control for the olive-fruit fly: Remote pest monitoring and optimized bait-sprays. In: CEUR Workshop Proceedings

In-field Experiments for Performance Evaluation of a New Low-Cost Active Multispectral Crop Sensor

Aristotelis C. Tagarakis, Marko Kostić, Natasa Ljubičić, Bojana Ivošević, Goran Kitić, and Miloš Pandžić

1 Introduction

Maize (*Zea mays* L.) is an important crop in the world with an estimated production area of 194 million hectares in 2017 and an average yield of 5.9 Mg ha^{-1} [1]. Obviously, there are a lot of possibilities for gaining higher quantities of harvested product, especially if considering the increasing trends of maize production during the past decades (Fig. 1). However, the yield increase was mainly achieved by genetically altering the crop, by producing improved hybrids, and by optimizing the cultivation practices [2]. To reach higher yields and higher economic feasibility of maize crop, fertilization management plays a critical role, and nitrogen (N) is considered as the most essential element in plant mineral nutrition [3].

Currently agricultural production is greatly depended on the use of N fertilizers, especially in cereal crops [4]. The physiological role of N in plants is of great importance, particularly in maize grown for grain which is highly demanding to nitrogen fertilizer [5], as N is essential element in grain protein and its sufficiency leads to

A. C. Tagarakis (✉)
BioSense Institute, University of Novi Sad, Novi Sad, Serbia

Institute for Bio-Economy and Agri-Technology (iBO), Centre for Research and Technology Hellas (CERTH), Thessaloniki, Greece
e-mail: a.tagarakis@certh.gr

M. Kostić
Faculty of Agriculture, University of Novi Sad, Novi Sad, Serbia
e-mail: marko.kostic@polj.uns.ac.rs

N. Ljubičić · B. Ivošević · G. Kitić · M. Pandžić
BioSense Institute, University of Novi Sad, Novi Sad, Serbia
e-mail: natasa.ljubicic@biosense.rs; gkitic@biosense.rs; milos.pandzic@biosense.rs

D. D. Bochtis et al. (eds.), *Information and Communication Technologies for Agriculture—Theme I: Sensors*, Springer Optimization and Its Applications 182,
https://doi.org/10.1007/978-3-030-84144-7_13

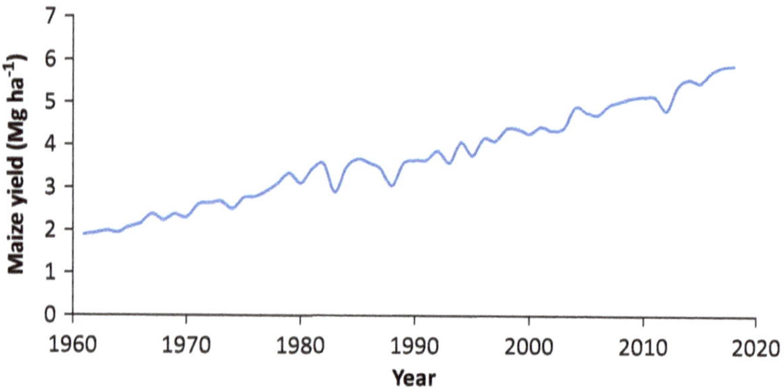

Fig. 1 Average yield trend for the years 1961–2018. (Source: Ref. [1])

improved grain yield and quality. Therefore, care must be taken to supply the plants with the appropriate rate of N fertilizer which would ensure sufficient uptake, while minimizing N losses avoiding groundwater and surface water contamination. According to Raun and Johnson [6], about 67% of the applied N fertilizer in the cereal production is lost as result of denitrification, surface runoff, volatilization, and leaching. Consequently, the assessment of the usage of the N applied to the crop in the form of fertilizer, is one of the most important aspects in plant nutrition and is referred in literature as the N use efficiency. One of the methods for the assessment of N use efficiency was defined by López-Bellido et al. [7] as the ratio of grain yield achieved to the amount of the N fertilizer applied. Many different approaches of N fertilizer rate calculation were established to increase N use efficiency which were based on potential yield, soil organic matter content, soil mineralization potential [8] and other aspects of crop production. Other actions for increasing the efficiency of the N fertilizer application lie on selecting the correct agronomic practices, such as choosing the appropriate N fertilizer source and the correct timing of N application and following the proper N application techniques for each specific crop.

During the last decades, the development of sensing technologies and sensing systems for agricultural use aims to overcome the limitations of traditional agriculture. Traditionally, farming was heavily influenced by the subjective point of view of individual actors with no true knowledge of the actual status of the dominant factors that influence crop growth, yield and quality. Site-specific field management is a higher level of decision-making strategy putting into focus information gathering and analyzing techniques in the context of a newly introduced farming management approach named Precision Agriculture (PA). It is often defined as a farming approach in which decisions are made at a high resolution according to the actual needs of the plants at each location. During the course of development of PA, several definitions were reported in literature. Gemtos et. al. [9] defined PA as *"the management of spatial and temporal variability in the fields using Information and Communications Technologies (ICT)"*. The International Society of Precision Agriculture (ISPA) has recently recognized the following as the official definition of PA: *"Precision*

Agriculture is a management strategy that gathers, processes and analyzes temporal, spatial and individual data and combines it with other information to support management decisions according to estimated variability for improved resource use efficiency, productivity, quality, profitability and sustainability of agricultural production" [10]. Robert [11] referred to the three "*R's*" of PA which correspond to the application of inputs in agricultural systems at the "*Right time*", at the "*Right amount*" and to the "*Right place*" while later on Khosla added another two "*R's*" corresponding to the use of "*the Right Source*", and the "*Right manner*"[12]. PA is commercially practiced since the early 1990's [13] when the Global Positioning Systems (GPS) and the first yield monitors with georeferencing capability became commercially available [14].

In practice, in order to address all the aspects of PA, fields and plants are monitored using a variety of sensing technologies and decisions concerning fertilizer, pesticide, irrigation and field operations are adjusted accordingly. PA management systems show significant advantages compared to traditional farming such as increased application efficiency, and minimal environmental footprint of agricultural applications [15–18]. Recent advances in technology provide an unprecedented opportunity for further development and advanced use of PA. Remote and proximal sensing using optical sensors are gaining popularity in crop production systems in the framework of PA applications. Nowadays, there is vast number of optical sensors recording the data about crops at different spatial, radiometric and temporal resolutions. Both remote and in-field sensors are used for monitoring plant health status, detecting deficiencies of nutrients and water, and soil condition.

Remote sensing includes different types of sensing devices which are mainly divided in two categories based on the measuring principles; (a) active sensors which carry their own source of light, emitting at certain ranges of the spectrum and measuring the reflectance from the targeted surface [5] and (b) passive sensors which use natural source of light and therefore measure the reflectance of the solar radiation from the plants. Passive sensors' measurements depend highly on the sun exposure, cloudiness, architecture and reflective characteristics of scanned objects that could jeopardize recording stability in time, especially if large fields are measured requiring a significant amount of time [19, 20]. The main advantage of active proximal sensors, as opposed to the passive systems, is their independence from illumination conditions, since they emit their own light, making feasible to operate under cloudy conditions or even at night. Furthermore, active sensing systems do not require calibration because the light source is known and constant [21].

Remote sensing systems can also be categorized based on the platform carrying the sensor; (a) satellite based remote sensing systems, (b) aerial vehicle or unmanned aerial vehicle (UAV) mounted proximal sensing systems and, (c) ground-based proximal sensing systems. The systems described in categories a and b are mainly passive while most of the sensing systems included in category c are active. This is mainly due to the close proximity to the canopy that the active sensing devices need to be positioned in order to adequately illuminate the sensing surface. Typically, the spatial resolution of proximal sensing is in the range from millimeters to centimeters, as opposed to remote sensing (satellite systems) that typically has resolution in the range from decimeters to hundreds of meters [22].

Site-specific crop management relies on cutting-edge sensing and decision-making technologies in which maximum productivity per unit of input and used land area are dominant criteria. In order to achieve the aforementioned, field activities must be spatio-temporal adaptive considering previously determined in-field conditions and specific requirements of the specific crop. Modern crop management is certainly related to the data collection methods hence information validity and usability in the decision-making process. The complexity of crop monitoring and modeling is the limiting factor for implementation of all the postulates of site-specific cropping concept. Remote sensing provides the opportunity to rapidly acquire data concerning crops' canopy characteristics. Remote sensing using satellite-derived data has been used in agriculture since 1970's when the first Landsat satellite was launched. However, the very low spatial resolution and the high cost of image acquisition restricted the use to only a few specialized applications. Over the period of nearly half a century, the spatial resolution, as well as the revisit frequency, increased dramatically [13]. In addition, during the last decade ESA and NASA have made available to general public certain satellite imagery at no cost [23, 24] leading to increased interest of the agricultural community towards satellite remote sensing. Therefore, having satellite imagery of sufficient resolution at greatly reduced cost, made remote sensing a popular approach for monitoring field and crop status. However, the satellite remote sensing approach is not very handy when more detailed spatial information is required particularly in tree and vegetable crops and in small sized fields. In addition, pre-processing and analysis of the data is prerequisite, making this source of data impractical for on-the-go in-field applications.

This study focused on the performance evaluation of a recently developed active multispectral sensor. Therefore, this chapter emphasizes on the use of ground-based proximal sensors in agriculture. Ground-based proximal sensing is performed by sensors at a relatively short distance from the object of interest. These types of sensing systems are hand-held, tractor-mounted or vehicle-mounted, or may facilitate both options. The main limitation of the ground-based proximal sensors is the small area coverage [25]. On the other hand, they show significant advantages, such as high spatial resolution and independent choice of the time of acquisition, since their measurements are not compromised by cloudiness or differences in illumination conditions, in the case of active sensors. Therefore, these sensors are ideal for practical applications such as on-the-go variable rate N applications [5, 26, 27]. To that end, numerous studies have been conducted with the aim to develop algorithms that employ multispectral crop sensing systems for nutrition management. The majority of these algorithms employ yield potential as a benchmark for calculation of the additional amount of fertilizer that needs to be side dressed. However, their robustness shows fluctuations due to the temporal variability that yield potential shows as it is highly dependable to the annual precipitation, temperature, relative humidity, and other climate parameters [28]. These algorithms are used to process the signal acquired by the sensors and calculate, in real-time, the amount of fertilizer that needs to be applied at the specific location being scanned. After prescribing the amount of fertilizer to be applied, the application system receives the signal to change the rate according to the calculations.

As already mentioned, the majority of the proximal sensors are active, having their own source of light emitting at certain wavelengths and measuring the reflectance from the target. Simple handling makes them suitable for a wide range of applications on different crops, and instant data can be collected following a non-destructive sampling method [29, 30]. Therefore, active proximal sensors have become very popular for crop status diagnostic. The results of numerous studies on such sensors offer a vast number of models which could be used in the prediction of maize maturity, yield potential, plant health estimation, etc. [31]. Spectral analysis of reflected waves from plant canopy is valuable in the recognition of spectral "fingerprints", which help identify some biotic or abiotic processes that are otherwise undetectable by human or machine. Furthermore, active multispectral sensors show great potential in rapid spatial assessment of nitrogen (N) status of growing plants in early season which provide adequate precondition for optimization of nitrogen management [32]. Therefore, active proximal sensing has been increasingly used in agriculture for assessing crop status and growth and has proved to be promising approach for end-of-season yield estimation in a large range of crops [33]. Most active proximal sensors, measure the reflectance of specific spectra of light, typically in the visible and the near-infrared, from the plant canopy providing a range of vegetation indices such as the Normalized Difference Vegetation Index (NDVI; [34]). The NDVI is the most widely used index for deriving yield estimates [35]. Apart from the in-season yield estimation, NDVI has also been related to nitrogen status, chlorophyll content, biomass, and leaf area, at micro and macro scale [27, 36].

Yield estimation from mid-season spectral canopy measurements is of particular importance since it is the first step in the development of an algorithm for real-time variable rate N applications [37]. The timing of sensing, in terms of growth stage, greatly impacts the accuracy of yield predictions from sensor data [38]. Previous studies in maize defined V7–V8 as the growth stages that provide the highest accuracy of end-of-season yield estimation and V6 as the stage with the highest variability in NDVI measurements [33, 38], important elements for maximizing the benefit of variable rate fertilization.

Raun et al. [39] introduced a new index approach to be used for end-of-season yield estimations from mid-season measurements and for on-the-go variable rate fertilization. This new index named Estimated Yield (EY) was defined as the sum of the NDVI measurements between two mid-season acquisition dates divided by the cumulative Growing Degree Days (GDD) from sensing day 1 to sensing day 2 (Eq. 1).

$$EY = \frac{\left(NDVI_{T1} + NDVI_{T2}\right)}{GDD} \tag{1}$$

Where: EY is the In Estimated Yield, $NDVI_{T1}$ and $NDVI_{T2}$ are the Normalized Difference Vegetation Index measured at sensing day 1 and sensing day 2 respectively, and GDD is defined as the cumulative growing degree days.

This modified index was an improvement in the use of vegetation indices for yield estimations [5]. This new index was further modified to derive a simplified and more applicable approach named In Season Estimated Yield (INSEY; [15]) (Eq. 2).

$$\mathrm{INSEY}_{\mathrm{DAP}} = \frac{\mathrm{NDVI}}{\mathrm{DAP}} \tag{2}$$

Where: $INSEY_{DAP}$ is the In Season Estimated Yield using the days after planting, NDVI is the Normalized Difference Vegetation Index, DAP is the number of Days After Planting for days with GDD>0, and GDD is defined as the growing degree days.

Another expression of INSEY is the $INSEY_{GDD}$ which uses the GDD to normalize the NDVI measurements (Eq. 3).

$$\mathrm{INSEY}_{\mathrm{GDD}} = \frac{\mathrm{NDVI}}{\mathrm{GDD}} \tag{3}$$

Where: $INSEY_{GDD}$ is the In Season Estimated Yield using GDD, NDVI is the Normalized Difference Vegetation Index, and GDD is defined as the cumulative growing degree days from planting to sensing.

Teal et al. [40] compared different models that used $INSEY_{GDD}$, and $INSEY_{DAP}$ showing similarly good results in the prediction of end-of-season yield. These new approaches normalize the NDVI measurements across time and various environmental conditions [40], accounting for the growing conditions from the time of planting to sensing and providing an estimate of the N uptake per day [41] and the biomass produced per day [38].

The timing of data acquisition plays an important role in the accuracy of the yield prediction form mid-season NDVI measurements. In studies conducted in maize crop, V6 to V8 (Ritchie scale; [42] were defined as the growth stages that provide good estimation of end-of-season yield and are early enough to perform corrective management practices [33, 43].

Over the years, various different proximal sensors found practical applications in the agricultural sector. The first devices developed, such as SPAD meter (Konica Minolta Inc., Osaka, Japan) had limited automations and therefore demanded manual work for data acquisition. With the rising of PA applications, new generation of proximal sensors were developed providing the option to perform continuous measurements in the field with limited input by the operator. Sensing systems such as Yara N-sensor (Yara International ASA, Oslo, Norway), GreenSeeker (Tribmle Inc., CA, USA), Crop Circle (Holland Scientific, NE, USA), CropScan (Next Instruments, Sydney, Australia) etc., are fully automated and provide the options of automated mapping and real time variable rate fertilizer applications. All these proximal sensors provide accurate assessment of the plants' growth, which can be linked to photosynthetic activity and chlorophyll content [44], level of evapotranspiration [45], crops' nitrogen status [5, 15, 46], and yield [43, 44].

Most active proximal sensors vary in central wavelengths or bandwidths for calculating indices such as the NDVI [47, 48]. Therefore, this study was conducted to evaluate the performance of the recently developed active multispectral proximal sensor Plant-O-Meter in real field conditions, and compare it with the GreenSeeker handheld, a widely accepted commercial crop sensor extensively used in literature. The main objectives of this study were to: (1) define the relationship between the NDVI measurements derived from the two sensors, (2) determine the specific growth stage at which the sensors provide more reliable end-of-season yield estimation under the specific climatic conditions of Vojvodina region, and (3) define the ability of Plant-O-Meter to estimate end-of-season yield from mid-season canopy measurements as compared to the GreenSeeker handheld sensor device.

2 Materials and Methods

2.1 Field Trials and Experimental Design

The present study was carried out at two experimental fields located approximately 80 km apart. The first trial was located in Ravno Selo (45.423614N, 19.618609E, Calcic Chernozem soil type; referred in the text as Field-1), in northern Vojvodina, while the second in Bajmok (45.962249N, 19.409621E, Gleyic Chernozem soil type; referred in the text as Field-2), in central part of Vojvodina region, in Serbia. The dominant soil types in Vojvodina region are characterized as Chernozems. These soils have good structure and stability of soil aggregates ensuring good permeability [19]. The total porosity of this type of soil is approximately 50% including about 20% of macropores and 30% of micropores. In general, these soils are fertile with high content in soil organic carbon [49] and are suited to optimally fulfill the soil functions, thus they consist ideal soils for growing arable and other crops, providing high yield potentials [50].

At each experimental field, three maize hybrids of different maturity classes and length of vegetation period were sown, namely P9537 (FAO 340), P9911 (FAO 450) and P0412 (FAO 530), here referred as Hybrid-1, Hybrid-2 and Hybrid-3 respectively. The fields were sown at 70 cm of inter-row spacing and 20 cm spacing between plants in the row. The study included five different N treatments (0, 50, 100, 150 and 200 kg N ha^{-1}) applied pre-plant by incorporating granular urea (46% N). These treatments were incorporated in the experiment in order to create variability in maize growth which subsequently corresponded to yield variability reflecting the effect of N availability according to the rate of N applied in each plot.

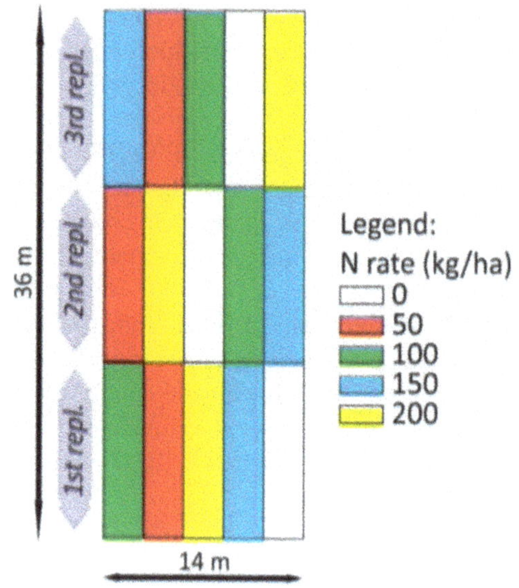

Fig. 2 Experimental design of a single maize hybrid

Henceforward, the two multispectral sensors were utilized to measure the crop canopy properties several times during maize crop vegetative stages. The aim was to capture the variability produced artificially by the N treatments, reflecting the effect of N sufficiency or deficiency on final yield. This information would help farmers make better decisions for in-field management.

The experiment was conducted following a randomized complete block design (RCBD) with three replications. Each plot contained four 12 m long rows of maize (Fig. 2). Only the central part, 6 m long, of the two middle rows was measured and harvested, while the plants located in the side rows and in the remaining 3 m at the beginning and at the end of the rows in the plots served as guard plants.

2.2 *Sensor Measurements and Sensor Description*

Two active proximal sensors were used to measure NDVI at V5, V6 and V8 growth stages of maize; the GreenSeeker hand-held (Trimble Inc., CA, USA) and a recently developed active multispectral optical sensor named Plant-O-Meter (BioSense Institute, Serbia).

GreenSeeker (Fig. 3a) is an active hand-held sensor, which emits light and measures the reflectance at 660 nm (R) and 770 nm (NIR) calculating the NDVI [51]. In-field reflectance measurements were taken by holding the GreenSeeker sensor approximately 60 cm above the crop canopy, with the sensing footprint perpendicular to the row direction, manually recording four average measurements from the measuring area in each plot.

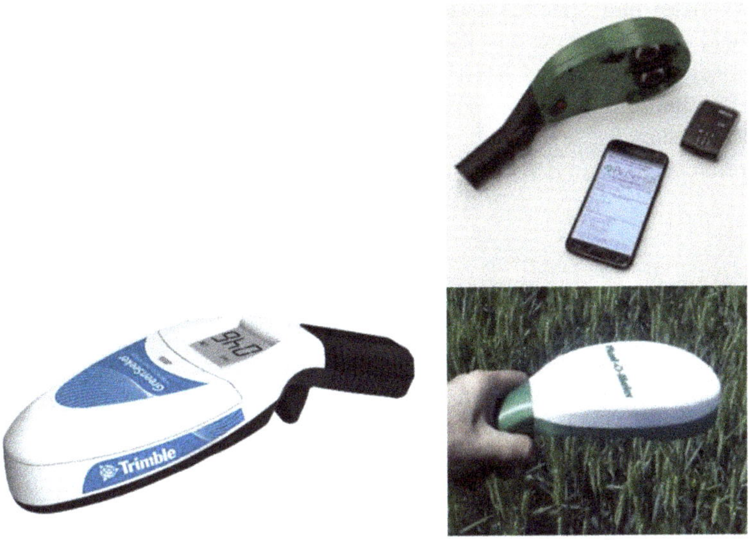

a b

Fig. 3 The two handheld active proximal sensors used in the study; (**a**) GreenSeeker handheld and (**b**) Plant-O-Meter

Plant-O-Meter (Fig. 3b) is an active sensor equipped with an integrated multispectral source that comprises light sources emitting at four indicative wavelengths (465 nm; Blue, 535 nm; Green, 630 nm; Red and 850 nm; Infrared) and senses the energy reflected from the plant canopy. The sensor records the reflectance for each band separately providing the ability to calculate more than 20 different indices. It connects to any Android device and uses its processing and storing capacity for data logging and processing. In addition, it uses the device's GPS antenna to georeference the measurements. The interface also supports connection to external Bluetooth GPS antenna maximizing the georeferencing accuracy. Detailed information concerning the working principles and specifications of the Plant-O-Meter device is provided in Kitić et al. [52]. In-field reflectance measurements were taken by holding the Plant-O-Meter approximately 60 cm above the crop canopy with the sensing footprint perpendicular to the row direction and scanning the whole length of the two middle rows in continuous mode which supports continuous mapping of the sensor's measurements. The data acquisition frequency was 1 Hz which corresponded to approximately 1 measurement every 0.9 m if travelling on average at walking speed. The central part, i.e., 6 m of each measured row (two middle rows per plot), was selected after processing the data using GIS software. The NDVI measurements with both instruments were made close to noon, between 11:00 a.m. and 1:00 p.m. for more consistent environmental conditions (illumination, temperature, relative humidity etc.) between the measurements.

In this study, the GreenSeeker hand-held device was used as a reference sensor in order to assess the operability and accuracy of the recently developed

Plant-O-Meter multispectral sensor. The main reason for selecting GreenSeeker hand-held for this purpose was that it works using the same operating principles, it is relatively low cost, and has similar measuring characteristics to the Plant-O-Meter. In addition, the GreenSeeker hand-held is a widely used, commercial, and well accepted by the scientific community, canopy sensor. Therefore, it serves as a good reference for comparison.

2.3 Harvest

Due to the small size of the experimental plots, it was impossible to use maize harvester equipped with yield monitor; thus, harvest was performed manually. At the stage of full maturity, the plants from the central part, 6 m long, of the two middle rows in each plot were hand-harvested by manually picking all developed ears and collecting in pre-labelled bags. The gross weight of each plot was measured using a handheld digital scale and the content of each bag was shelled to calculate the net grain weight. A GAC® 2500-INTL Grain Analysis Computer (Dickey-John, IL, U.S.A.) was used to measure grain moisture content and the final yield was normalized at 14% moisture content.

2.4 Data Analysis

Descriptive statistics and analysis of variance (ANOVA) were performed on the yield datasets to find possible differences between the three different varieties used in the study. Pearson's correlation was used to define the relationship between yield and NDVI derived from Plant-O-Meter and GreenSeeker crop sensors measurements acquired at V5, V6 and V8 growth stages of maize. The datasets from both fields were joined based on the growth stage, regardless the date of data acquisition. Then, the unified dataset was used for the correlation analysis. The statistical analyses were performed using the Statistica 12 software (Dell Software, TX, USA).

Further in the study, the sensor measurements were transformed to $INSEY_{DAP}$ dividing the NDVI by the days after planting as recommended by previous studies [15, 33]. The use of INSEY adjusted the sensor measurements to the specific growing conditions of each field from planting until sensing. Additionally, it facilitated the joining of data from both fields into a unified dataset for the final regression analysis. Linear regression analysis was used to define the relationships between the INSEY and end-of-season yield for each growth stage. This analysis was very useful to define the growth stage when the most reliable estimation of end-of-season yield could be acquired from the mid-season crop canopy measurements using the two multispectral sensors.

Linear regression models were also used to define the relationship between the GreenSekeer and Plant-O-Meter NDVI measurements. Since the purpose of this analysis was the direct comparison between the two sensors scanning, almost simultaneously, the same canopy surfaces, all the scanning data from all growth stages were used.

3 Results and Discussion

3.1 Descriptive Statistics and Analysis of Variance

According to the descriptive statistics of the yield datasets (Table 1), variability (Variance and Coefficient of Variance) of average end-of-season yield was considerably higher in Field-1 as compared to Field-2. This indicates that the different N fertilization rates had greater impact on yield in Field-1 which was more responsive to N fertilizer application. In addition, Field-2 showed considerably higher yields compared to Field-1 for all maize hybrids used in the study. The higher fertility in Field-2 explains the lower response of the maize plants to N fertilization. Hybrid-3 provided the highest average yield followed by Hybrid-2, while Hybrid-1 had the lowest yield. These results were consistent for both fields and according to ANOVA results the yield differences between hybrids within each field were statistically significant ($p < 0.05$) with the exception of the yield differences between Hybrid-2 and Hybrid-3 in Field-1. In that specific case, yield of Hybrid-3 was higher than Hybrid-2, consistent with the results in Field-2, but the difference was not statistically significant as result of the lowest response to N application in Field11. The results in Table 1 revealed relatively weak impact of applied N on the final grain yield. This would be followed by enough N which had been mineralized from soil organic matter or that deposited in the rainfall to meet all of the plant N needs. In open field crops, great N losses (over 50%) are often caused by well-known soil processes such as denitrification, volatilization or fixation depending on soil conditions, fertilizer application method and weather conditions [53].

Table 1 Descriptive statistics and analysis of variance (ANOVA) results of end-of-season yield ($Mg\ ha^{-1}$) for the three maize varieties used in the study and for each field

	Hybrid	Min	Max	Mean	Std. Dev.	Var.	C.V. (%)
Field-1	1	9.5	13.8	12.10a†	1.3	1.6	10.4
	2	11.9	15.5	13.32b	1.1	1.2	8.5
	3	12.3	15.5	13.75b	1.0	1.1	7.7
Field-2	1	10.4	14.1	13.17a	1.0	1.1	7.1
	2	13.1	15.8	14.49b	0.8	0.6	5.7
	3	15.0	17.5	16.10c	0.8	0.6	4.7

†Within each field, means followed by the same letter are not significantly different ($p < 0.05$)

3.2 Weather

Concerning the climatic and weather conditions, minimum temperature (T_{min}; Fig. 4a) in Field-1 was somewhat lower for the growing season of maize. On the other hand, the average maximum temperature (T_{max}) was quite similar at both fields. The highest average min and max temperatures were recorded, as expected, in August.

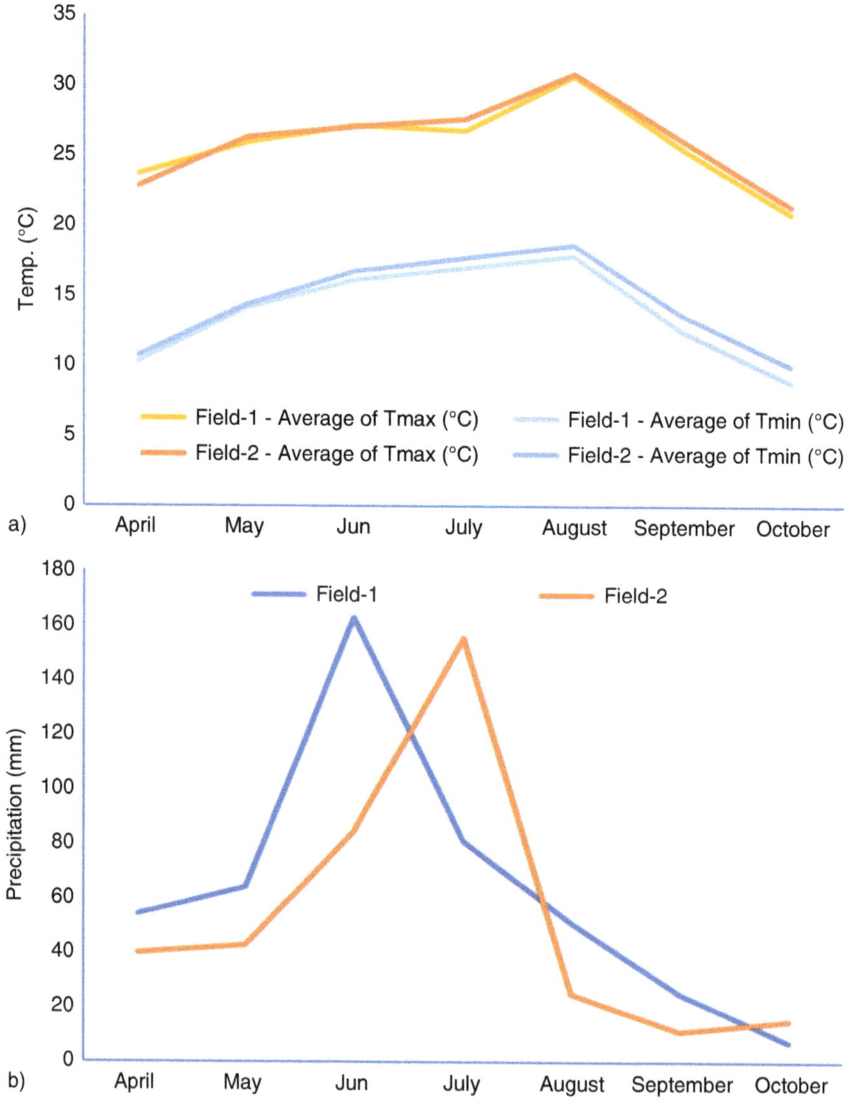

Fig. 4 Time charts of minimum and maximum monthly average temperature and cumulated precipitation (monthly) for the two experimental locations during maize growing season in 2018

In addition, precipitation showed a quite favorable trend for maize production throughout the 2018 growing season. After sowing, there was adequate precipitation for emergence and for the initial stages of maize growth. Then, precipitation increased during the rapid growth stages, when there is high water demand, and decreased during the reproductive stages reaching the minimum close to maturity. Distribution of precipitation was probably better in Field-2 with the maximum cumulative precipitation occurring during July contrary to Field-1 where the wettest month was June. This fact, in conjunction with the differences in soil properties are the main factors leading to increased yield potential in Field-1.

3.3 Correlation Analysis

Pearson's correlation analysis revealed the relationship between NDVI derived from Plant-O-Meter and GreenSeeker measurements, and end-of-season yield (Table 2). The canopy reflectance measurements took place at three different stages of maize growth, V5, V6 and V8, in both fields. These stages were selected as they proved, in several studies, to provide adequate estimation of end-of-season yield whilst being early enough to facilitate corrective actions [33, 40]. According to the results, all Plant-O-Meter measurements were significantly correlated to yield ($p < 0.01$) showing an increasing trend as the growth stage proceeded. Similarly, the GreenSeeker measurements showed stronger correlation at V8 stage. In this case, the relationship between GreenSeeker NDVI acquired at V6 and V8 stages and Yield was statistically significant ($p < 0.01$) but this was not the case for the measurements acquired at V5.

In addition, the strongest correlation between Plant-O-Meter and GreenSeeker NDVI measurements were acquired at V8 ($r = 0.853$; $p < 0.01$) followed by V6 ($r = 0.780$; $p < 0.01$) growth stages while at V5 the correlation was non-significant. These results suggest that the level of uncertainty introduced by random detection of soil surface had significant effect on the measurements. During the early growth

Table 2 Relationship between NDVI derived from Plant-O-Meter and GreenSeeker measurements, acquired at V5, V6, and V8 growth stages of maize, and end-of-season yield (Pearson's correlation)

			Plant-O-Meter			GreenSeeker		
		Yield	**V5**	**V6**	**V8**	**V5**	**V6**	**V8**
	Yield	1						
Plant-O-meter	**V5**	0.719[a]	1					
	V6	0.867[a]	0.792[a]	1				
	V8	0.892[a]	0.773[a]	0.883[a]	1			
GreenSeeker	**V5**	0.359	0.171	0.454[b]	0.301	1		
	V6	0.741[a]	0.780[a]	0.799[a]	0.743[a]	0.549[a]	1	
	V8	0.863[a]	0.808[a]	0.767[a]	0.853[a]	0.225	0.677[a]	1

[a]Correlation is significant at the 0.01 level
[b]Correlation is significant at the 0.05 level

stages when the plants' leaf area was small, the proportion of soil in the sensor's field of view introduced noise to the system affecting the NDVI readings. This effect was higher for the GreenSeeker sensor due to the significantly lower spatial resolution of the measurements. The increased noise in data acquisition of the GreenSeeker sensor at V5 stage was also reflected on the relationship with yield.

3.4 *Linear Regression Analysis*

Regression analysis between INSEY and end of season yield showed that for both sensors, good yield estimation can be achieved at V6 – V8 growth stages. This finding is in agreement with the results of Tagarakis and Keterings [33] who defined V6 as the earliest growth stage for accurate yield estimations. In our study, the most accurate estimation of yield for both sensors were achieved at V8 growth stage; coefficient of determination (R^2) was 0.8 and 0.75 for Plant-O-Meter and GreenSeeker respectively (Fig. 5). This is consistent with the findings of previous studies [40]. In general, Plant-O-Meter provided better estimation of end-of-season yield for all three maize growth stages. However, the measurements with the GreenSeeker sensor were performed manually, providing only four measurements in each experimental plot, unlike the Plant-O-Meter which, sensing using the

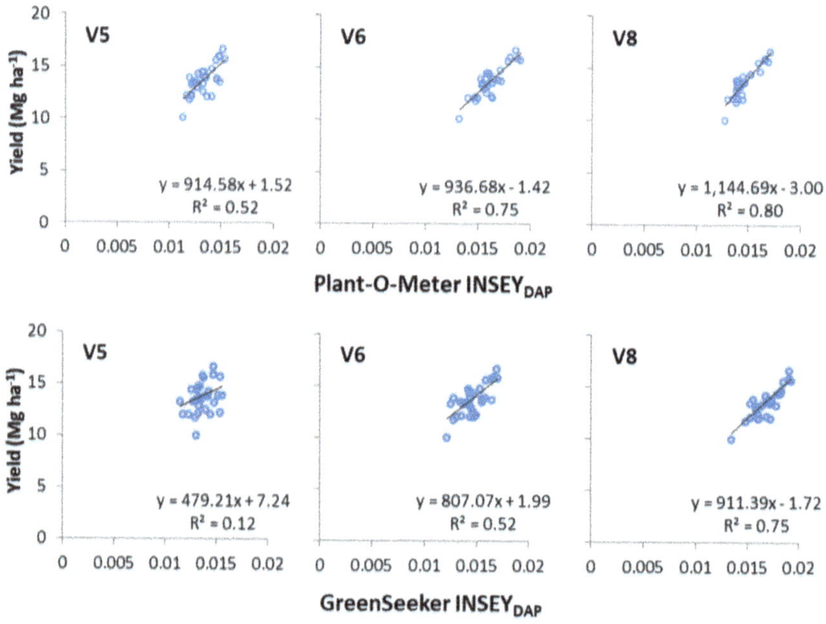

Fig. 5 Relationship between end-of-season yield and In-Season Estimated Yield ($INSEY_{DAP}$) measured using the Plant-O-Meter (**a**) and the GreenSeeker (**b**) at V5, V6 and V8 maize growth stages

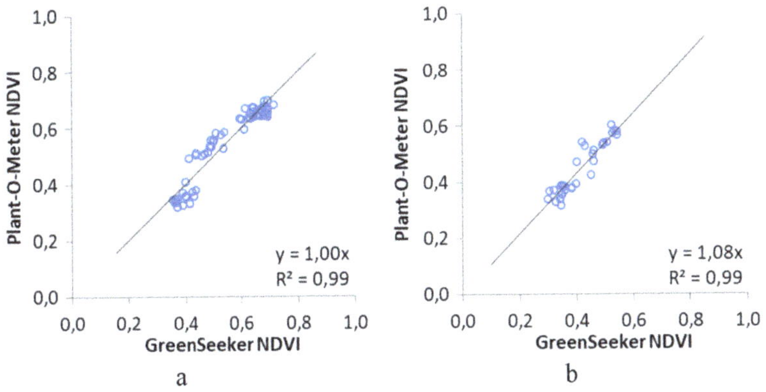

Fig. 6 Relationship between NDVI measured using the Plant-O-Meter (y axis) and NDVI measured using the GreenSeeker (x axis). Data derived from (**a**) Field-1, and (**b**) Field-2

mapping mode, provided a significantly larger number of measurements per plot leading to greater accuracy. Therefore, the current results need to be further investigated in various environmental and climatic conditions and at more detailed temporal and spatial resolution for both sensors.

Regression analysis was also used to directly compare the Plant-O-Meter and GreenSeeker sensors' measurements. Since the only purpose of this analysis was the direct comparison of the results derived by the two sensors, all the datasets, from each field of study for all the timings of measuring, were combined and analyzed together as one unified dataset. This way, the number of samples in the statistical analysis increased significantly leading to more reliable results. According to the analysis, the results confirmed a strong relationship between the NDVI measurements from the two sensors (R^2 = 0.89), for both Field-1 and Field-2 datasets (Fig. 6). This finding suggests that the relationship between the NDVI measurements derived from Plant-O-Meter, were similar to the ones derived from GreenSeeker regardless the location where the measurements were taken from.

In a second phase, the datasets from the two fields were combined into a unified dataset and the regression analysis was repeated. The results were similar to the analysis from each field separately, confirming the strong relationship between the measurements of the two sensors (Fig. 7), despite the fact that they measure canopy reflectance at different wavelengths. GreenSeeker is one of the most widely used commercial active proximal sensors with a vast number of scientific studies proving its reliability for in-field canopy measurements and for real-time applications in a range of crops. Considering that Plant-O-Meter performed equally well to the GreenSeeker, the results of this study suggest that the new low-cost sensor shows great potential to be used for on-the-go variable rate applications as it performs similarly to the GreenSeeker in real field conditions.

This was the first trial for in-field testing of the Plant-O-Meter multispectral active proximal sensor. The results revealed its potential to estimate end-of-season yield form mid-season canopy measurements which is the first step in the

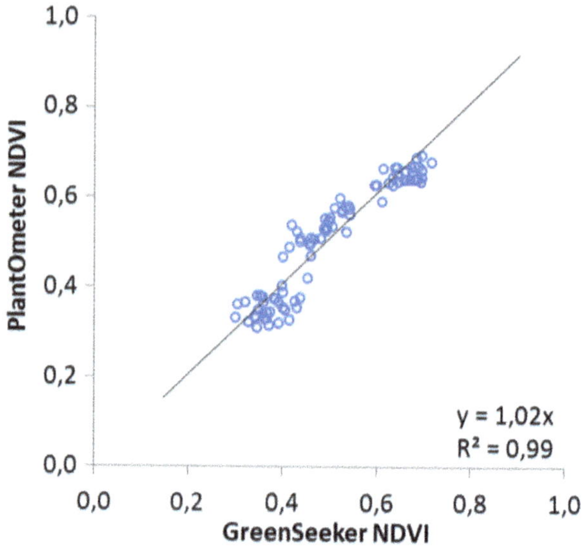

Fig. 7 Relationship between NDVI measured using the Plant-O-Meter (y axis) and NDVI measured using the GreenSeeker (x axis). Data derived from both fields combined

development of an algorithm for variable rate nitrogen applications [37]. Further studies in a range of crops are needed to ensure the Plant-O-Meter's reliability for extended commercial use. In addition, the sensor's operability and reliability need to be further evaluated in different environmental conditions to ensure its stability in adverse environments.

4 Future Prospects for Development

Active proximal sensors, such as the GreenSeeker and the Plant-O-Meter presented here, are promising tools for predicting end-of-season yield and N requirements in arable crops. So far, the main drawbacks limiting the adoption of such sensing systems by the farmers were the high cost and the complexity of use. Therefore, our effort was focused on the development of a low-cost sensing system using a farmer-friendly interface in order to maximize the adoption by the farmers. Towards that direction, Plant-O-Meter crop sensor offers a wide range of possibilities for plant status detection and could potentially work as a useful tool for decision making and for variable rate N fertilizer applications. Additional N rate trials are set in order to evaluate Plant-O-Meter's usage in different crops and to develop algorithms for variable rate N application. The studies are initially focused in the two main arable crops, maize and wheat, with the intention to expand to other crops as well.

After finalizing the development and evaluation stage, the system will be available for commercial use. The expectations are to receive wide acceptance by the farmers due to simplification of the data acquisition process, the extended modes of operation, and the affordable cost pushing up the farmers' enthusiasm for modernization regarding the engagement of sophisticated tools in crop production.

During the performance evaluation of Plant-O-Meter in laboratory and field conditions several directions for potential improvements of the device's performance have emerged. In that sense, future research and development will focus on the following aspects:

- Automatic calibration. Currently the device is calibrated by manually adjusting potentiometers controlling the current supplied to the LEDs. This calibration can only be performed by trained personnel. Future plans include using digital potentiometers which can be set by the microcontroller making the calibration process automated, minimizing the maintenance costs;
- Measurement of the emitted radiation from the LEDs. This feature would provide the ability to measure absolute values of emitted and reflected signals, increasing the number of supported vegetation indices;
- Adjusting the Plant-O-Meter to tractors or other agricultural machinery. For this upgrade, a mechanical module for attaching the sensor to the tractor should be designed. In addition, an appropriate communication module, ideally following the ISOBUS standards, will be developed.

5 Conclusions and Outlook

In this study, Plant-O-Meter hand-held device was tested in real, outdoor environment for measuring maize canopy properties, as compared to the GrenSeeker handheld commercial crop sensor. Optimal field conditions (precipitation, temperature, and soil fertility) during the maize growing season reduced the influence of applied N on the final yield. However, useful information was extracted by the datasets acquired from the two fields of the study. Based on the present findings reliable end-of-season yield estimation was attained measuring the mid-season NDVI between V7 and V8 growth stage of maize. The overall results indicated that NDVI obtained using GreenSeeker were quite similar to the NDVI measured by the Plant-O-Meter showing an almost 1:1 relationship. In addition, both sensors provided good estimation of end-of-season yield. The precision of yield estimation was maximized for the measurements acquired at V8 growth stage of maize crop. This applies to both sensors and for both locations. The Plant-O-Meter provided slightly better estimation of end-of-season yield especially for the measurements performed earlier in the season (V5 and V6 stages). However, this needs to be further investigated since the data attained with the Plant-O-Meter had significantly higher spatial resolution compared to the GreenSeeker data which were recorded manually. Nevertheless, the results indicated that Plant-O-Meter exhibits strong potential for accurate plant canopy measurements and for real time variable rate fertilization applications in maize.

Considering that the optimal stage for measuring NDVI depends on the environmental conditions, further studies in more diverse conditions are needed to test and evaluate Plant-O-Meter's performance in all possible field conditions. This will enrich the database resulting to the final equation for end-of-season yield estimation using mid-season crop canopy measurements. Defining the equation for yield estimation is the first step for developing algorithm for real time variable rate applications. In addition, similar studies should be performed in various crops in order to increase the applicability of the system. Since it is an active sensor, it is not dependent on illumination sources of the measuring environment (e.g. sunlight for outdoor measurements, artificial light for indoor measurements, etc.). Therefore, it can be a useful tool for indoor environments such as greenhouses, while it is suitable for night-time measurements. Furthermore, the low cost and the ease of use of the Plant-O-Meter sensor are expected to make it a reliable and affordable solution for everyone, even for small and medium size farmers, that wish to take active part in smart farming. In large-scale agricultural production, Plant-O-Meter measurements can serve as auxiliary data, or as ground-truth, for remote sensing data applications.

Acknowledgments This study was supported by the project "Development of the device for measurement and mapping of nitrogen as the most important parameter in sustainable agriculture", contract no. 114-451-2794/2016-03, funded by Provincial Secretariat for Higher Education and Scientific Research of Autonomous Province of Vojvodina, Republic of Serbia and by the project "Improvement of the quality of tractors and mobile systems with the aim of increasing competitiveness and preserving soil and environment", contract no. TR-31046, funded by the Ministry of Education, Science and Technological Development of the Republic of Serbia.

References

1. FAOstat (2020) 'Production quantities of Maize. Average for the years 1961 - 2018'. http://www.fao.org/faostat/en/#data/QC/visualize. (Date accessed 18/02/2020).
2. Hammer, G. L., Dong, Z., McLean, G., Doherty, A., Messina, C., Schussler, J., Zinselmeier, C., Paszkiewicz, S. and Cooper, M. (2009) 'Can Changes in Canopy and/or Root System Architecture Explain Historical Maize Yield Trends in the U.S. Corn Belt?', Crop Sci. 49, pp. 299–312. https://doi.org/10.2135/cropsci2008.03.0152.
3. Shapiro, C. A. and Wortmann, C. S. (2006) 'Corn response to nitrogen rate, row spacing and plant density in Eastern Nebraska', Agron. J., 98, pp. 529–535.
4. Ladha, K.J., Pathak, H., Krupnik, T.J., Six, J. and van Kessel, C. (2005) 'Efficiency of Fertilizer nitrogen in cereal production: Retrospects and prospects', Adv. Agron., 87, pp. 85–156.
5. Tagarakis, A. C. and Ketterings, Q. M. (2018) 'Proximal sensor-based algorithm for variable rate nitrogen application in maize in northeast U.S.A.', Computers and Electronics in Agriculture, 145, pp. 373-378. https://doi.org/10.1016/j.compag.2017.12.031.
6. Raun, W., and Johnson, G. (1999). 'Improving nitrogen use efficiency for cereal production', Agronomy Journal, 91, pp. 357–363.
7. López-Bellido, R. and López-Bellido, L. (2001) 'Efficiency of nitrogen in wheat under Mediterranean conditions: Effect of tillage, crop rotation and N fertilization', Field Crop. Res., 71, pp. 31–46.

8. Setiyono, T. D., Yang, H., Walters, D. T., Dobermann, A., Ferguson, R. B., Roberts, D. F., Lyon, D. J., Clay, D. E. and Cassaman, K. G. (2011) 'Maize-N: A decision tool for nitrogen management in maize', Agron. J.,103, pp. 1276–1283.
9. Gemtos, T., Fountas, S., Tagarakis, A. and Liakos, V. (2013) 'Precision agriculture application in fruit crops:experience in handpicked fruits', Procedia Technology. 8, pp. 324–332.
10. International Society of Precision Agriculture – ISPA (2018) 'Official definition of Precision Agriculture', https://www.ispag.org/about/definition (accessed 28 January 2020).
11. Robert, P., Rust, R. and Larson, W. (1994) 'Site-specific Management for Agricultural Systems', Proceedings of the 2nd International Conference on Precision Agriculture, 1994, Madison, WI.
12. Khosla, R. (2008) 'The 9th International Conference on Precision Agriculture opening ceremony presentation', July 20-23rd, 2008. ASA/CSSA/SSSA.
13. Mulla, D. J. (2013) 'Twenty five years of remote sensing in precision agriculture: Key advances and remaining knowledge gaps', Biosystems Engineering, Special Issue: Sensing in Agriculture, pp. 358–371. https://doi.org/10.1016/j.biosystemseng.2012.08.009.
14. Yang, C., Everitt, J. H., Du, Q., Luo, B. and Chanussot, J. (2013) 'Using High-Resolution Airborne and Satellite Imagery to Assess Crop Growth and Yield Variability'. Proceedings of the IEEE, 101 (3), pp. 582-592. https://doi.org/10.1109/JPROC.2012.2196249.
15. Raun, W. R., Solie, J. B., Johnson, G. V., Stone, M. L., Mullen, R. W., Freeman, K. W., Thomason, W. E, and Lukina, E. V. (2002) 'Improving nitrogen use efficiency in cereal grain production with optical sensing and variable rate application', Agron. J. 94:815–820. https://doi.org/10.2134/agronj2002.8150.
16. Tagarakis, A. C., Ketterings, Q. M., Lyons, S. and Godwin, G. (2017) 'Proximal sensing to estimate yield of brown midrib forage sorghum', Agronomy Journal, 109(1), pp. 107–114. https://doi.org/10.2134/agronj2016.07.0414.
17. Auernhammer, H. (2001) 'Precision farming — the environmental challenge', Computers and Electronics in Agriculture, 30 (1–3), pp. 31–43.
18. Tagarakis, A., Liakos, V., Fountas, S., Koundouras, S. and Gemtos, T. A. (2013) Management zones delineation using fuzzy clustering techniques in grapevines. Precision Agriculture, 14, pp. 18–39.
19. Oberti, R., Marchi, M., Tirelli, P., Calcante, A., Iriti, M. and Borghese, A. N. (2014) 'Automatic detection of powdery mildew on grapevine leaves by image analysis: Optimal view angle range to increase the sensitivity', Computers and Electronics in Agriculture, 104, pp. 1-8.
20. Whetton, R., Waine, T., Mouazen, A. (2017) 'Optimising configuration of a hyperspectral imager for on-line field measurement of wheat canopy', Biosystems Engineering, 155, pp. 84-95.
21. Fitzgerald, G. J. (2010) 'Characterizing vegetation indices derived from active and passive sensors', International Journal of Remote Sensing, 31:16, pp. 4335-4348. https://doi.org/10.1080/01431160903258217.
22. Oerke, E.C., Mahlein, A.K. and Steiner, U. (2014) 'Proximal sensing of plant diseases' In: Gullino, M.L., Bonants, P.J.M. (eds) 'Detection and diagnostics of plant pathogens', Springer, Dordrecht, p.p. 55–68. https://doi.org/10.1007/978-94-017-9020-8_4.
23. Aschbacher, J. and Milagro-Pérez, M. P. (2012) 'The European Earth monitoring (GMES) programme: Status and perspectives', Remote Sensing of Environment, 120, pp. 3-8. https://doi.org/10.1016/j.rse.2011.08.028.
24. Woodcock, C. E., Allen, R., Anderson, M., Belward, A., Bindschadler, R., Cohen, W., Gao, F., Goward, S. N., Helder, D., Helmer, E., Nemani, R., Oreopoulos, L., Schott, J., Thenkabail, P. S., Vermote, E. F., Vogelmann, J., Wulder, M. A. and Wynne, R. (2008) 'Free access to Landsat imagery', Science, 320, pp. 1011. https://doi.org/10.1126/science.320.5879.1011a.
25. Jackson, R. D. (1986) 'Remote Sensing of Biotic and Abiotic Plant Stress', Annual review of Phytopathology, 24, pp. 265–287. https://doi.org/10.1146/annurev.py.24.090186.001405.

26. Shanahan, J. F., Kitchen, N. R., Raun, W. R. and Schepers, J. S. (2008) 'Responsive in-season nitrogen management for cereals', Computers and Electronics in Agriculture, 61, pp. 51-62. https://doi.org/10.1016/j.compag.2007.06.006.
27. Solari, F., Shanahan, J., Ferguson, R. B., Schepers, J. S. and Gitelson, A. A. (2008) 'Active sensor reflectance measurements to corn nitrogen status and yield potential', Agronomy Journal, 100, pp. 571–579. https://doi.org/10.2134/agronj2007.0244.
28. Girma, K., Holtz, S. L., Arnall, D. B., Fultz, L. M., Hanks, T. L., Lawles, K. D., Mack, C. J., Owen, K. W., Reed, S. D., Santillano, J., Walsh, O., White, M. J. and Raun, W. R. (2007). 'Weather, fertilizer, previous year grain yield and fertilizer response level affect ensuing year grain yield and fertilizer response of winter wheat', Agronomy Journal, 99, pp. 1607–1614.
29. Kostić, M., Rakić, D., Savin, L., Dedović, N. and Simikić, M. (2016) 'Application of an original soil tillage resistance sensor in spatial prediction of selected soil properties', Computers and Electronics in Agriculture, 127, pp. 615–624. https://doi.org/10.1016/j.compag.2016.07.027.
30. Magney, S. T., Eitel, J. U. H., Huggins, D. R. and Vierling, L. A. (2016) 'Proximal NDVI derived phenology improves in-season predictions of wheat quantity and quality', Agricultural and Forest Meteorology, 217, pp. 46–60. https://doi.org/10.1016/j.agrformet.2015.11.009.
31. Zecha, C. W., Peteinatos, G. G., Link, J. and Claupein, W. (2018) 'Utilisation of ground and airborne optical sensors for nitrogen level identification and yield prediction in wheat', Agriculture, 8(6) pp. 79. https://doi.org/10.3390/agriculture8060079.
32. Bean, G. M., Kitchen, N. R., Camberato, J. J., Ferguson, R. B., Fernandez, F. G., Franzen, D. W., Laboski, C. A. M., Nafziger, E. D., Sawyer, J. E., Scharf, P. C., Schepers, J. and Shanahan, J. S. (2018) 'Active-optical reflectance sensing corn algorithms evaluated over the United States Midwest corn belt', Agronomy Journal, 110, pp. 2552–2565.
33. Tagarakis, A. C. and Ketterings, Q. M. (2017) 'In-season estimation of corn yield potential using proximal sensing', Agronomy Journal, 109(4), pp. 1323–1330. https://doi.org/10.2134/agronj2016.12.0732.
34. Rouse, J. W., Haas, R. H., Schell, J. A. and Deering, D. W. (1973) 'Monitoring vegetation systems in the Great Plains with ERTS', NASA. Goddard Space Flight Center 3d ERTS-1 Symp., 1, pp. 309–317.
35. Hatfield, J. L., Gitelson, A. A., Schepers, J. S. and Walthall, C. L.(2008) 'Application of spectral remote sensing for agronomic decisions', Agronomy Journal, 100, pp. 117–131. https://doi.org/10.2134/agronj2006.0370c.
36. Wang, R., Cherkauer, K. A. and Bowling, L. C. (2016) 'Corn Response to Climate Stress Detected with Satellite-Based NDVI Time Series', Remote Sensing, 8(4), pp. 269. https://doi.org/10.3390/rs8040269.
37. Moges, S. M., Girma, K., Teal, R. K., Freeman, K. W., Zhang, H. and Arnall, D. B. (2007) 'In-season estimation of grain sorghum yield potential using a hand-held optical sensor', Arch. of Agron. and Soil Sci., 53(6), pp. 617–628. https://doi.org/10.1080/03650340701597251.
38. Raun, W. R., Solie, J. B., Martin, K. L., Freeman, K. W., Stone, M. L., Johnson, G. V. and Mullen, R. W. (2005) 'Growth stage, development, and spatial variability in corn evaluated using optical sensor readings', J. Plant Nutr., 28, pp. 173–182. https://doi.org/10.1081/PLN-200042277.
39. Raun,W. R., Johnson, G.V., Stone, M.L., Solie, J.B., Lukina, E.V., Thomason, W.E., and Schepers, J.S. (2001) 'In-season prediction of potential grain yield in winter wheat using canopy reflectance', Agronomy Journal, 93, pp. 131–138.
40. Teal, R. K., Tubana, B., Girma, K., Freeman, K. W., Arnall, D. B., Walsh, O. and Raun, W. R. (2006) In-season prediction of corn grain yield potential using normalized difference vegetation index', Agron. J., 98, pp. 1488–1494. https://doi.org/10.2134/agronj2006.0103.
41. Lukina, E. V., Freeman, K. W., Wynn, K. J., Thomason, W. E., Mullen, R. W., Stone, M. L., Solie, J. B., Klatt, A. R., Johnson, G. V., Elliott, R. L. and Raun, W. R. (2001) 'Nitrogen fertilization optimization algorithm based on in-season estimates of yield and plant nitrogen uptake', Journal of Plant Nutrition, 24(6), pp. 885-898. https://doi.org/10.1081/PLN-100103780.

42. Ritchie, S. W., Hanway, J. J. and Benson, G. O. (1997) 'How a Corn Plant Develops', Special Report No. 48, Iowa State University Cooperative Extension Service: Ames, IA, USA, 1997.
43. Rogers, N. G. (2016) 'Sensor Based Nitrogen Management for Corn Production in Coastal Plain Soils', All Theses. 2579.
44. Adamsen, F.J., Pinter Jr., P.J., Barnes, E.M., LaMorte, R.L., Wall, G.W., Leavitt, S.W. and Kimball, B.A. (1999) 'Measuring wheat senescence with a digital camera', Crop Science, 39, pp. 719-724. https://doi.org/10.2135/cropsci1999.0011183X003900030019x.
45. Helman, D., Bonfil, D. J. and Lensky, I. M. (2019) 'Crop RS-Met: A biophysical evapotranspiration and root-zone soil water content model for crops based on proximal sensing and meteorological data', Agricultural Water Management, 211, pp. 210–219. https://doi.org/10.1016/j.agwat.2018.09.043.
46. Stone, M. L., Solie, J. B., Raun, W. R., Whitney, R. W., Taylor, S. L. and Ringer, J. D. (1996) 'Use of spectral radiance for correcting in-season fertilizer nitrogen deficiencies in winter wheat', Trans. ASAE 39, pp. 1623–1631. https://doi.org/10.13031/2013.27678.
47. Kim, Y., Huete, A., Miura, T. and Jiang, Z. (2010) 'Spectral compatibility of vegetation indices across sensors: band decomposition analysis with Hyperion data', Journal of Applied Remote Sensing 4(1) 043520. https://doi.org/10.1117/1.3400635.
48. Yao, X., Yao, X., Jia, W., Tian, Y., Ni, J., Cao, W. and Zhu, Y. (2013) 'Comparison and intercalibration of vegetation indices from different sensors for monitoring above-ground plant nitrogen uptake in winter wheat', Sensors, 13(3), pp. 3109-3130. https://doi.org/10.3390/s130303109.
49. Belic, M., Manojlivic, M., Nesic, L., Ciric, V., Vasin, J., Benka, P. and Seremesic S. (2013) 'Pedo-Ecological significance of Soil Organic Carbon stock in South-Eastern Pannonian basin', Carpathian Journal of Earth and Environmental Sciences, 8 (1), pp. 171 – 178.
50. Altermann, M., Rinklebe, J., Merbach, I., Körschens, M., Langer, U. and Hofmann, B. (2005) 'Chernozem—Soil of the Year 2005', J. Plant Nutr. Soil Sci. 2005, 168, pp. 725–740. https://doi.org/10.1002/jpln.200521814.
51. Tremblay N., Wang Z., Ma, B. L., Belec, C. and Vigneault, P. (2009) 'A comparison of crop data measured by two commercial sensors for variable-rate nitrogen application', Precision Agriculture, 10, pp. 145-161. https://doi.org/10.1007/s11119-008-9080-2.
52. Kitić, G., Tagarakis, A., Cselyuszka, N., Panić, M., Birgermajer, S., Sakulskia, D. and Matović, J. (2019) 'A new low-cost portable multispectral optical device for precise plant status assessment', Computers and Electronics in Agriculture, 162, pp. 300–308.
53. Johnson, G. V. and Raun, W. R. (2003) 'Nitrogen response index as a guide to fertilizer management', Journal of Plant Nutrition, 26, pp. 249–262.

www.ingramcontent.com/pod-product-compliance
Ingram Content Group UK Ltd.
Pitfield, Milton Keynes, MK11 3LW, UK
UKHW021304100626
13040UKWH00003B/26
* 9 7 8 3 0 3 0 8 4 1 4 3 0 *